# Pro Apache Hadoop

## Second Edition

Sameer Wadkar
Madhu Siddalingaiah

Apress®

**Pro Apache Hadoop**



ISBN-13 (pbk): 978-1-4302-4863-7

ISBN-13 (electronic): 978-1-4302-4864-4



Publisher: Heinz Weinheimer
Lead Editor: Jonathan Gennick
Technical Reviewer: Vimlesh Om Mittal
Editorial Board: Steve Anglin, Mark Beckner, Ewan Buckingham, Gary Cornell, Louise Corrigan, Jim DeWolf, Jonathan Gennick, Jonathan Hassell, Robert Hutchinson, Michelle Lowman, James Markham, Matthew Moodie, Jeff Olson, Jeffrey Pepper, Douglas Pundick, Ben Renow-Clarke, Dominic Shakeshaft, Gwenan Spearing, Matt Wade, Steve Weiss
Coordinating Editor: Jill Balzano
Copy Editor: Nancy Sixsmith
Compositor: SPi Global
Indexer: SPi Global
Artist: SPi Global
Cover Designer: Anna Ishchenko

Distributed to the book trade worldwide by Springer Science+Business Media New York, 233 Spring Street, 6th Floor, New York, NY 10013. Phone 1-800-SPRINGER, fax (201) 348-4505, e-mail orders-ny@springer-sbm.com, or visit www.springeronline.com. Apress Media, LLC is a California LLC and the sole member (owner) is Springer Science + Business Media Finance Inc (SSBM Finance Inc). SSBM Finance Inc is a Delaware corporation.

For information on translations, please e-mail rights@apress.com, or visit www.apress.com.

Apress and friends of ED books may be purchased in bulk for academic, corporate, or promotional use. eBook versions and licenses are also available for most titles. For more information, reference our Special Bulk Sales–eBook Licensing web page at www.apress.com/bulk-sales.

Any source code or other supplementary material referenced by the author in this text is available to readers at www.apress.com. For detailed information about how to locate your book's source code, go to www.apress.com/source-code/.

*This book is dedicated to the Open Source Community whose contributions have made Software Development such as vibrant and exciting profession.*

*—Sameer*

*To Sasha, Eashan, and Shilpa.*

*—Madhu*

# Contents at a Glance

# Contents

# About the Authors

**Sameer Wadkar** has more than 16 years of experience in software architecture and development. He has a bachelor's degree in electrical engineering and an MBA in finance from Mumbai University, a postgraduate diploma in software engineering from the National Center of Software Technology (now Center for Development of Advanced Computing), and a master's degree in applied and computational mathematics from Johns Hopkins University. He has implemented distributed systems and high-traffic web sites for a wide variety of clients ranging from federal agencies to investment banking companies. Sameer has been actively working on Hadoop/HBase implementations since 2011 and is also an open-source contributor. His GitHub page is `https://github.com/sameerwadkar`.

Sameer's open-source contributions include a version of the popular text mining algorithm known as Latent Dirichlett Allocation, which scales to millions of documents on a single machine. He is an avid chess player and plays actively on `www.chessclub.com`.

**Madhu Siddalingaiah** is a technology consultant with 25 years of experience in a variety of business domains, including aerospace, health care, financial, energy, defense, and scientific research. Over the years, he has specialized in electrical engineering, Internet technologies, and Big Data. More recently, Madhu has delivered several high-profile Big Data systems and solutions.

He earned his physics degree from the University of Maryland. Outside of his profession, Madhu is a private helicopter pilot, enjoys travel, and participates in a constantly growing list of hobbies.

# About the Technical Reviewer

**Vimlesh Om Mittal** has more than 14 years of technology implementation experience and is a Cloudera Certified Developer for Apache Hadoop CDH4. He has a very broad technology background, and his experience includes projects involving modernizing business processes, custom application development, ETL development, digital application development, business intelligence, database design, and data conversions.

Vimlesh's experience is a strong blend of architecting robust information management systems, solution architecture, and software development in several development frameworks such as Big Data using the Hadoop ecosystem; data ingestion/acquisition; ETL; reporting/analytics; .NET; J2EE; and scripting languages in different vendor technologies such as Cloudera, SAS, IBM, Oracle, Microsoft, iOS, and BEA.

Vimlesh's focus also includes solutions for financial services data, mobile application design, and development. He is also well-versed in all aspects of the software development life cycle and has extensive experience in project management disciplines.

# Acknowledgments

Hadoop has come a long way over the past decade with a large number of vendors and independent developers contributing to it. Hadoop in turn depends on a large number of open source libraries. We would like to acknowledge the effort of all the open source contributors who have contributed to Hadoop. We would also like to acknowledge the blog writers and professionals who contribute to answering Hadoop-related queries on forums. Their participation contributes to making a complex product such as Hadoop mainstream and easier to use.

We want to thank the Apress staff members who have applied their expertise to make this book into something readable.

We also want to thank our family, friends, and colleagues for supporting us through the writing of this book.

# Introduction

This book is designed to be a concise guide to using the Hadoop software. Despite being around for more than half a decade, Hadoop development is still a very stressful yet very rewarding task. The documentation has come a long way since the early years, and Hadoop is growing rapidly as its adoption is increasing in the Enterprise. Hadoop 2.0 is based on the YARN framework, which is a significant rewrite of the underlying Hadoop platform. It has been our goal to distill the hard lessons learned while implementing Hadoop for clients in this book. As authors, we like to delve deep into the Hadoop source code to understand why Hadoop does what it does and the motivations behind some of its design decisions. We have tried to share this insight with you. We hope that not only will you learn Hadoop in depth but also gain fresh insight into the Java language in the process.

This book is about Big Data in general and Hadoop in particular. It is not possible to understand Hadoop without appreciating the overall Big Data landscape. It is written primarily from the point of view of a Hadoop developer and requires an intermediate-level ability to program using Java. It is designed for practicing Hadoop professionals. You will learn several practical tips on how to use the Hadoop software gleaned from our own experience in implementing Hadoop-based systems.

This book provides step-by-step instructions and examples that will take you from just beginning to use Hadoop to running complex applications on large clusters of machines. Here's a brief rundown of the book's contents:

Chapter 1 introduces you to the motivations behind Big Data software, explaining various Big Data paradigms.

Chapter 2 is a high-level introduction to Hadoop 2.0 or YARN. It introduces the key concepts underlying the Hadoop platform.

Chapter 3 gets you started with Hadoop. In this chapter, you will write your first MapReduce program.

Chapter 4 introduces the key concepts behind the administration of the Hadoop platform.

Chapters 5, 6, and 7, which form the core of this book, do a deep dive into the MapReduce framework. You learn all about the internals of the MapReduce framework. We discuss the MapReduce framework in the context of the most ubiquitous of all languages, SQL. We emulate common SQL functions such as `SELECT`, `WHERE`, `GROUP BY`, and `JOIN` using MapReduce. One of the most popular applications for Hadoop is ETL offloading. These chapters enable you to appreciate how MapReduce can support common data-processing functions. We discuss not just the API but also the more complicated concepts and internal design of the MapReduce framework.

Chapter 8 describes the testing frameworks that support unit/integration testing of MapReduce frameworks.

Chapter 9 describes logging and monitoring of the Hadoop Framework.

Chapter 10 introduces the Hive framework, the data warehouse framework on top of MapReduce.

Chapter 11 introduces the Pig and Crunch frameworks. These frameworks enable users to create data-processing pipelines in Hadoop.

Chapter 12 describes the HCatalog framework, which enables Enterprise users to access data stored in the Hadoop file system using commonly known abstractions such as databases and tables.

Chapter 13 describes how Hadoop can used for streaming log analysis.

Chapter 14 introduces you to HBase, the NoSQL database on top of Hadoop. You learn about use-cases that motivate the use of Hbase.

Chapter 15 is a brief introduction to data science. It describes the main limitations of MapReduce that make it inadequate for data science applications. You are introduced to new frameworks such as Spark and Hama that were developed to circumvent MapReduce limitations.

Chapter 16 is a brief introduction to using Hadoop in the cloud. It enables you to work on a true production-grade Hadoop cluster from the comfort of your living room.

Chapter 17 is a whirlwind introduction to the key addition to Hadoop 2.0: the capability to develop your own distributed frameworks such as MapReduce on top of Hadoop. We describe how you can develop a simple distributed download service using Hadoop 2.0.

CHAPTER 1

# Motivation for Big Data

The computing revolution that began more than 2 decades ago has led to large amounts of digital data being amassed by corporations. Advances in digital sensors; proliferation of communication systems, especially mobile platforms and devices; massive scale logging of system events; and rapid movement toward paperless organizations have led to a massive collection of data resources within organizations. And the increasing dependence of businesses on technology ensures that the data will continue to grow at an even faster rate.

Moore's Law, which says that the performance of computers has historically doubled approximately every 2 years, initially helped computing resources to keep pace with data growth. However, this pace of improvement in computing resources started tapering off around 2005.

The computing industry started looking at other options, namely parallel processing to provide a more economical solution. If one computer could not get faster, the goal was to use many computing resources to tackle the same problem in parallel. Hadoop is an implementation of the idea of multiple computers in the network applying MapReduce (a variation of the single instruction, multiple data [SIMD] class of computing technique) to scale data processing.

The evolution of cloud-based computing through vendors such as Amazon, Google, and Microsoft provided a boost to this concept because we can now rent computing resources for a fraction of the cost it takes to buy them.

This book is designed to be a practical guide to developing and running software using Hadoop, a project hosted by the Apache Software Foundation and now extended and supported by various vendors such as Cloudera, MapR, and Hortonworks. This chapter will discuss the motivation for Big Data in general and Hadoop in particular.

## What Is Big Data?

In the context of this book, one useful definition of Big Data is any dataset that cannot be processed or (in some cases) stored using the resources of a single machine to meet the required service level agreements (SLAs). The latter part of this definition is crucial. It is possible to process virtually any scale of data on a single machine. Even data that cannot be stored on a single machine can be brought into one machine by reading it from a shared storage such as a network attached storage (NAS) medium. However, the amount of time it would take to process this data would be prohibitively large with respect to the available time to process this data.

Consider a simple example. If the average size of the job processed by a business unit is 200 GB, assume that we can read about 50 MB per second. Given the assumption of 50 MB per second, we will need 2 seconds to read 100 MB of data from the disk sequentially, and it would take us approximately 1 hour to read the entire 200 GB of data. Now imagine that this data was required to be processed in under 5 minutes. If the 200 GB required per job could be evenly distributed across 100 nodes, and each node could process its own data (consider a simplified use-case such as simply selecting a subset of data based on a simple criterion: SALES_YEAR>2001), discounting the time taken to perform the CPU processing and assembling the results from 100 nodes, the total processing can be completed in under 1 minute.

This simplistic example shows that Big Data is context-sensitive and that the context is provided by business need.

---

■ **Note** Dr. Jeff Dean Keynote discusses parallelism in a paper you can find at `www.cs.cornell.edu/projects/ladis2009/talks/dean-keynote-ladis2009.pdf`. To read 1 MB of data sequentially from a local disk requires 20 million nanoseconds. Reading the same data from a 1 Gbps network requires about 250 million nanoseconds (assuming that 2 KB needs 250,000 nanoseconds and 500,000 nanoseconds per round-trip for each 2 KB). Although the link is a bit dated, and the numbers have changed since then, we will use these numbers in the chapter for illustration. The proportions of the numbers with respect to each other, however, have not changed much.

---

# Key Idea Behind Big Data Techniques

Although we have made many assumptions in the preceding example, the key takeaway is that we can process data very fast, yet there are significant limitations on how fast we can read the data from persistent storage. Compared with reading/writing node local persistent storage, it is even slower to send data across the network.

Some of the common characteristics of all Big Data methods are the following:

- Data is distributed across several nodes (Network I/O speed << Local Disk I/O Speed).
- Applications are distributed to data (nodes in the cluster) instead of the other way around.
- As much as possible, data is processed local to the node (Network I/O speed << Local Disk I/O Speed).
- Random disk I/O is replaced by sequential disk I/O (Transfer Rate << Disk Seek Time).

The purpose of all Big Data paradigms is to parallelize input/output (I/O) to achieve performance improvements.

## Data Is Distributed Across Several Nodes

By definition, Big Data is data that cannot be processed using the resources of a single machine. One of the selling points of Big Data is the use of commodity machines. A typical commodity machine would have a 2–4 TB disk. Because Big Data refers to datasets much larger than that, the data would be distributed across several nodes.

Note that it is not really necessary to have tens of terabytes of data for processing to distribute data across several nodes. You will see that Big Data systems typically process data in place on the node. Because a large number of nodes are participating in data processing, it is essential to distribute data across these nodes. Thus, even a 500 GB dataset would be distributed across multiple nodes, even if a single machine in the cluster would be capable of storing the data. The purpose of this data distribution is twofold:

- Each data block is replicated across more than one node (the default Hadoop replication factor is 3). This makes the system resilient to failure. If one node fails, other nodes have a copy of the data hosted on the failed node.
- For parallel processing reasons, several nodes participate in the data processing. Thus, 50 GB of data shared within 10 nodes enables all 10 nodes to process their own subdataset, achieving 5–10 times improvement in performance. The reader may well ask why all the data is not on the network file system (NFS), in which each node can read its portion. The answer is that reading from a local disk is significantly faster than reading from the network. Big Data systems make the local computation possible because the application libraries are copied to each data node before a job (an application instance) is started. We discuss this in the next section.

## Applications Are Moved to the Data

For those of us who rode the J2EE wave, the three-tier architecture was drilled into us. In the three-tier programming model, the data is processed in the centralized application tier after being brought into it over the network. We are used to the notion of data being distributed but the application being centralized.

Big Data cannot handle this network overhead. Moving terabytes of data to the application tier will saturate the networks and introduce considerable inefficiencies, possibly leading to system failure. In the Big Data world, the data is distributed across nodes, but the application moves to the data. It is important to note that this process is not easy. Not only does the application need to be moved to the data but all the dependent libraries also need to be moved to the processing nodes. If your cluster has hundreds of nodes, it is easy to see why this can be a maintenance/deployment nightmare. Hence Big Data systems are designed to allow you to deploy the code centrally, and the underlying Big Data system moves the application to the processing nodes prior to job execution.

## Data Is Processed Local to a Node

This attribute of data being processed local to a node is a natural consequence of the earlier two attributes of Big Data systems. All Big Data programming models are distributed- and parallel-processing based. Network I/O is orders of magnitude slower than disk I/O. Because data has been distributed to various nodes, and application libraries have been moved to the nodes, the goal is to process the data in place.

Although processing data local to the node is preferred by a typical Big Data system, it is not always possible. Big Data systems will schedule tasks on nodes as close to the data as possible. You will see in the sections to follow that for certain types of systems, certain tasks require fetching data across nodes. At the very least, the results from every node have to be assimilated on a node (the famous reduce phase of MapReduce or something similar for massively parallel programming models). However, the final assimilation phases for a large number of use-cases have very little data compared with the raw data processed in the node-local tasks. Hence the effect of this network overhead is usually (but not always) negligible.

## Sequential Reads Preferred Over Random Reads

First, you need to understand how data is read from the disk. The disk head needs to be positioned where the data is located on the disk. This process, which takes time, is known as the *seek* operation. Once the disk head is positioned as needed, the data is read off the disk sequentially. This is called the *transfer* operation. Seek time is approximately 10 milliseconds; transfer speeds are on the order of 20 milliseconds (per 1 MB). This means that if we were reading 100 MB from separate 1 MB sections of the disk, it would cost us 10 (seek time) * 100 (seeks) = 1 second, plus 20 (transfer rate per 1MB) * 100 = 2 seconds. This is a total of 3 seconds to read 100 MB. However, if we were reading 100 MB sequentially from the disk, it would cost us 10 (seek time) * 1 (seek) = 10 milliseconds + 20*100=2 seconds, for a total of 2.01 seconds.

Note that we have used the numbers based on the Dr. Jeff Dean's address, which is from 2009. Admittedly, the numbers have changed; in fact, they have improved since then. However, relative proportions between numbers have not changed, so we will use it for consistency.

Most throughput-oriented Big Data programming models exploit this feature. Data is swept sequentially off the disk and filtered in the main memory. Contrast this with a typical relational database management system (RDBMS) model that is much more random-read-oriented.

## An Example

Suppose that you want to get the total sales numbers for the year 2000 ordered by state, and the sales data is distributed randomly across multiple nodes. The Big Data technique to achieve this can be summarized in the following steps:

1. Each node reads in the entire sales data and filters out sales data that is not for the year 2000. Data is distributed randomly across all nodes and read in sequentially on the disk. The filtering happens in main memory, not on the disk, to avoid the cost of seek times.
2. Each node process proceeds to create groups for each state as they are discovered and adds the sales numbers for a given state bucket. (The application is present on all nodes, and data is processed local to a node.)
3. When all the nodes have completed the process of sweeping the sales data from the disk and computing the total sales by state numbers, they send their respective number to a designated node (we call this node the *assembler node*), which has been agreed upon by all nodes at the beginning of the process.
4. The designated assembler node assembles all the total sales by state number from each node and adds up the values received from each node per state.
5. The assembler node sorts the final numbers by state and delivers the results.

This process demonstrates typical features of a Big Data system: focusing on maximizing throughput (how much work gets done per unit time) over latency (how fast a request is responded to, one of the critical aspects based on which transactional systems are judged because we want the fastest possible response).

# Big Data Programming Models

The major types of Big Data programming models you will encounter are the following:

- *Massively parallel processing (MPP) database system*: EMC's Greenplum and IBM's Netezza are examples of such systems.
- *In-memory database systems*: Examples include Oracle Exalytics and SAP HANA.
- *MapReduce systems*: These systems include Hadoop, which is the most general-purpose of all the Big Data systems.
- *Bulk synchronous parallel (BSP) systems*: Examples include Apache HAMA and Apache Giraph.

## Massively Parallel Processing (MPP) Database Systems

At its core, MPP systems employ some form of splitting data based on values contained in a column or a set of columns. For example, in the earlier example in which sales for the year 2000 ordered by state were computed, we could have partitioned the data by state, so certain nodes would contain data for certain states. This method of partitioning would enable each node to compute the total sales for the year 2000.

The limitation of such a system should be obvious. You need to decide how the data will be split at design time. The splitting criteria chosen will often be driven by the underlying use-case. As such, it is not suitable for *ad hoc* querying. Certain queries will execute at a blazing fast speed because they can take advantage of how the data is split between nodes. Others will operate at a crawl speed because the data is not distributed in a manner consistent with how it is accessed to execute the query resulting in data needed to be transferred to the nodes over the network.

To handle this limitation, it is common for such systems to store the data multiple times, split by different criteria. Depending on the query, the appropriate dataset is picked.

Following is the way in which the MPP programming model meets the attributes defined earlier for Big Data systems (consider the sales ordered by the state example):

- Data is split by state on separate nodes.
- Each node contains all the necessary application libraries to work on its own subset of the data.
- Each node reads data local to itself. An exception is when you apply a query that does not respect how the data is distributed; in this case, each task needs to fetch its own data from other nodes over the network.
- Data is read sequentially for each task. All the sales data is co-located and swept off the disk. The filter (year = 2000) is applied in memory.

## In-Memory Database Systems

From an operational perspective, in-memory database systems are identical to MPP systems. The implementation difference is that each node has a significant amount of memory, and most data is preloaded into memory. SAP HANA operates on this principle. Other systems, such as Oracle Exalytics, use specialized hardware to ensure that multiple hosts are housed in a single appliance. At the core, an in-memory database is like an in-memory MPP database with a SQL interface.

One of the major disadvantages of the commercial implementations of in-memory databases is that there is a considerable hardware and software lock-in. Also, given that the systems use proprietary and very specialized hardware, they are usually expensive. Trying to use commodity hardware for in-memory databases increases the size of the cluster very quickly. Consider, for example, a commodity server that has 25 GB of RAM. Trying to host 1 TB in-memory databases will need more than 40 hosts (accounting for other activities that need to be performed on the server). 1 TB is not even that big, and we are already up to a 40-node cluster.

The following describes how the in-memory database programming model meets the attributes we defined earlier for the Big Data systems:

- Data is split by state in the earlier example. Each node loads data into memory.
- Each node contains all the necessary application libraries to work on its own subset.
- Each node reads data local to its nodes. The exception is when you apply a query that does not respect how the data is distributed; in this case, each task needs to fetch its own data from other nodes.
- Because data is cached in memory, the Sequential Data Read attribute does not apply except when the data is read into memory the first time.

## MapReduce Systems

MapReduce is the paradigm on which this book is based. It is by far the most general-purpose of four methods. Some of the important characteristics of Hadoop's implementation of MapReduce are the following:

- It uses commodity scale hardware. Note that commodity scale does not imply laptops or desktops. The nodes are still enterprise scale, but they use commonly available components.
- Data does not need to be partitioned among nodes based on any predefined criteria.
- The user needs to define only two separate processes: map and reduce.

We will discuss MapReduce extensively in this book. At a very high level, a MapReduce system needs the user to define a map process and a reduce process. When Hadoop is being used to implement MapReduce, the data is typically distributed in 64 MB–128 MB blocks, and each block is replicated twice (a replication factor of 3 is the default in Hadoop). In the example of computing sales for the year 2000 and ordered by state, the entire sales data would be loaded into the Hadoop Distributed File System (HDFS) as blocks (64 MB–128 MB in size). When the MapReduce process is launched, the system would first transfer all the application libraries (comprising the user-defined map and reduce processes) to each node.

Each node will schedule a map task that sweeps the blocks comprising the sales data file. Each Mapper (on the respective node) will read records of the block and filter out the records for the year 2000. Each Mapper will then output a record comprised of a key/value pair. *Key* will be the state and *value* will be the sales number from the given record if the sales record is for the year 2000.

Finally, a configurable number of Reducers will receive the key/value pairs from each of the Mappers. Keys will be assigned to specific Reducers to ensure that a given key is received by one and only one Reducer. Each Reducer will then add up the sales value number for all the key/value pairs received. The data format received by the Reducer is key (state), and a list of values for that key (sales records for the year 2000). The output is written back to the HDFS. The client will then sort the result by states after reading it from the HDFS. The last step can be delegated to the Reducer because the Reducer receives its assigned keys in the sorted order. In this example, we need to restrict the number of Reducers to one to achieve this, however. Because communication between Mappers and Reducers causes network I/O, it can lead to bottlenecks. We will discuss this issue in detail later in the book.

This is how the MapReduce programming model meets the attributes defined earlier for the Big Data systems:

- Data is split into large blocks on HDFS. Because HDFS is a distributed file system the data blocks are distributed across all the nodes redundantly.
- The application libraries, including the map and reduce application code, are propagated to all the task nodes.
- Each node reads data local to its nodes. Mappers are launched on all the nodes and read the data blocks local to themselves (in most cases, the mapping between tasks and disk blocks is up to the scheduler, which may allocate remote blocks to map tasks to keep all nodes busy).
- Data is read sequentially for each task on large block at a time (blocks are typically of size 64 MB–128 MB)

One of the important limitations of the MapReduce paradigm is that it is not suitable for iterative algorithms. A vast majority of data science algorithms are iterative by nature and eventually converge to a solution. When applied to such algorithms, the MapReduce paradigm requires each iteration to be run as a separate MapReduce job, and each iteration often uses the data produced by its previous iteration. But because each MapReduce job reads fresh from the persistent storage, the iteration needs to store its results in persistent storage for the next iteration to work on. This process leads to unnecessary I/O and significantly impacts the overall throughput. This limitation is addressed by the BSP class of systems, described next.

## Bulk Synchronous Parallel (BSP) Systems

The BSP class of systems operates very similarly to the MapReduce approach. However, instead of the MapReduce job terminating at the end of its processing cycle, the BSP system is composed of a list of processes (identical to the map processes) that synchronize on a barrier, send data to the Master node, and exchange relevant information. Once the iteration is completed, the Master node will indicate to each processing node to resume the next iteration.

Synchronizing on a barrier is a commonly used concept in parallel programming. It is used when many threads are responsible for performing their own tasks, but need to agree on a checkpoint before proceeding. This pattern is needed when all threads need to have completed a task up to a certain point before the decision is made to proceed or abort with respect to the rest of the computation (in parallel or in sequence). Synchronization barriers are used all the time in the real world processes. Example, carpool mates often meet at a designated place before proceeding in a single car. The overall process is only as fast as the last person (or thread) arriving at the barrier.

The BSP method of execution allows each map-like process to cache its previous iteration's data significantly improving the throughput of the overall process. We will discuss BSP systems in the Data Science chapter of this book. They are relevant to iterative algorithms.

# Big Data and Transactional Systems

It is important to understand how the concept of transactions has evolved in the context of Big Data. This discussion is relevant to NoSQL databases. Hadoop has HBase as its NoSQL data store. Alternatively, you can use Cassandra or NoSQL systems available in the cloud such as Amazon Dynamo.

Although most RDBMS users expect ACID properties in databases, these properties come at a cost. When the underlying database needs to handle millions of transactions per second at peak time, it is extremely challenging to respect ACID features in their purest form.

---

■ **Note** *ACID* is an acronym for *atomicity, consistency, isolation, and durability.* A detailed discussion can be found at the following link: `http://en.wikipedia.org/wiki/ACID`.

---

Some compromises are necessary, and the motivation behind these compromises is encapsulated in what is known as the CAP theorem (also known as Brewer's theorem). *CAP* is an acronym for the following:

- *Consistency*: All nodes see the same copy of the data at all times.
- *Availability*: A guarantee that every request receives response about success and failure within a *reasonable and well-defined* time interval.
- *Partition tolerance*: The system continues to perform despite failure of its parts.

The theorem goes on to prove that in any system only two of the preceding features are achievable, not all three. Now, let's examine various types of systems:

- *Consistent and available*: A single RDBMS with ACID properties is an example of a system that is consistent and available. It is not partition-tolerant; if the RDBMS goes down, users cannot access the data.
- *Consistent and partition-tolerant*: A clustered RDBMS is such as system. Distributed transactions ensure that all users will always see the same data (consistency), and the distributed nature of the data will ensure that the system remains available despite loss of nodes. However, by virtue of distributed transactions, the system will be unavailable for durations of time when two-phase commits are being issued. This limits the number of simultaneous transactions that can be supported by the system, which in turn limits the availability of the system.
- *Available and partition-tolerant*: The type of systems classified as "eventually consistent" fall into this category. Consider a very popular e-commerce web site such as `Amazon.com`. Imagine that you are browsing through the product catalogs and notice that two units of a certain item are available for sale. By nature of the buying process, you are aware that between you noticing that a certain number of items are available and issuing the buy request, someone could come in first and buy the items. So there is little incentive for always showing the most updated value because inventory changes. Inventory changes will be propagated to all the nodes serving the users. Preventing the users from browsing inventory while this propagation is taking place in order to provide the most current value of the inventory will limit the availability of the web site, resulting in lost sales. Thus, we have sacrificed consistency for

availability, and partition tolerance allows multiple nodes to display the same data (although there may be a small window of time in which each user sees different data, depending on the nodes they are served by).

These decisions are very critical when developing Big Data systems. MapReduce, which is the main topic of the book, is only one of the components of the Big Data ecosystem. Often it exists in the context of other products such as HBase, in which making the trade-offs discussed in this section are critical to developing a good solution.

# How Much Can We Scale?

We made several assumptions in our examples earlier in the chapter. For example, we ignored CPU time. For a large number of business problems, computational complexity does not dominate. However, with the growth in computing capability, various domains became practical from an implementation point of view. One example is data mining using complex Bayesian statistical techniques. These problems are indeed computationally expensive. For such problems, we need to increase the number of nodes to perform processing or apply alternative methods.

---

▒ **Note** The paradigms used in Big Data computing such as MapReduce have also been extended to other parallel computing methods. For example, *general-purpose computation on graphics programming units (GPGPU)* computing achieves massive parallelism for compute-intensive problems.

---

We also ignored network I/O costs. Using 50 compute nodes to process data also requires the use of a distributed file system and communication costs for assembling data from 50 nodes in the cluster. In all Big Data solutions, I/O costs will dominate. These costs introduce serial dependencies in the computational process.

## A Compute-Intensive Example

Consider processing 200 GB of data with 50 nodes, in which each node processes 4 GB of data located on a local disk. Each node takes 80 seconds to read the data (at the rate of 50 MB per second). No matter how fast we compute, we cannot finish in under 80 seconds. Assume that the result of the process is a total dataset of size 200 MB, and each node generates 4 MB of this result. which is transferred over a 1 Gbps (1 MB per packet) network to a single node for display. It will take about 3 milliseconds (each 1 MB requires 250 microseconds to transfer over the network, and the network latency per packet is assumed to be 500 microseconds (based on the previously referenced talk by Dr. Jeff Dean) to transfer the data to the destination node. Ignoring computational costs, the total processing time cannot be under 40.003 seconds.

Now imagine that we have 4000 nodes, and magically each node reads its own 500 MB of data from a local disk and produces 0.1 MB of result set. Notice that we cannot go faster than 1 second if data is read in 50 MB blocks. This translates to maximum performance improvement by a factor of about 4000. In other words for a certain class of problems, if it takes 4000 hours to complete the processing, we cannot do better than 1 hour, no matter how many nodes are thrown at the problem. A factor of 4000 might sound like a lot, but there is an upper limit to how fast we can get. In this simplistic example, we have made many simplifying system assumptions. We also assumed that there are no serial dependencies in the application logic, which is usually a false assumption. Once we add those costs, the maximum performance gain possibly falls drastically.

Serial dependencies, which are the bane of all parallel computing algorithms, limit the degree of performance improvement. The limitation is well known and documented as the Amdhal's Law.

## Amdhal's Law

Just as the speed of light defines the theoretical limit of how fast we can travel in our universe, Amdhal's Law defines the limits of performance gain we can achieve by adding more nodes to clusters.

---

■ **Note** See `http://en.wikipedia.org/wiki/Amdahl's_law` for a full discussion of Amdhal's Law.

---

In a nutshell, the law states that if a given solution can be made perfectly parallelizable up to a proportion P (where P ranges from 0 to 1), the maximum performance improvement we can obtain given an infinite number of nodes (a fancy way of saying a lot of nodes in the cluster) is 1/(1-P). Thus, if we have even 1 percent of the execution that cannot be made, parallel the best improvement we can get is 100 fold. All programs have some serial dependencies, and disk I/O and network I/O will add more. There are limits to how many improvements we can achieve regardless of the methods we use.

# Business Use-Cases for Big Data

Big Data and Hadoop have several applications in the business world. At the risk of sounding cliché, the three big attributes of Big Data are considered to be these:

- Volume
- Velocity
- Variety

*Volume* relates the size of the data processed. If your organization needs to extract, load, and transform 2 TB of data in 2 hours each night, you have a volume problem.

*Velocity* relates to speed at which large data arrives. Organizations such as Facebook and Twitter encounter the velocity problem. They get massive amounts of tiny messages per second that need to be processed almost immediately, posted to the social media sites, propagated to related users (family, friends, and followers), events generated, and so on.

*Variety* is related to an increasing number of formats that need to be processed. Enterprise search systems have become commonplace in organizations. Open-source software such as Apache Solr has made search-based systems ubiquitous. Most unstructured data is not stand-alone; it has considerable structured data associated with it. For example, consider a simple document such as an e-mail. E-mail has considerable metadata associated with it. Examples include sender, receivers, order of receivers, time sent/received, organizational information about the senders/receivers (for example, a title at the time of sending), and so on.

Some of this information is even dynamic. For example, if you are analyzing years of e-mail (Area of Legal Practice has several use-cases around this), it is important to know what the title of senders or receivers were when the e-mail was first sent. This feature of dynamic master data is commonplace and leads to several interesting challenges.

Big Data helps solve everyday problems such as large-scale extract, transform, load (ETL) issues by using commodity software and hardware. In particular, open-source Hadoop, which runs on commodity servers and can scale by adding more nodes, enables ETL (or *ELT*, as it is commonly called in the Big Data domain) to be performed significantly faster at commodity costs.

Several open-source products have evolved around Hadoop and the HDFS to support velocity and variety use-cases. New data formats have evolved to manage the I/O performance around massive data processing. This book will discuss the motivations behind such developments and the appropriate use-cases for them.

Storm (which evolved at Twitter) and Apache Flume (designed for large-scale log analysis) evolved to handle the velocity factor. The choice of which software to use depends on how close to "real time" the processes need to be. Storm is useful for tackling problems that require "more real-time" processing than Flume.

The key message is this: Big Data is an ecosystem of various products that work in concert to solve very complex business problems. Hadoop is often at the center of such solutions. Understanding Hadoop enables you to develop a strong understanding of how to use the other entrants in the Big Data ecosystem.

# Summary

Big Data has now become mainstream, and the two main drivers behind it are open-source Hadoop software and the advent of the cloud. Both of these developments allowed the mass-scale adoption of Big Data methods to handle business problems at low cost. Hadoop is the cornerstone of all Big Data solutions. Although other programming models, such as MPP and BSP, have sprung up to handle very specific problems, they all depend on Hadoop in some form or other when the scale of data to be processed reaches a multiterabyte scale. Developing a deep understanding of Hadoop enables users of other programming models to be more effective. The goal of this book is to you achieve that.

The chapters to come will guide you through the specifics of using the Hadoop software as well as offer practical methods for solving problems with Hadoop.

CHAPTER 2

# Hadoop Concepts

Applications frequently require more resources than are available on an inexpensive (commodity) machine. Many organizations find themselves with business processes that no longer fit on a single, cost-effective computer. A simple but expensive solution has been to buy specialty machines that cost a lot of memory and have many CPUs. This solution scales as far as what is supported by the fastest machines available, and usually the only limiting factor is your budget. An alternative solution is to build a high-availability cluster, which typically attempts to look like a single machine and usually requires very specialized installation and administration services. Many high-availability clusters are proprietary and expensive.

A more economical solution for acquiring the necessary computational resources is cloud computing. A common pattern is to have bulk data that needs to be transformed, in which the processing of each data item is essentially independent of other data items; that is, by using a single-instruction, multiple-data (SIMD) scheme. Hadoop provides an open-source framework for cloud computing, as well as a distributed file system.

This book is designed to be a practical guide to developing and running software using Hadoop, a project hosted by the Apache Software Foundation. This chapter introduces you to the core Hadoop concepts. It is meant to prepare you for the next chapter, in which you will get Hadoop installed and running.

## Introducing Hadoop

Hadoop is based on the Google paper on MapReduce published in 2004, and its development started in 2005. At the time, Hadoop was developed to support the open-source web search engine project called Nutch. Eventually, Hadoop separated from Nutch and became its own project under the Apache Foundation.

Today Hadoop is the best-known MapReduce framework in the market. Currently, several companies have grown around Hadoop to provide support, consulting, and training services for the Hadoop software.

At its core, Hadoop is a Java-based MapReduce framework. However, due to the rapid adoption of the Hadoop platform, there was a need to support the non-Java user community. Hadoop evolved into having the following enhancements and subprojects to support this community and expand its reach into the Enterprise:

- *Hadoop Streaming*: Enables using MapReduce with any command-line script. This makes MapReduce usable by UNIX script programmers, Python programmers, and so on for development of *ad hoc* jobs.
- *Hadoop Hive*: Users of MapReduce quickly realized that developing MapReduce programs is a very programming-intensive task, which makes it error-prone and hard to test. There was a need for more expressive languages such as SQL to enable users to focus on the problem instead of low-level implementations of typical SQL artifacts (for example, the `WHERE` clause, `GROUP BY` clause, `JOIN` clause, etc.). Apache Hive evolved to provide a data warehouse (DW) capability to large datasets. Users can express their queries in Hive Query Language, which is very similar to SQL. The Hive engine converts these queries to low-level MapReduce jobs

transparently. More advanced users can develop user-defined functions (UDFs) in Java. Hive also supports standard drivers such as ODBC and JDBC. Hive is also an appropriate platform to use when developing Business Intelligence (BI) types of applications for data stored in Hadoop.

- *Hadoop Pig*: Although the motivation for Pig was similar to Hive, Hive is a SQL-like language, which is declarative. On the other hand, Pig is a procedural language that works well in data-pipeline scenarios. Pig will appeal to programmers who develop data-processing pipelines (for example, SAS programmers). It is also an appropriate platform to use for extract, load, and transform (ELT) types of applications.

- *Hadoop HBase*: All the preceding projects, including MapReduce, are batch processes. However, there is a strong need for real-time data lookup in Hadoop. Hadoop did not have a native key/value store. For example, consider a Social Media site such as Facebook. If you want to look up a friend's profile, you expect to get an answer immediately (not after a long batch job runs). Such use-cases were the motivation for developing the HBase platform.

We have only just scratched the surface of what Hadoop and its subprojects will allow us to do. However the previous examples should provide perspective on why Hadoop evolved the way it did. Hadoop started out as a MapReduce engine developed for the purpose of indexing massive amounts of text data. It slowly evolved into a general-purpose model to support standard Enterprise use-cases such as DW, BI, ELT, and real-time lookup cache. Although MapReduce is a very useful model, it was the adaptation to standard Enterprise use-cases of the type just described (ETL, DW) that enabled it to penetrate the mainstream computing market. Also important is that organizations are now grappling with processing massive amounts of data.

For a very long time, Hadoop remained a system in which users submitted jobs that ran on the entire cluster. Jobs would be executed in a First In, First Out (FIFO) mode. However, this lead to situations in which a long-running, less-important job would hog resources and not allow a smaller yet more important job to execute. To solve this problem, more complex job schedulers in Hadoop, such as the Fair Scheduler and Capacity Scheduler were created. But Hadoop 1.x (prior to version 0.23) still had scalability limitations that were a result of some deeply entrenched design decisions.

Yahoo engineers found that Hadoop had scalability problems when the number of nodes (`http://developer.yahoo.com/blogs/hadoop/scaling-hadoop-4000-nodes-yahoo-410.html`) increased to an order of a few thousand. As these problems became better understood, the Hadoop engineers went back to the drawing board and reassessed some of the core assumptions underlying the original Hadoop design; eventually this lead to a major design overhaul of the core Hadoop platform. Hadoop 2.x (from version 0.23 of Hadoop) is a result of this overhaul.

This book will cover version 2.x with appropriate references to 1.x, so you can appreciate the motivation for the changes in 2.x.

# Introducing the MapReduce Model

Hadoop supports the MapReduce model, which was introduced by Google as a method of solving a class of petascale problems with large clusters of commodity machines. The model is based on two distinct steps, both of which are custom and user-defined for an application:

- *Map*: An initial ingestion and transformation step in which individual input records can be processed in parallel

- *Reduce*: An aggregation or summarization step in which *all associated* records must be processed together by a single entity

The core concept of MapReduce in Hadoop is that input can be split into logical chunks, and each chunk can be initially processed independently by a map task. The results of these individual processing chunks can be physically partitioned into distinct sets, which are then sorted. Each sorted chunk is passed to a reduce task. Figure 2-1 illustrates how the MapReduce model works.

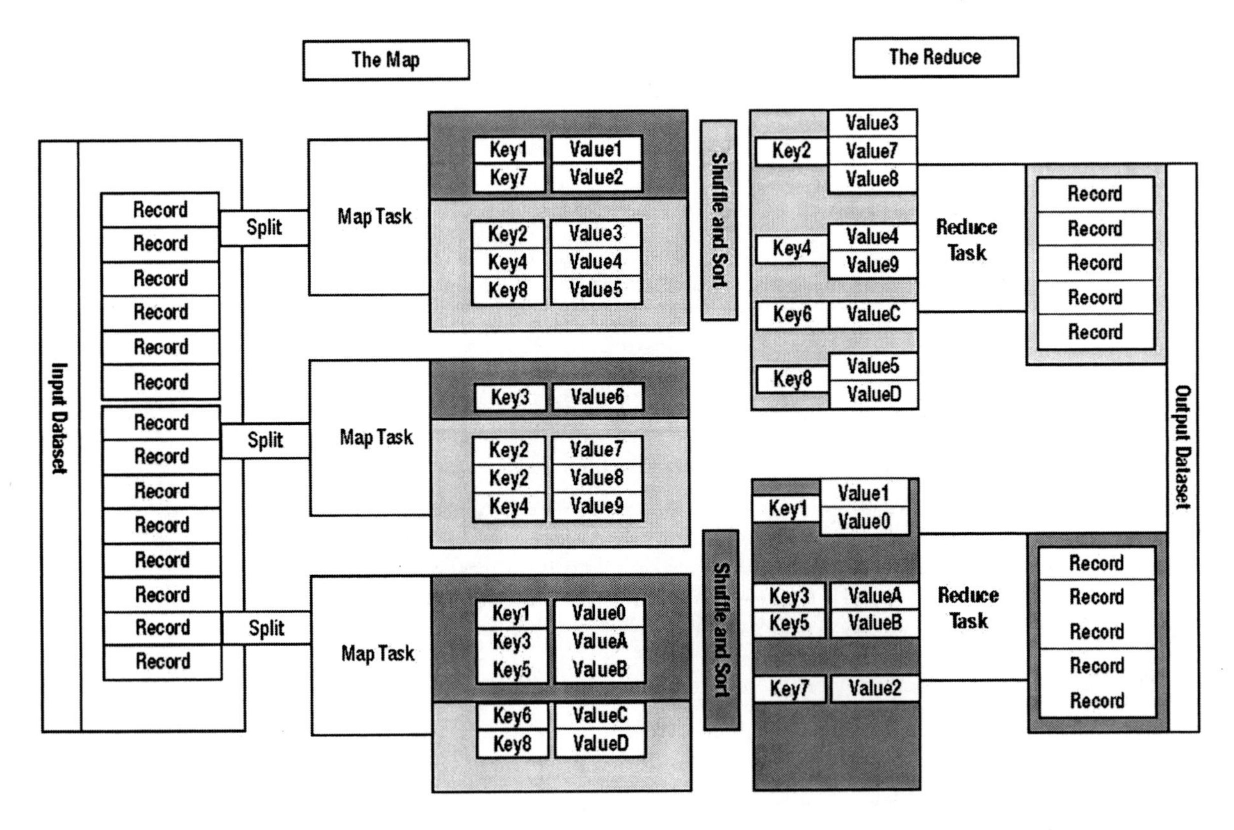


***Figure 2-1.*** *MapReduce model*

A map task can run on any compute node in the cluster, and multiple map tasks can run in parallel across the cluster. The map task is responsible for transforming the input records into key/value pairs. The output of all the maps will be partitioned, and each partition will be sorted. There will be one partition for each reduce task. Each partition's sorted keys and the values associated with the keys are then processed by the reduce task. There can be multiple reduce tasks running in parallel on the cluster.

Typically, the application developer only to provide only four items to the Hadoop framework: the class that will read the input records and transform them into one key/value pair per record, a Mapper class, a Reducer class, and a class that will transform the key/value pairs that the reduce method outputs into output records.

Let's illustrate the concept of MapReduce using what has now become the "Hello-World" of the MapReduce model: the word-count application.

Imagine that you have a large number of text documents. Given the increasing interest in analyzing unstructured data, this situation is now relatively common. These text documents could be Wikipedia pages downloaded from the following web site `http://dumps.wikimedia.org/`. Or they could be a large organization's e-mail archive analyzed for legal purposes (for example, the Enron Email Dataset: `www.cs.cmu.edu/~enron/`). There are many interesting analyses you can perform on text (for example, information extraction, document clustering based on content, and document classification based on sentiment). However, most such analyses begin with getting a count of each word in the document corpus (a collection of documents is often referred to as a *corpus*). One reason is to compute the term-frequency/inverse-document -frequency (TF/IDF) for a word/document combination.

---

■ **Note** A good discussion on TF/IDF and some related applications can be found at the following link: `http://en.wikipedia.org/wiki/Tf-idf`.

---

Intuitively, it should be easy to do so. Assume for simplicity that each document comprises words separated by spaces. A straightforward solution is this:

1. Maintain a hashmap whose key is a "word," and value is the count of the word.
2. Load each document in memory.
3. Split each document into words.
4. Update the global hashmap for every word in the document.
5. After each document is processed, we have the count for all words.

Most corpora have unique word counts that run into a few million, so the previous solution is logically workable. However, the major caveat is the size of the data (after all, this book is about Big Data). When the document corpus is of terabyte scale, it can take hours or even a few days to complete the process on a single node.

Thus, we use MapReduce to tackle the problem when the scale of data is large. Take note; this is the usual scenario you will encounter; you have a pretty straightforward problem that simply will not scale on a single machine. You should use MapReduce.

The MapReduce implementation of the above solution is the following:

1. A large cluster of machine is provisioned. We assume a cluster size of 50, which is quite typical in a production scenario.
2. A large number of map processes run on each machine. A reasonable assumption is there will be as many map processes as there are files. This assumption will be relaxed in the later sections (when we talk about compression schemes and alternative file formats such as sequence files) but let's go with it for now. Assume that there are 10 million files; there will be 10 million map processes started. At a given time, we assume that there are as many map processes running as there are CPU cores. Given a dual quad-core CPU machine, we assume that eight Mappers run simultaneously, so each machine is responsible for running 200,000 map processes. Thus there are 25,000 iterations (each iteration runs eight Mappers per iteration, one on each of its cores) of eight Mappers running on each machine during the processing.
3. Each Mapper processes a file, extracts words, and emits the following key/value pair: `<{WORD},1>`. Examples of Mapper output are these:
    - `<the,1>`
    - `<the,1>`
    - `<test,1>`
4. Assume that we have a single Reducer. Again, this is not a requirement; it is the default setting. This default needs to be changed frequently in practical scenarios, but it is appropriate for this example.

5. The Reducer receives key/value pairs that have the following format: <{WORD},[1,....1]>. That is, key/value pairs received by the Reducer is such that the key is a word emitted from any of the Mappers (<WORD>, and the value is a list of values ([1,....1]) emitted by any of the Mappers for the word. Examples of Reducer input key/values are these:
    - <the,[1,1,1,...,1]>
    - <test,[1,1]>
6. The Reducer simply add up the 1s to provide a final count of the {WORD} and send the result to the output as the following key/value pair: <{WORD},{COUNT OF WORD}>. Examples of the Reducer output are these:
    - <the,1000101>
    - <test,2>

The key to receiving a list of values for a key in the reduce phase is a phase known as the sort/shuffle phase in MapReduce. All the key/value pairs emitted by the Mapper are sorted by the key in the Reducer. If multiple Reducers are allocated, a subset of keys will be allocated to each Reducer. The key/value pairs *for a given* Reducer are sorted by key, which ensures that all the values associated with one key are received by the Reducer together.

---

■ **Note** The Reducer phase does not actually create a list of values before the reduce operation begins for each key. This would require too much memory for typical stop words in the English language. Suppose that 10 million documents have 20 occurrences of the word *the* in our example. We would get a list of 200 million *1*s for the word *the*. This would easily overwhelm the Java Virtual Machine (JVM) memory for the Reducer. Instead, the sort/shuffle phase accumulates the *1*s together for the word *the* in a local file system of the Reducer. When the Reducer operation initiates for the word *the*, the *1*s simply stream out through the Java iterator interface.

---

Figure 2-2 shows the logical flow of the process just described.

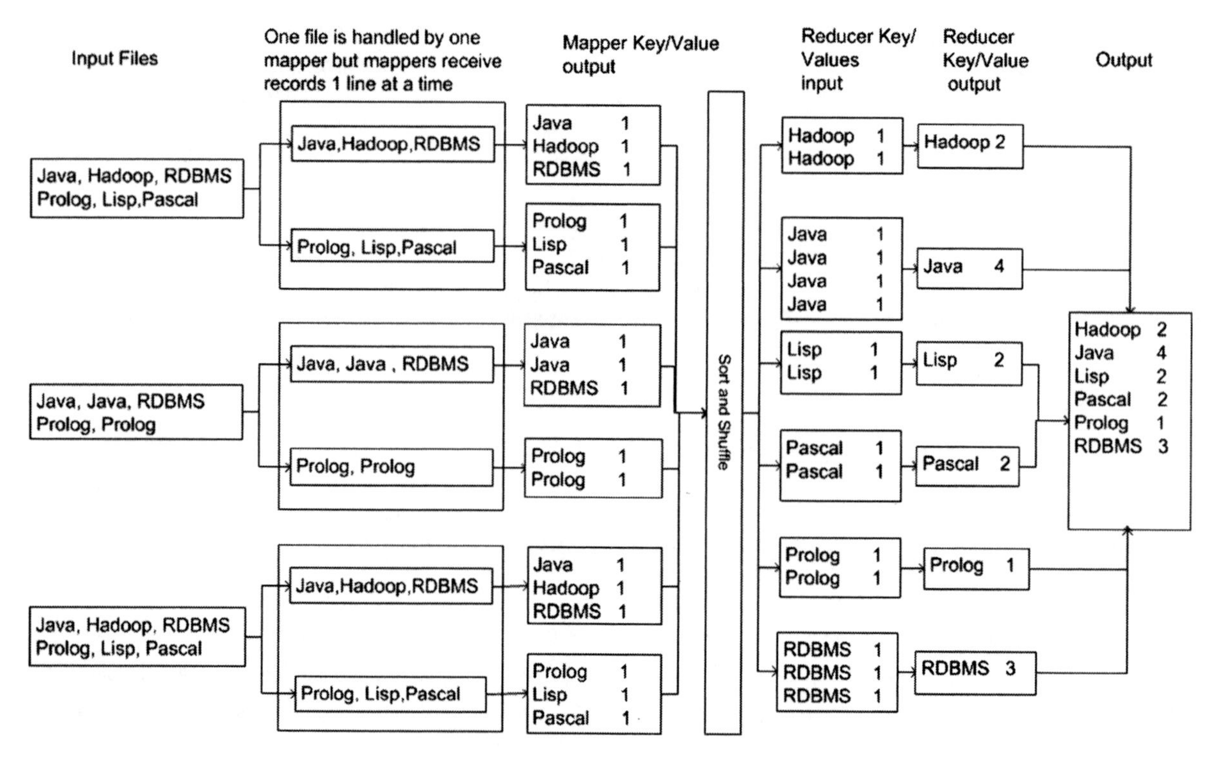


***Figure 2-2.*** *Word count MapReduce application*

At this point you are probably wondering how each Mapper accesses its file. Where is the file stored? Does each Mapper get it from a network file system (NFS)? It does not! Remember from Chapter 1 that reading from the network is an order of magnitude slower than reading from a local disk. Thus, the Hadoop system is designed to ensure that most Mappers read the file from a local disk. This means that the entire corpus of documents in our case is distributed across 50 nodes. However, the MapReduce system sees a unified single file system, although the overall design of the HDFS allows each file to be network-switch-aware to ensure that work is effectively scheduled to disk local processes. This is the famous Hadoop Distributed File System (HDFS) We will discuss the HDFS in more detail in the following sections.

# Components of Hadoop

We will begin a deep dive into various components of Hadoop in this section. We will begin with Hadoop 1.x components and eventually discuss the new 2.x components. At a very high level, Hadoop 1.x has following daemons:

- *NameNode*: Maintains the metadata for each file stored in the HDFS. Metadata includes the information about blocks comprising the file as well their locations on the DataNodes. As you will soon see, this is one of the components of 1.x that becomes a bottleneck for very large clusters.
- *Secondary NameNode*: This is not a backup NameNode. In fact, it is a poorly named component of the Hadoop platform. It performs some housekeeping functions for the NameNode.
- *DataNode*: Stores the actual blocks of a file in the HDFS on its own local disk.

- *JobTracker*: One of the master components, it is responsible for managing the overall execution of a job. It performs functions such as scheduling child tasks (individual Mapper and Reducer) to individual nodes, keeping track of the health of each task and node, and even rescheduling failed tasks. As we will soon demonstrate, like the NameNode, the Job Tracker becomes a bottleneck when it comes to scaling Hadoop to very large clusters.
- *TaskTracker*: Runs on individual DataNodes and is responsible for starting and managing individual Map/Reduce tasks. Communicates with the JobTracker.

Hadoop 1.x clusters have two types of nodes: master nodes and slave nodes. Master nodes are responsible for running the following daemons:

- NameNode
- Secondary NameNode
- JobTracker

Slave nodes are distributed across the cluster and run the following daemons:

- DataNode
- TaskTracker

Although only one instance of each of the master daemons runs on the entire cluster, there are multiple instances of the DataNode and TaskTracker. On a smaller or development/test cluster, it is typical to have all the three master daemons run on the same machine. For production systems or large clusters, however, it is more prudent to keep them on separate nodes.

# Hadoop Distributed File System (HDFS)

The HDFS is designed to support applications that use very large files. Such applications write data once and read the same data many times.

The HDFS is a result of the following daemons acting in concert:

- NameNode
- Secondary NameNode
- DataNode

The HDFS has master/slave architecture. The NameNode is the master node, and the DataNodes are the slave nodes. Usually, a DataNode daemon runs on each slave node. It manages the storage attached to each DataNode. The HDFS exposes a file system namespace and allows data to be stored on a cluster of nodes while providing the user a single system view of the file system. The NameNode is responsible for managing the metadata for the files.

## Block Storage Nature of Hadoop Files

First, you should understand how files are physically stored in the cluster. In Hadoop, each file is broken into multiple blocks. A typical block size is 64 MB, but it is not atypical to configure block sizes of 32 MB or 128 MB. Block sizes can be configured per file in the HDFS. If the file is not an exact multiple of the block size, the space is not wasted, and the last block is just smaller than the total block size. A large file will be broken up into multiple blocks.

Each block is stored on a DataNode. It is also replicated to ensure against failure. The default replication factor in Hadoop is 3. A rack-aware Hadoop system stores one block on one node in the local rack (assuming that the Hadoop client is running on one of the DataNodes; if not, the rack is chosen randomly). The second replica is placed on a node of a different remote rack, and the last node is placed on a node in the same remote rack. A Hadoop system is made rack-aware by configuring the rack to node Domain Name System (DNS) name mapping in a separate network topology file, the path of which is referenced through the Hadoop configuration files.

---

■ **Note** Some Hadoop systems can drop the replication factor to 2. One example is Hadoop running on the EMC Isilon hardware. The underlying rationale is that the hardware uses RAID 5, which provides a built-in redundancy, enabling a drop in replication factor. Dropping the replication factor has obvious benefits because it enables faster I/O performance (writing 1 replica less). The following white paper illustrates the design of such systems: `www.emc.com/collateral/software/white-papers/h10528-wp-hadoop-on-isilon.pdf`.

---

Why not just place all three replicas on different racks? After all, it would only increase the redundancy. It would further ensure against rack failure as well as improve rack throughput. However, the possibility of rack failures over node failure is far less, and attempting to save replicas to multiple racks only degrades the write performance. Hence, a trade-off is made to save two replicas to nodes on the same remote rack in return for improved performance. Such subtle design decisions motivated by performance constraints are common in the Hadoop system.

## File Metadata and NameNode

When a client requests a file or decides to store a file in HDFS, it needs to know which DataNodes to access. Given this information, the client can directly write to the individual DataNodes. The responsibility for maintaining this metadata rests with the NameNode.

The NameNode exposes a file system namespace and allows data to be stored on a cluster of nodes while allowing the user a single system view of the file system. HDFS exposes a hierarchical view of the file system with files stored in directories, and directories can be nested. The NameNode is responsible for managing the metadata for the files and directories.

The NameNode manages all the operations such as file/directory open, close, rename, move, and so on. The DataNodes are responsible for serving the actual file data. This is an important distinction! When a client requests or sends data, the data does not physically go through NameNode. This would be huge bottleneck. Instead, the client simply gets the metadata about the file from the NameNode and fetches the file blocks directly from the nodes.

Some of the metadata stored by the NameNode includes these:

- File/directory name and its location relative to the parent directory.
- File and directory ownership and permissions.
- File name of individual blocks. Each block is stored as a file in the local file system of the DataNode in the directory that can be configured by the Hadoop system administrator.

It should be noted that the NameNode does not store the location (DataNode identity) for each block. This information is obtained from each of the DataNodes at the time of the cluster startup. The NameNode only maintains information about which blocks (the file names of each block on the DataNode) which makes up the file in the HDFS.

The metadata is stored on the disk but loaded in memory during the cluster operation for fast access. This aspect is critical to fast operation of Hadoop, but also results in one of its major bottlenecks that inspired Hadoop 2.x.

Each item of metadata consumes about 200 bytes of RAM. Consider a 1 GB file and block size of 64 MB. Such a file requires 16 x 3 (including replicas) = 48 blocks of storage. Now consider 1,000 files of 1 MB each. This system of files requires 1000 x 3 = 3,000 blocks for storage. (Each block is only 1 MB large, but multiple files cannot be stored in a single block). Thus, the amount of metadata has increased significantly. This will result in more memory usage on the NameNode. This example should also serve to explain why Hadoop systems prefer large files over small files. A large number of small files will simply overwhelm the NameNode.

The NameNode file that contains the metadata is `fsimage`. Any changes to the metadata during the system operation are stored in memory and persisted to another file called `edits`. Periodically, the `edits` file is merged with the `fsimage` file by the Secondary NameNode. (We will discuss this process in detail when we discuss the Secondary NameNode.) These files do not contain the actual data; the actual data is stored on individual blocks in the slave nodes running the DataNode daemon. As mentioned before, the blocks are just files on the slave nodes. The block stores only the raw content, no metadata. Thus, losing the NameNode metadata renders the entire system unusable. The NameNode metadata enables clients to make sense of the blocks of raw storage on the slave nodes.

The DataNode daemons periodically send a heartbeat message to the NameNode. This enables the NameNode to remain aware of the health of each DataNode and not direct any client requests to a failed node.

## Mechanics of an HDFS Write

An HDFS write operation relates to file creation. From a client perspective, HDFS does not support file updates. (This is not entirely true because the file append feature is available for HDFS for HBase purposes. However, it is not recommended for general-purpose client use.) For the purpose of the following discussion, we will assume a default replication factor of 3.

Figure 2-3 depicts the HDFS write process in a diagram form, which is easier to take in at a glance.

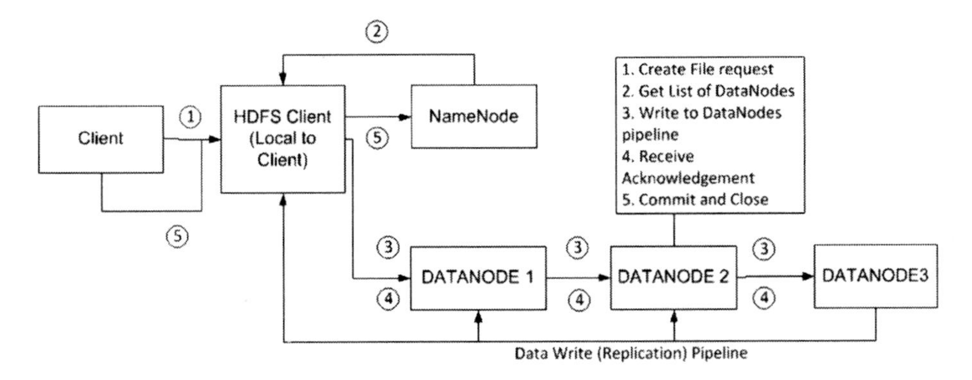


***Figure 2-3.*** *HDFS write process*

The following steps allow a client to write a file to the HDFS:

1. The client starts streaming the file contents to a temporary file in its local file system. It does this before contacting the NameNode.
2. When the file data size reaches the size of a block, the client contacts the NameNode.
3. The NameNode now creates a file in the HDFS file system hierarchy and notifies the client about the block identifier and location of the DataNodes. This list of DataNodes also contains the list of replication nodes.
4. The client uses the information from the previous step to flush the temporary file to a data block location (first DataNode) received from the NameNode. This results in the creation of an actual file on the local storage of the DataNode.
5. When the file (HDFS file as seen by the client) is closed, the NameNode commits the file and it becomes visible in the system. If the NameNode goes down before the commit is issued, the file is lost.

Step 4 deserves some added attention. The flushing process in that step operates as follows:

1. The first DataNode receives the data from the client in smaller packets (typically 4 KB in size). Although this portion is being written to the disk on the first DataNode, it starts to stream it to the second DataNode.
2. The second DataNode starts writing the streaming data block to its own disk and at the same time starts streaming the packets of the data block to the third DataNode.
3. The third DataNode now writes data to its own disk. Thus, data is written and replicated through a DataNodes in a pipelined manner.
4. Acknowledgment packets are sent back from each DataNode to the previous one in the pipeline. The first DataNode eventually sends the acknowledgment to the client node.
5. When the client receives the acknowledgment for a data block, the block is assumed to be persisted to all nodes, and it sends the final acknowledgment to the NameNode.
6. If any DataNode in the pipeline fails, the pipeline is closed. The data will still be written to the remaining DataNodes. The NameNode is made aware that the file is under-replicated and takes steps to re-replicate the data on a good DataNode to ensure adequate replication levels.
7. A checksum is also computed for each block and it is used to verify the integrity of the block. These checksums are stored in a separate hidden file in the HDFS and are used to verify the integrity of the block data when it is read back.

## Mechanics of an HDFS Read

Now we will discuss how the file is read from HDFS. The HDFS read process is depicted in Figure 2-4.

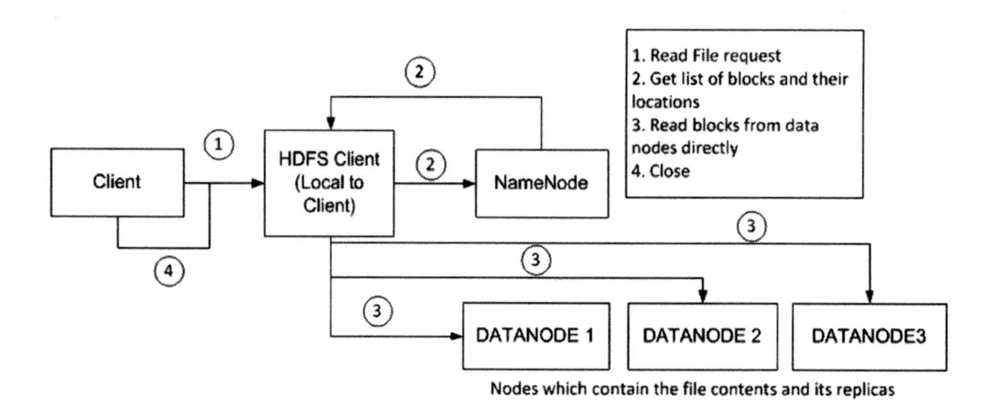


***Figure 2-4.*** *HDFS read process*

The following steps enable the file to be read by a client:

1. The client contacts the NameNode that returns the list of blocks and their locations (including replica locations).
2. The client initiates reading the block directly by contacting the DataNode. If the DataNode fails, the client contacts the DataNode hosting the replica.
3. As the block is being read, the checksum is calculated and compared with the checksum calculated at the time of the file write. If the checksum fails, the block is retrieved from the replica.

## Mechanics of an HDFS Delete

To delete a file from HDFS, follow these steps:

1. The NameNode merely renames the file path to indicate that the file is moved into the `/trash` directory. Note that the only operation occurring here is a metadata update operation linked to renaming the file path. This is a very fast process. The file stays in the `/trash` directory for a predefined interval of time (6 hours is the current setting and it is currently *not* configurable). During this time, the file can be restored easily by moving it from the `/trash` directory.
2. Once the time interval for which the file should be maintained in `/trash` directory expires, the NameNode deletes the file from the HDFS namespace.
3. The blocks making up the deleted file are freed up, and the system shows an increased available space.

The replication factor of a file is not static. It can be reduced. This information is transferred over to the NameNode via the next heartbeat message. The DataNode then actively removes the block from its local storage, which makes more space available to the cluster. Thus, the NameNode actively maintains the replication factor of each file.

## Ensuring HDFS Reliability

Hadoop and HDFS are designed to be resilient to failure. Data loss can occur in two ways:

- *DataNodes can fail*: Each DataNode periodically sends heartbeat messages to the NameNode (the default is 3 seconds). If the NameNode does not receive heartbeat messages within a predefined interval, it assumes that the DataNode has failed. At this point, it actively initiates replication of blocks stored in the lost node (obtained from one of its replicas) to a healthy node. This enables proactive maintenance of the replication factor.
- *Data can get corrupted due to a phenomenon called* bit rot: This is defined as an event in which the small electric charge that represents a "bit" disperses, resulting in loss of data. This condition can be detected only during an HDFS read operation due to "checksum" mismatch. If the checksum of the block does not match, re-replication is initiated because the block is considered corrupted, and the NameNode actively tries to restore the replication count for the block.

## Secondary NameNode

We are now ready to discuss the role of the Secondary NameNode. This component probably takes the cake for being the most misnamed component in the Hadoop platform. The Secondary NameNode is *not* a failover node.

You learned earlier that the NameNode maintains all its metadata in memory. It first reads it from the fsimage file stored in the local file system of the NameNode. During the course of the Hadoop system operation, the updates to the NameNode contents are applied in memory. However, to ensure against data loss, these edits are also applied to a local file called edits.

---

■ **Note** The fsimage file does not actually store the location of the blocks. It gets this information from each DataNode during the system startup from the DataNodes and keeps it in memory.

---

The goal of the edits file is to accumulate the changes during the system operation. If the system is restarted, the contents of the edits file can be rolled into fsimage during the restart. However, this would slow down Hadoop restarts. The Secondary NameNode was created to handle this problem.

The role of the Secondary NameNode is to periodically merge the contents of the edits file in the fsimage file. To this end, the Secondary NameNode periodically executes the following sequence of steps:

1. It asks the Primary to roll over the edits file, which ensures that new edits go to a new file. This new file is called edits.new.
2. The Secondary NameNode requests the fsimage file and the edits file from the Primary.
3. The Secondary NameNode merges the fsimage file and the edits file into a new fsimage file.
4. The NameNode now receives the new fsimage file from the Secondary NameNode with which it replaces the old file. The edits file is now replaced with the contents of the edits.new file created in the first step.
5. The fstime file is updated to record when the checkpoint operation took place.

It should now be clear why the NameNode is the single point of failure in Hadoop 1.x. If the fsimage and edits files get corrupted, all the data in the HDFS system is lost. So although the DataNodes can simply be commodity machines with JBOD (which means "just a bunch of disks"), the NameNode and the Secondary NameNode must be connected to more reliable storage (RAID-based) to ensure against the loss of data. The two files mentioned previously must also be regularly backed up. If they need to be restored on backups, all the updates between now and up until the backup was taken are lost. Table 2-1 summarizes the key files that enable the NameNode to support the HDFS.

***Table 2-1.*** *Key NameNode files*

| File Name | Description |
|---|---|
| fsimage | Contains the persisted state of the HDFS metadata as of the last checkpoint |
| edits | Contains the state changes to the HDFS metadata since the last checkpoint |
| fstime | Contains the timestamp of the last checkpoint |

## TaskTracker

The TaskTracker daemon, which runs on each compute node of the Hadoop cluster, accepts requests for individual tasks such as Map, Reduce, and Shuffle operations. Each TaskTracker is configured with a set of slots that is usually set up as the total number of cores available on the machine. When a request is received (from the JobTracker) to launch a task, the TaskTracker initiates a new JVM for the task. JVM reuse is possible, but actual usage examples of this feature are hard to come by. Most users of the Hadoop platform simply turn it off. The TaskTracker is assigned a task depending on how many free slots it has (total number of tasks = actual tasks running). The TaskTracker is responsible for sending heartbeat messages to the JobTracker. Apart from telling the JobTracker that it is healthy, these messages also tell the JobTracker about the number of available free slots.

## JobTracker

The JobTracker daemon is responsible for launching and monitoring MapReduce jobs. When a client submits a job to the Hadoop system, the sequence of steps shown in Figure 2-5 is initiated.

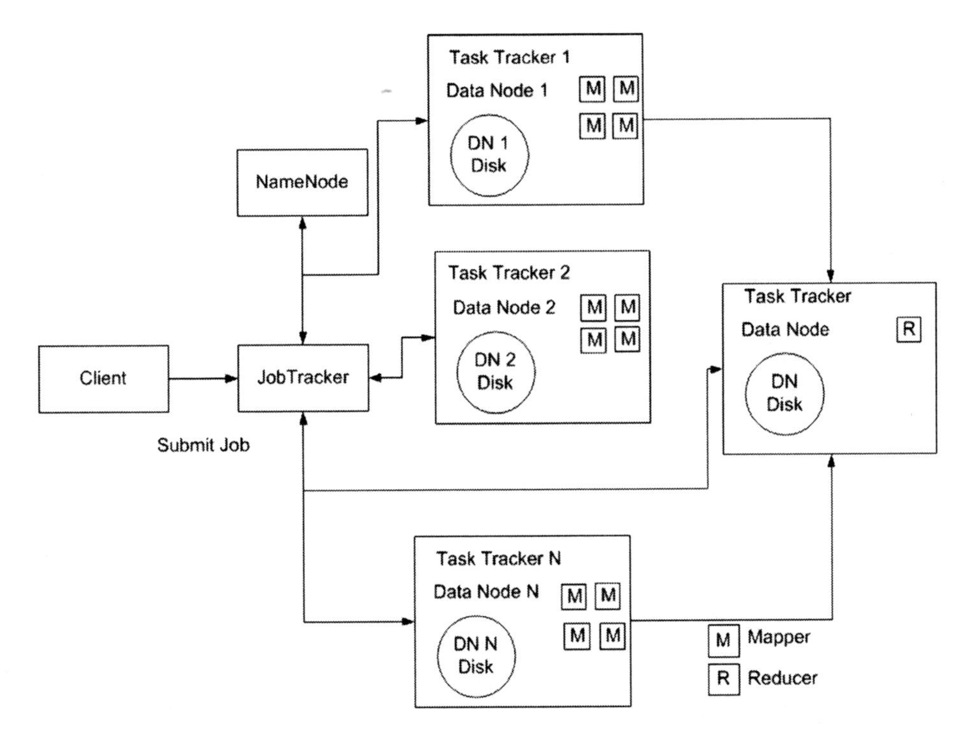


*Figure 2-5. Job submission process*

The process is detailed in the following steps:

1. The job request is received by the Job Tracker.
2. Most MapReduce jobs require one or more input directories. The Job Tracker requests the NameNode for a list of DataNodes hosting the blocks for the files contained in the list of input directories.

3. The JobTracker now plans for the job execution. During this step, the JobTracker determines the number of tasks (Map tasks and Reduce tasks) needed to execute the job. It also attempts to schedule the tasks as close to the data blocks as possible.

4. The JobTracker submits the tasks to each TaskTracker node for execution. The TaskTracker nodes are monitored for their health. They send heartbeat messages to the JobTracker node at predefined intervals. If heartbeat messages are not received for a predefined duration of time, the TaskTracker node is deemed to have failed, and the task is rescheduled to run on a separate node.

5. Once all the tasks have completed, the JobTracker updates the status of the job as successful. If a certain number of tasks fail repeatedly (the exact number is specified through configuration in the Hadoop configuration files), the JobTracker announces a job as failed.

6. Clients poll the JobTracker for updates about the Job progress.

The discussion so far on Hadoop 1.x components should have made it clear that even the JobTracker is a single point of failure. If the JobTracker goes down, so does the entire cluster with the running jobs. Also there is only a single JobTracker, which increases the load on a single JobTracker in an environment of multiple jobs running simultaneously.

# Hadoop 2.0

MapReduce has undergone a complete overhaul. The result is Hadoop 2.0, which is sometimes called MapReduce 2.0 (MR v2) or YARN. This book will often reference the version as 2.x because the point releases are not expected to change behavior and architecture in any fundamental way.

MR v2 is application programming interface (API)-compatible with MR v1, with just a recompile step. However, the underlying architecture has been rehauled completely. In Hadoop 1.x, the JobScheduler has two major functions:

- Resource management
- Job scheduling/job monitoring

YARN aims to separate these functions into separate daemons. The idea is to have a global Resource Manager and a per-application Application Master. Note, we mentioned *application*, not *job*. In the new Hadoop 2.x, an application can either be a single job in the sense of the classical MapReduce job or a Directed Acyclic Graph (DAG) of jobs. A DAG is a graph whose nodes are connected so that no cycles are possible. That is, regardless of how you traverse a graph, you cannot reach a node again in the process of traversal. In plain English, a DAG of jobs implies jobs with hierarchical relationships between each other.

YARN also aims to expand the utility of Hadoop beyond MapReduce. We discover various limitations of the MapReduce framework in the following chapters. Newer frameworks have evolved to address these limitations. For example, Apache Hive arrived to bring SQL features on top of Hadoop, Apache PIG addresses script-based, data-flow style processing. Even newer frameworks such as Apache HAMA address iterative computation, which is very typical in machine learning-style use-cases.

Spark/Shark frameworks from Berkley are a cross between Hive and HAMA, providing low-latency SQL access as well some in-memory computations. Although these frameworks are all designed to work on top of HDFS, not all are first-class citizens of the Hadoop Framework. What is needed is an over-arching framework that enables newer frameworks with varying computing philosophies (not just the MapReduce model), such as the bulk synchronous parallel (BSP) model on which HAMA is based or an in-memory caching and computation model on which Shark/Spark are based, to be integrated into the Hadoop framework.

The new framework should be designed from the ground up to support newer types of applications while still operating within the overall Hadoop System. This will enable system-wide policies around security and resource management to be applied consistently, even though all systems share the same underlying HDFS.

The YARN system has the following components:

- Global Resource Manager
- Node Manager
- Application-specific Application Master
- Scheduler
- Container

A container includes a subset of the total number of CPU cores and size of the main memory. An application will run in set of containers. An Application Master *instance* will request resources from the Global Resource Manager. The Scheduler will allocate the resources (containers) through the per-node Node Manager. The Node Manager will then report the usage of the individual containers to the Resource Manager.

The Global Resource Manager and the per-node Node Manager form the management system for the new MapReduce framework. The Resource Manager is the ultimate authority for allocating resources. Each application type has an Application Master. (For example, MapReduce is a type, and each MapReduce job is an instance of the MapReduce type, similar to the class and object relationship in object-oriented programming). For each application of the application type, an Application Master instance is instantiated. The Application Master *instance* negotiates with the Resource Manager for containers to execute the jobs. The Resource Manager utilizes the scheduler (global component) in concert with the per-node Node Manager to allocate these resources. From a system perspective, the Application Master also runs in a container.

The overall architecture for YARN is depicted in Figure 2-6.

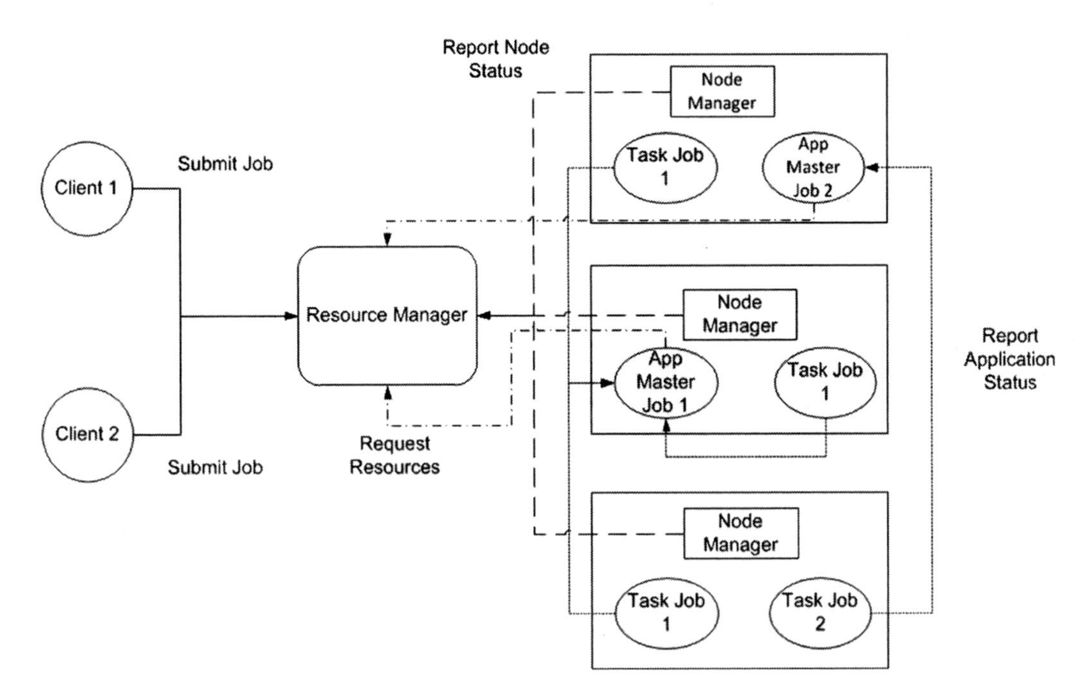


***Figure 2-6.*** *YARN architecture*

The MapReduce v1 Framework has been reused without any major modifications, which will enable backward compatibility with existing MapReduce programs.

# Components of YARN

Let's discuss each component in more detail. At a high level, we have a bunch of commodity machines set up in a Hadoop cluster. Each machine is called a *node.*

## Container

The *container* is a computational unit in the YARN framework. It is a subsystem in which a unit of work occurs. Or, in the language of MapReduce v1, it is a component in which the equivalent of a *task* executes. The relationship between a container and a node is this: one node runs several containers, but a container cannot cross a node boundary.

A container is a set of allocated system resources. Currently only two types of system resources are supported:

- CPU core
- Memory in MB

The container comprising the resources will execute on a certain node, so implicit in a container is the notion of the "resource name" that is the name of the rack and the node on which the container runs. When a container is requested, it is requested on a specific node. Thus, a container is a right conferred upon an application to use a specific number of CPU cores and a specific amount of memory on a specific host.

Any job or application (single job or DAG of jobs) will essentially run in one or more containers. The YARN framework entity that is ultimately responsible for physically allocating a container is called a Node Manager.

## Node Manager

A *Node Manager* runs on a single node in the cluster, and each node in the cluster runs its own Node Manager. It is a slave service: it takes requests from another component called the Resource Manager and allocates containers to applications. It is also responsible for monitoring and reporting usage metrics to the Resource Manager. Together with the Resource Manager, the Node Manager forms the framework responsible for managing the resource allocation on the Hadoop cluster. While the Resource Manager is a global component, the Node Manager is a per-node agent responsible for managing the health of individual nodes in the Hadoop cluster. Tasks of the Node Manager include the following:

- Receives requests from the Resource Manager and allocates containers on behalf of the job.
- Exchanges messages with the Resource Manager to ensure the smooth functioning of the overall cluster. The Resource Manager keeps track of global health based on reports received from each Node Manager, which is delegated the task of monitoring and managing its own health.
- Manages the life cycle of each launched container.
- Logs management on each node.
- Executes auxiliary services exploited by various YARN applications. For example, the Shuffle service in MapReduce is implemented as an auxiliary service in the current Hadoop implementation.

When a node starts up, it registers with the Resource Manager and tells the Resource Manager how many of resources (eventually allocated to form containers) are available. During runtime, this information is constantly updated as the Node Manager and Resource Manager work together to ensure a fully functional and optimally utilized cluster.

The Node Manager is responsible for managing only the abstract notion of containers; it does not contain any knowledge of the individual application or the application type. This responsibility is delegated to a component called the Application Master. But before we discuss the Application Master, let's briefly visit the Resource Manager.

## Resource Manager

The *Resource Manager* is primarily a scheduler: it arbitrates resources among competing applications to ensure optimal cluster utilization. The Resource Manager has a *pluggable* scheduler that is responsible for allocating resources to the various running applications, subject to familiar constraints of capacities and queues. Examples of schedulers include the Capacity Scheduler and Fair Scheduler in Hadoop, both of which you will encounter in subsequent chapters.

The actual task of creating, provisioning, and monitoring resources is delegated to the per-node Node Manager. This separation of concerns enables the Resource Manager to scale much more than the traditional JobScheduler.

## Application Master

The *Application Master* is the key differentiator between the older MapReduce v1 framework and YARN. The Application Master is an instance of a *framework-specific library*. It negotiates resources from the Resource Manager and works with the Node Manager to acquire those resources and execute its tasks. The Application Master is the component that negotiates resource containers from the Resource Manager.

The key benefits the Application Master brings to the YARN framework are these:

- Improved scalability
- A more generic framework

In MapReduce v1, the responsibility for managing the task failover rested with the JobTracker. The JobTracker also had the responsibility for allocating resources to jobs. Scalability is improved in v2 because the Resource Manager (the replacement for JobTracker) is now responsible only for scheduling. The task of managing the jobs or application rests with the Application Master. If a task fails, the Application Master will negotiate resources from the Resource Manager and attempt to re-execute the task.

In MapReduce v1, the Hadoop Framework supported only MapReduce-type jobs; it was not a generic framework. The main reason is that the key components such as JobTracker and TaskTracker were developed with the notions of Map and Reduce tasks deeply entrenched in their design. As MapReduce got more traction, people discovered that certain types of computations are not practical using MapReduce. So new frameworks, such as the BSP frameworks on which Apache HAMA and Apache Giraph are based, were developed. They did graph computations well, and they also worked well with the HDFS. As of this writing, in-memory frameworks such as Shark/Spark are gaining traction. Although they also work well with the HDFS, they do not fit into Hadoop 1.x because they are designed using a very different computational philosophy.

Introducing the Application Master approach in v2 as a part of YARN changes all that. Enabling the individual design philosophies to be embedded into an Application Master enables several frameworks to coexist in a single managed system. So while Hadoop/HAMA/Shark ran on separately managed systems on the same HDFS in Hadoop 1.x, resulting in unintended system and resource conflicts, they can now run in the same Hadoop 2.x system. They will all arbitrate resources from the Resource Manager. YARN will enable the Hadoop system to become more pervasive. Hadoop will now support more than just MapReduce-style computations, and it gets more pluggable: if new systems are discovered to work better with certain types of computations, their Application Masters can be developed and plugged in to the Hadoop system. The Application Master concept now allows Hadoop to extend beyond MapReduce and enables MapReduce to coexist and cooperate with other frameworks.

## Anatomy of a YARN Request

When a user submits a job to the Hadoop 2.x framework, the underlying YARN framework handles the request (see Figure 2-7).

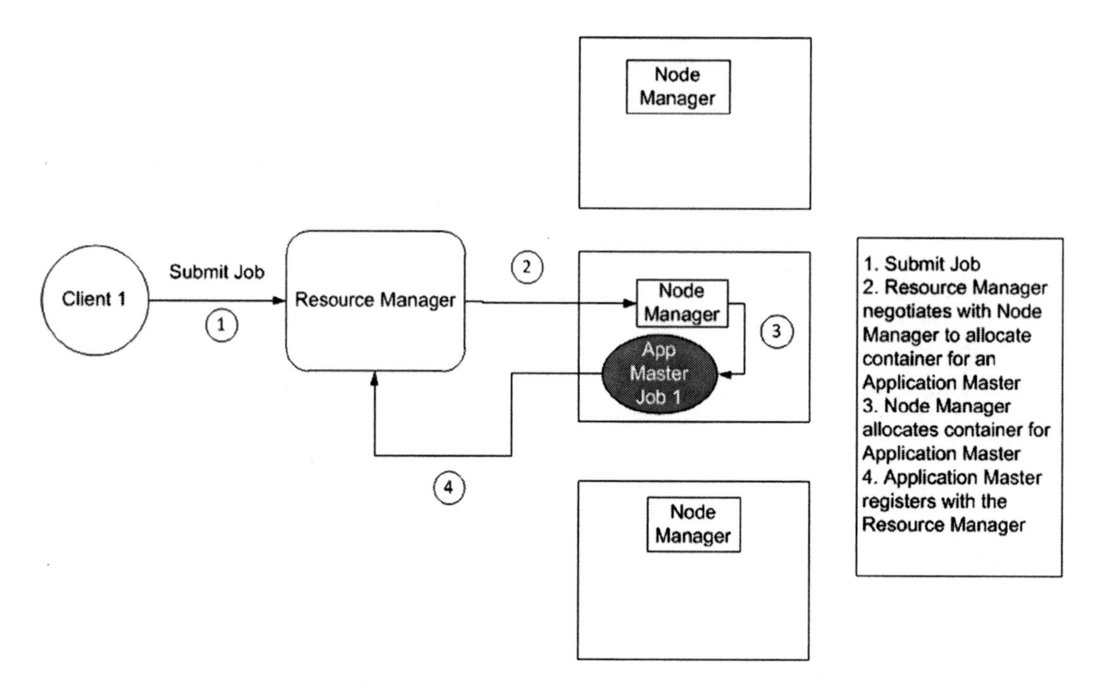


***Figure 2-7.*** *Application master startup*

Here are the steps used:

1. A client program submits the application. The application type that in turn determines the Application Master is also specified.
2. The Resource Manager negotiates resources to acquire a container on a node to launch an instance of the Application Master.
3. The Application Master registers with the Resource Manager. This registration enables the client to query the Resource Manager for details about the Application Master. Thus the client will communicate with the Application Master it has launched through its own Resource Manager.
4. During its operation, the Application Master negotiates resources from the Resource Manager through resource requests. A resource request contains, among other things, the node on which containers are requested and the specifications of the container (CPU code and memory specifications).
5. The application code executing in the launched container reports its progress to the Application Master (possibly remote) through an application-specific protocol.
6. The client program communicates with the Application Master through the application-specific protocol. The client references the Application Master through querying the Resource Manager it registered with in step 3.

The preceding steps are shown in Figure 2-8.

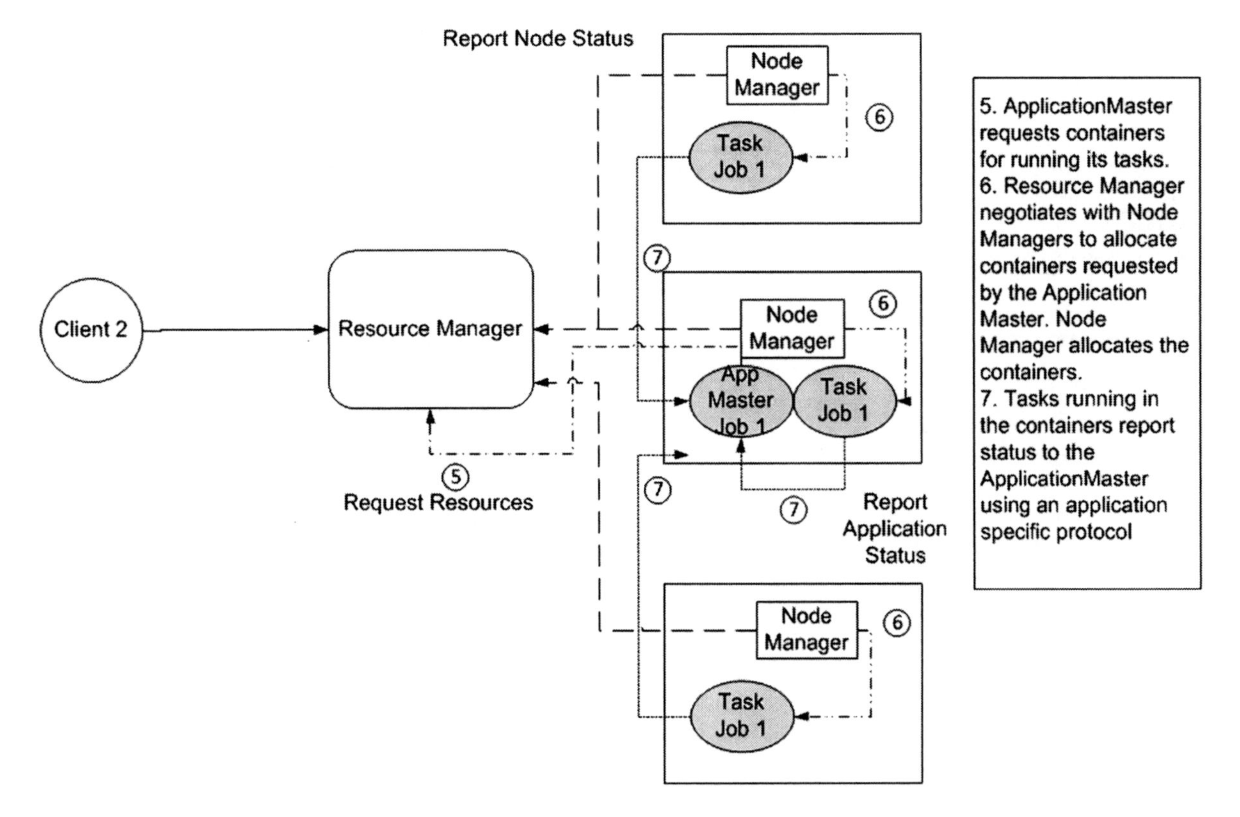


***Figure 2-8.*** *Job resource allocation and execution*

Once the application completes execution, the Application Master deregisters with the Resource Manager, and the containers used are released back to the system.

# HDFS High Availability

The earlier discussion on HDFS made it clear that in Hadoop 1.x, the NameNode is a single point of failure. The Hadoop 1.x system has a single NameNode, and if the machine hosting the NameNode service becomes unavailable, the entire cluster becomes inaccessible unless the NameNode is restarted and brought up on a separate machine. Apart from accidental NameNode losses, there are also constraints from a maintenance point of view. If the node running the NameNode needs to be restarted, the entire cluster will be unavailable during the period in which the NameNode is not running.

Hadoop 2.x introduces the notion of a High Availability NameNode, which is discussed here only from a conceptual perspective. Consult the Hadoop web site for evolving details of how to implement a High Availability NameNode.

The core idea behind the High Availability NameNode is that two similar NameNodes are used: one is in *active mode*, and the other is in *standby mode*. The active node supports clients in the system; the standby node needs to be synchronized with the active NameNode data to allow for a rapid failover operation. To ensure this in the current design, both NameNodes must share a storage device (through an NFS). Any modification to the active NameNode space is applied to the `edits` log file in the shared storage device. The standby node will keep applying these changes to its own namespace. In the event of a failure, the standby first ensures that all the edits are applied and takes over the responsibility of the active NameNode.

Remember that NameNodes do not contain metadata about blocks' locations; it is obtained by the NameNode from the DataNode during startup. To ensure that the standby NameNode starts up quickly, the DataNodes know the location of both NameNodes and send this information to both at startup. The heartbeat messages are also exchanged with both NameNodes.

## Summary

This chapter introduced the various concepts of the Hadoop system. It started with a canonical word count example and proceeded to explore various key features in Hadoop. You learned about the Hadoop Distributed File System (HDFS) and saw how jobs are managed in Hadoop 1.x using JobTracker and TaskTracker daemons. Using the knowledge of the way these daemons limit scalability, you were introduced to YARN, the feature of Hadoop 2.x that addresses these limitations. You then explored High Availability NameNode.

The next chapter will explore the installation of Hadoop software, and you will write and execute your first MapReduce program.

CHAPTER 3

# Getting Started with the Hadoop Framework

Previous chapters discussed the motivation for Big Data, followed by a high-level introduction to Hadoop, the most important Big Data framework in the market. In this chapter, you actually use Hadoop. It guides you through the process of setting up your Hadoop development environment and provides general guidelines for installing Hadoop on the operating system of your choice. You can then write your first few Hadoop programs, which introduce you to the deeper concepts underlying the Hadoop architecture.

## Types of Installation

Although installing Hadoop is often a task for experienced system administrators, and installation details can be found on the Apache web site for Hadoop, it is important to have a basic idea about installing Hadoop on various platforms, for two reasons:

- To enable unit-testing of Hadoop programs, Hadoop needs to be installed in stand-alone mode. This process is relatively straightforward for Linux-based systems, but it is more involved for Windows-based systems.
- To enable simulation of Hadoop programs in a real cluster, Hadoop provides a pseudo-distributed cluster mode of operation.

This chapter covers various modes in which Hadoop can be used. The configuration of the Hadoop development environment is discussed in the context of using VMs from vendors that come equipped with a development environment. We demonstrate Hadoop installation in stand-alone mode on Windows and Linux (the pseudo-cluster installation of Linux is discussed as well). Hadoop is an evolving software and its installation is very complex. Appendix A describes the installation steps for Windows and Linux platforms. These steps must be viewed as a set of general guidelines for installation. Your mileage may vary. We recommend that you use the VM method to install a development environment, described in this chapter, for performing development for the Hadoop 2.x platform.

### Stand-Alone Mode

Stand-alone is the simplest mode of operation and most suitable for debugging. In this mode, the Hadoop processes run in a single JVM. Although this mode is obviously the least efficient from a performance perspective, it is the most efficient for development turnaround time.

## Pseudo-Distributed Cluster

In this mode, Hadoop runs on a single node in a pseudo-distributed manner, and all the daemons run in a separate Java process. This mode is used to simulate a clustered environment.

## Multinode Node Cluster Installation

In this mode, Hadoop is indeed set up on a cluster of machines. It is the most complex to setup and is often a task for an experienced Linux system administrator. From a logical perspective, it is identical to the pseudo-distributed cluster.

## Preinstalled Using Amazon Elastic MapReduce

Another method you can use to quickly get started on a real Hadoop cluster is the Amazon Elastic MapReduce (EMR) service. This service now supports both 1.x and 2.x versions of Hadoop. It also supports various distributions of Hadoop such as the Apache version and the MapR distribution.

EMR enables users to spin up a Hadoop cluster with a few simple clicks on a web page. The main idea behind EMR is as follows:

1. The user loads the data on the Amazon S3 service, which is a simple storage service. Amazon S3 is a distributed file storage system offered by Amazon Web Services. It supports storage via Web Services interfaces. Hadoop can be configured to treat S3 as a distributed file system. In this mode, the S3 service acts like the HDFS.
2. The user also loads the application libraries in the Amazon S3 service.
3. The user starts the EMR job by indicating the location of the libraries and the input files, as well as the output directory on S3 in which the job will write its output.
4. A Hadoop cluster launches on the Amazon cloud, the job is executed, and the output is placed persistently in the output directory specified in the earlier step.

In its default behavior, the cluster is shut down automatically, and the user stops paying. However, there is an option (now available on the web page that launches the EMR) that enables you to indicate that you want to keep the cluster alive: the Auto-terminate option. When No is selected for this option, the cluster does not shut down after the job is complete.

You can choose to enter into any of the nodes using a Secure Shell (SSH) client. After users are connected to a physical mode through an SSH client, they can continue to use Hadoop as a fully functional cluster. Even the HDFS is available to the user.

The user could use one of the sample and tiny jobs to launch the cluster, which executes and keeps the cluster running. The user can run more jobs by connecting to one of the nodes. A simple two-node cluster costs about $1.00 per hour (depending on the server type chosen; the price can rise as high as $14.00 per hour if high-end servers are chosen). After the users finish their work, they can shut down the cluster and stop paying for it. So for a small price, users can experience running real-world jobs on a real-world production grade Hadoop cluster. (Chapter 16 discusses Hadoop in the cloud.)

---

■ **Caution** Even $1.00 per hour can add up over a month's time. Pay careful attention to the status of services you run in Amazon's cloud offering. Shut them down when you're not using them and make sure that you understand precisely what you are being billed for. At least one person we know was caught out to the tune of several hundred dollars by leaving a server running all month when he wasn't using it. Don't let that happen to you.

---

# Setting up a Development Environment with a Cloudera Virtual Machine

This book is primarily focused on Hadoop development, and Hadoop installation is a complex task that is often simplified by using tools provided by vendors. For example, Cloudera provides Cloudera Manager, which simplifies Hadoop installation. As a developer, you want to have a reliable development environment that can be installed and set up quickly. Cloudera has released CDH 5.0 for both VMware and VirtualBox. If you do not have these VM players installed, download their latest versions first. Next, download the Cloudera 5 QuickStart VM from this link:

`www.cloudera.com/content/support/en/downloads/download-components/download-products.html?productID=F6mO278Rvo`

Note that the Cloudera 5 VM requires 8 GB of memory. Ensure that your machine has adequate memory to execute the VM. Alternatively, follow the steps in the subsequent section to install your own development environment.

When you launch the VM, you see the screen shown in Figure 3-1. The figure points to the Eclipse icon on the desktop inside the VM. You can simply open Eclipse and begin development of the Hadoop code because the environment is configured to run jobs directly from the Eclipse environment in local mode.

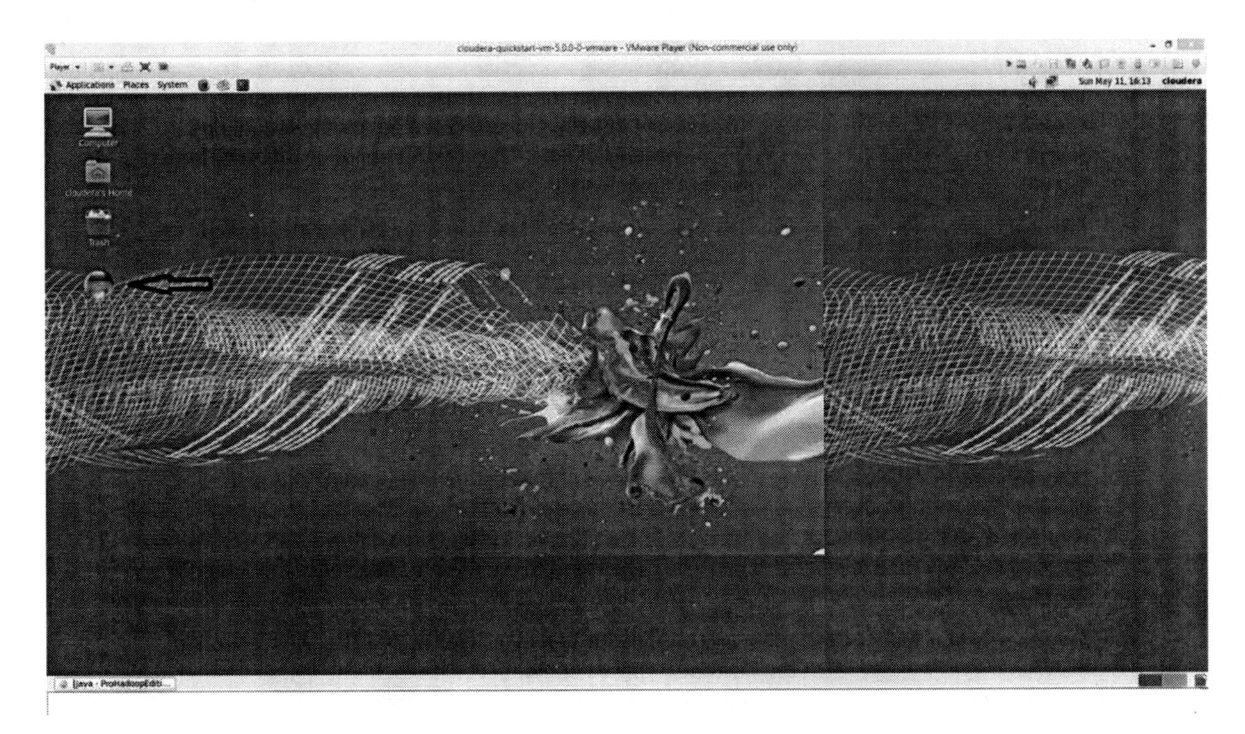

***Figure 3-1.*** *Cloudera 5 VM*

This is all you need to get started with Hadoop 2.0. The environment also enables the user to execute jobs in pseudo-distributed mode to simulate testing on a real cluster. As such, it is a complete environment for development, unit testing, and integration testing. The environment is also configured to allow the use of Cloudera Manager, a user-friendly GUI tool to monitor and manage your jobs. You are encouraged to become familiar with this tool because it greatly simplifies the tasks of job management and tracking.

We highly recommend this approach to have your Hadoop 2.0 development environment set up quickly.

---

■ **Note** If you intend to use the Cloudera VM mentioned in this section, it is not required to read about installing Hadoop. However, we have described the installation process for Hadoop on Windows and Linux in Appendix A, and you should follow the steps described in Appendix A to install Hadoop in the pseudo-cluster mode.

---

# Components of a MapReduce program

This section describes the various components that make up a MapReduce program in Java. The following list describes each of these components:

- *Client Java program*: A Java program that is launched from the client node (also referred to as the *edge node*) in the cluster. This node has access to the Hadoop cluster. It can also sometimes (not always) be one of the data nodes in the cluster. It is merely a machine in the cluster that has access to the Hadoop installation.
- *Custom Mapper class*: Includes a Mapper class that is often a custom class, except in the simplest cases. Instances of this class are executed on remote task nodes except in the case of executing jobs in the pseudo-cluster. These nodes are often different from the node in which the Client Java program launches the job.
- *Custom Reducer class*: Includes a Reducer class that is often a custom class, except in the simplest cases. Similar to Mapper, instances of this class are executed on remote task nodes, except in the case of executing jobs in the pseudo-cluster. These nodes are often different from the node in which the Client Java program launches the job.
- *Client-side libraries*: Libraries separate from the standard Hadoop libraries that are needed during the runtime execution of the client. The Hadoop libraries needed by the client are already installed and configured into the `CLASSPATH` by the Hadoop `Client` command (which is different from the Client program). It is found in the `$HADOOP_HOME/bin/` folder and is called `hadoop`. Just as the `java` command is used to execute a Java program, the `hadoop` command is used to execute the Client program that launches the Hadoop job. These libraries are configured by setting the environment variable `HADOOP_CLASSPATH`. Similar to the `CLASSPATH` variable, it is a colon-separated list of libraries.
- *Remote libraries*: Libraries needed for the execution of the custom Mapper and Reducer classes. They exclude the Hadoop set of libraries because the Hadoop libraries are already configured on the DataNodes. For example, if the Mapper is using a specialized XML parser, the libraries including the parser have to be transferred to the remote DataNodes that execute the Mapper.
- *Java Application Archive (JAR) files*: Java applications are packaged in JAR files, which contain the Client Java class as well as the Custom Mapper and Reducer classes. It also includes other custom dependent classes used by the Client and Mapper/Reducer classes.

# Your First Hadoop Program

In this section, you explore your first Hadoop program. The environment you use for this application is Eclipse with the Maven plugin. If you are using the Cloudera VM, it is preinstalled for you. The process of creating a Maven project in Eclipse is documented in Appendix B.

First create an empty Maven project and the necessary dependencies. The Project Object Model (POM) looks like the one shown in Listing 3-1. Create a file named `pom.xml` and put the code from Listing 3-1 into that file.

***Listing 3-1.*** pom.xml

```
<project xmlns="http://maven.apache.org/POM/4.0.0"
xmlns:xsi="http://www.w3.org/2001/XMLSchema-instance"
    xsi:schemaLocation="http://maven.apache.org/POM/4.0.0
http://maven.apache.org/xsd/maven-4.0.0.xsd">
    <modelVersion>4.0.0</modelVersion>
    <groupId>prohadoop</groupId>
    <artifactId>prohadoop</artifactId>
    <version>0.0.1-SNAPSHOT</version>
    <name>ProHadoop</name>
    <dependencies>
        <dependency>
            <groupId>org.apache.hadoop</groupId>
            <artifactId>hadoop-client</artifactId>
            <version>2.2.0</version>
        </dependency>
    </dependencies>
</project>
```

Now you can develop your first MapReduce program and run it locally. However, it is confusing because there are two MapReduce APIs: the old API and the new API. The old API has been deprecated, but it is still widely used. We run the jobs using both APIs so that you can become familiar with both.

---

■ **Tip** You do not need to type the example listings by hand. Go to the Apress.com catalog page for this book. You can find a tab on that page from which to download a zip archive containing all the listings.

---

## Prerequisites to Run Programs in Local Mode

It is important to be able to run Hadoop programs in local mode; it enables rapid development because readers can unit-test programs. The only prerequisites are that the `HADOOP_HOME` environment variable should be set correctly and that `{HADOOP_HOME}/bin` should be included in the `PATH` variable. For the Linux environment, ensuring these two settings is adequate.

However, in the Windows environment, the entire build process needs to be applied to ensure that all the required dynamic link libraries (DLLs) are built on the machine. They are not included in the `tar` file, which is downloaded from the Apache web site. The process of installing Hadoop on Windows for the pseudo-cluster or local mode is described in Appendix A.

If you are using the VM described in an earlier section, these prerequisites are already configured in the side VM.

Assuming that the prerequisites are in place, the example programs to follow can now be executed like any other Java program from the integrated development environment (IDE) of your choice.

The sample data for running all the jobs described in this book in the local mode is maintained in the `${project.basedir}/src/main/resources/input` folder. The `${project.basedir}` directory is the base directory for the project. The sample text file for running the WordCount application is maintained in the `wordcount` subfolder of the previously mentioned folder.

## WordCount Using the Old API

The `WordCountOldAPI.java` program in Listing 3-2 shows the code for the WordCount application using the old API, which has been deprecated. However, it is still widely used, and there is a lot of legacy code written there. The old API is not going away, and it is important to know how to use it because the reader may encounter it frequently when maintaining code.

***Listing 3-2.*** WordCountOldAPI.java

```
package org.apress.prohadoop.c3;

import java.io.IOException;
import java.util.Iterator;

import org.apache.hadoop.fs.Path;
import org.apache.hadoop.io.IntWritable;
import org.apache.hadoop.io.LongWritable;
import org.apache.hadoop.io.Text;
import org.apache.hadoop.mapred.FileInputFormat;
import org.apache.hadoop.mapred.FileOutputFormat;
import org.apache.hadoop.mapred.JobClient;
import org.apache.hadoop.mapred.JobConf;
import org.apache.hadoop.mapred.MapReduceBase;
import org.apache.hadoop.mapred.Mapper;
import org.apache.hadoop.mapred.OutputCollector;
import org.apache.hadoop.mapred.Reducer;
import org.apache.hadoop.mapred.Reporter;
import org.apache.hadoop.mapred.TextInputFormat;
import org.apache.hadoop.mapred.TextOutputFormat;

public class WordCountOldAPI {

    public static class MyMapper extends MapReduceBase implements
            Mapper<LongWritable, Text, Text, IntWritable> {
        public void map(LongWritable key, Text value,
                OutputCollector<Text, IntWritable> output, Reporter reporter)
                throws IOException {
            output.collect(new Text(value.toString()), new IntWritable(1));
        }
    }
    public static class MyReducer extends MapReduceBase implements
            Reducer<Text, IntWritable, Text, IntWritable> {
        public void reduce(Text key, Iterator<IntWritable> values,
                OutputCollector<Text, IntWritable> output, Reporter reporter)
                throws IOException {
            int sum = 0;
            while (values.hasNext()) {
                sum += values.next().get();
            }
            output.collect(key, new IntWritable(sum));
        }
    }
```

```
    public static void main(String[] args) throws Exception {
        JobConf conf = new JobConf(WordCountOldAPI.class);
        conf.setJobName("wordcount");

        conf.setOutputKeyClass(Text.class);
        conf.setOutputValueClass(IntWritable.class);

        conf.setMapperClass(MyMapper.class);
        conf.setCombinerClass(MyReducer.class);
        conf.setReducerClass(MyReducer.class);
        conf.setNumReduceTasks(1);
        conf.setInputFormat(TextInputFormat.class);
        conf.setOutputFormat(TextOutputFormat.class);
        FileInputFormat.setInputPaths(conf, new Path(args[0]));
        FileOutputFormat.setOutputPath(conf, new Path(args[1]));

        JobClient.runJob(conf);
    }
}
```

Begin by examining the `main()` method. The important concepts to grasp are described in the following list:

- `JobConf` is the primary interface through which the MapReduce job is configured for the Hadoop Framework. The framework executes the job as described in the `JobConf` object.
- `TextInputFormat` declares to the Hadoop Framework that the input is in the form of text. It is the subclass of `InputFormat`. (The various `InputFormat` classes are discussed in detail in Chapter 7.) For now, it is sufficient to know that the `TextInputFormat` class reads each line of the input file as a single record.
- `TextOutputFormat` specifies the output specification of the job. For example, it verifies that the output directory does not already exist. Hadoop does not allow the job to proceed if the output directory already exists. Hadoop jobs operate on massive amounts of data, and it often takes a few minutes to hours to execute the jobs, and it would be unfortunate to lose all that work if the job were accidentally rerun. Various types of specialized `OutputFormat` classes are described in Chapter 7, but for now know that the `TextOutputFormat` class is used to produce text output files from MapReduce programs.
- `FileInputFormat.setInputPaths(conf, new Path(args[0]))` tells the Hadoop Framework the input directory from which the files will be picked up. The directory can have one to many files, and each file has one word per line. Notice the use of the plural form in `setInputPaths`. This method can take an array of directories that serve as the input paths to the Hadoop program. The files in all these paths form the input specification of the program.
- `FileOutputFormat.setOutputPath(conf, new Path(args[1]))` specifies the output directory of the program. The final output is placed in this directory.
- `conf.setOutputKeyClass` and `conf.setOutputValueClass` specify the output Key and Value classes. They should match the specification of the Reducer class. It might seem redundant—and it is. However, it needs to be consistent with the Reducer specification or else there will be `RuntimeException` at the time of execution.
- The default value of the number of Reducers is 1. The value can be changed and often is to improve the efficiency of the program. The call on the `JobConf.class` instance that enables you to control it is the `setNumReduceTasks(int n)` method.

Now you understand how the entire program works in the Hadoop Framework. When the program begins, the Mapper reads designated blocks from the input files in the input folders. The stream of bytes that represents a file is then converted to a record (Key/Value format) and supplied as input to the Mapper. The key is the byte offset of the current line (instance of `LongWritable.class`), and the value (instance of `Text.class`) is a line of text from the file. In this simplified example, using the sample file in the books source code, one line is one word.

The Mapper emits the word and the integer 1. Notice that Hadoop requires you to use its own version of `Integer.class` called `IntWritable.class`. The reason lies in the I/O concepts underlying Hadoop (they are discussed in upcoming chapters). For now, accept that the input and output of Hadoop Mappers and Reducers need to be instances of the type `Writable.class`.

Next comes the Shuffle phase of Hadoop, which is not obvious in the preceding listing. All the keys (words, in our example) are sorted by the Shuffle/Sort phase and presented to the Reducer. Although the Reducer sees an iterator of `IntWritable.class` instances, it is only a logical view. In reality, the same `IntWritable.class` instance is being used in the Reducer. As the Reducer iterates to the values for the key (a list of 1s in the example), the same instance of the `IntWritable.class` is reused. Internally, the `IntWritable.set` method is called by the Hadoop Framework on the `values.next()` call to the iterator. Beware of this when you are collecting values and reusing them as references. Due to this feature of Hadoop (an optimization to save memory usage), reusing the references of the `values.next()` instances leads to unexpected results.

Finally, as the Reducer emits values, they are written to the output folder designated by the `FileOutputFormat.setOutputPath(conf, new Path(args[1]))` call in the Client `WordCountOldAPI` program. The Reducer sends its output Key and Value instances to the output file. The default separator for the output Key and Value instances in the output file is a `TAB` that can be changed by setting a configuration parameter: `mapreduce.textoutputformat.separator`. For example, the call `conf.set("mapreduce.textoutputformat.separator",",")` in the `main(...)` method would have converted the separator between the Key/Value instances in the output to become a ",".

In the case of a real cluster, there are as many files as there are Reducers configured. In this example, the file contains output as a word followed by the count of words.

The command line to execute the previous program on a cluster is similar to the command line used in the next section. This is discussed when we explore the new API in the next section.

# Building the Application

Next, you build your application using Maven. This process is described in Appendix B, so if you are not familiar with Maven, you should start there.

The build process produces the `prohadoop-0.0.1-SNAPSHOT.jar` file. The various components of the application are described in Table 3-1.

***Table 3-1.*** *Components of WordCount Using the Old API*

| Component | Name |
|---|---|
| Client class | `WordCountOldAPI.java` |
| Mapper class | `WordCountOldAPI.MyMapper.java` |
| Reducer class | `WordCountOldAPI.MyReducer.java` |
| Application JAR | `prohadoop-0.0.1-SNAPSHOT.jar` |
| Client-side libraries | None for this simple program |
| Remote libraries | None for this simple program |

## Running WordCount in Cluster Mode

You can start MapReduce jobs in cluster (or pseudo-cluster) mode. Following are two differences when executing in cluster mode:

- The Map and Reduce tasks each run on their own nodes that are usually not the same as the Client node.
- The Map and Reduce tasks execute in their own JVMs.

The first step of executing the WordCount program (including the WordCount program from the earlier section) in cluster mode is to bundle the application class files into a JAR file. The program can then be executed as follows:

```
export HADOOP_CLASSPATH=$HADOOP_CLASSPATH:<DEPENDENT_JARS_USED_BY_ CLIENT_CLASS>

hadoop jar prohadoop-0.0.1-SNAPSHOT.jar  \
org.aspress.prohadoop.c3.WordCountOldAPI \
<INPUT_PATH> <OUTPUT_PATH>
```

Although we do not have any specialized libraries that we use in the Client program, the correct way to make them accessible is by making the following call:

```
export HADOOP_CLASSPATH=$HADOOP_CLASSPATH:<DEPENDENT_JARS_USED_BY_ CLIENT_CLASS>
```

This HADOOP_CLASSPATH environment variable is used by the hadoop command to configure the client-side CLASSPATH before executing the Client program.

The `hadoop` command is also responsible for sending the `prohadoop-0.0.1-SNAPSHOT.jar` file to the remote nodes and ensuring that the `CLASSPATH` is configured to prepare for the execution of the Mapper and Reducer instances on them. This is an example of moving the application to the data that was mentioned in Chapter 1, "Motivation for Big Data."

The `<INPUT_PATH>` and `<OUTPUT_PATH>` folders refer to folders in the HDFS. The former is the input folder in which the files are consumed by the Mapper instances. The latter is the output folder to which the Reducer output is written.

## WordCount Using the New API

This section explores the new API, which was designed to offer better control to the programmer. The new API replaces `JobConf.class` with `Job.class`. It enables the user to configure the job, submit the job, control its execution, and monitor its progress. The `set()` methods of `Job.class` are similar to the `set()` methods of `JobConf.class`, and they can be invoked only until the job is submitted. If they are invoked after the job is submitted, they throw an instance of `IllegalStateException`. Methods of the `Job.class` such as `getStatus()` enable the user to monitor job progress. Listing 3-3 shows the WordCount program using the new API. Note that the `main()` method now uses `Job.class` instead of `JobConf.class`.

***Listing 3-3.*** WordCountNewAPI.java

```
package org.apress.prohadoop.c3;

import java.io.IOException;

import org.apache.hadoop.conf.Configuration;
import org.apache.hadoop.fs.Path;
import org.apache.hadoop.io.IntWritable;
```

```
import org.apache.hadoop.io.LongWritable;
import org.apache.hadoop.io.Text;
import org.apache.hadoop.mapreduce.Job;
import org.apache.hadoop.mapreduce.Mapper;
import org.apache.hadoop.mapreduce.Reducer;
import org.apache.hadoop.mapreduce.lib.input.FileInputFormat;
import org.apache.hadoop.mapreduce.lib.input.TextInputFormat;
import org.apache.hadoop.mapreduce.lib.output.FileOutputFormat;
import org.apache.hadoop.mapreduce.lib.output.TextOutputFormat;

public class WordCountNewAPI {
    public static class MyMapper extends
            Mapper<LongWritable, Text, Text, IntWritable> {
        public void map(LongWritable key, Text value, Context context)
                throws IOException, InterruptedException {
            String w = value.toString();
            context.write(new Text(w), new IntWritable(1));
        }
    }

    public static class MyReducer extends
            Reducer<Text, IntWritable, Text, IntWritable> {

        public void reduce(Text key, Iterable<IntWritable> values,
                Context context) throws IOException, InterruptedException {
            int sum = 0;
            for (IntWritable val : values) {
                sum += val.get();
            }
            context.write(key, new IntWritable(sum));
        }
    }

    public static void main(String[] args) throws Exception {
        Job job = Job.getInstance(new Configuration());
        job.setJarByClass(WordCountNewAPI.class);
        job.setOutputKeyClass(Text.class);
        job.setOutputValueClass(IntWritable.class);
        job.setMapperClass(MyMapper.class);
        job.setReducerClass(MyReducer.class);
        job.setInputFormatClass(TextInputFormat.class);
        job.setOutputFormatClass(TextOutputFormat.class);
        FileInputFormat.setInputPaths(job, new Path(args[0]));
        FileOutputFormat.setOutputPath(job, new Path(args[1]));
        boolean status = job.waitForCompletion(true);
        if (status) {
            System.exit(0);
        } else {
            System.exit(1);
        }
    }
}
```

## Building the Application

The application can be built again to produce the prohadoop-0.0.1-SNAPSHOT.jar file. The various components of the application are described in Table 3-2.

***Table 3-2.** Components of WordCount Using the New API*

| Component | Name |
|---|---|
| Client class | WordCountNewAPI.java |
| Mapper class | WordCountNewAPI.MyMapper.java |
| Reducer class | WordCountNewAPI.MyReducer.java |
| Application JAR | prohadoop-0.0.1-SNAPSHOT.jar |
| Client-side libraries | None for this simple program |
| Remote libraries | None for this simple program |

## Running WordCount in Cluster Mode

We will now execute the MapReduce program like we did in the previous section when we executed the MapReduce program based on the old API. The command to run the MapReduce program in cluster mode is the following:

```
hadoop jar prohadoop-0.0.1-SNAPSHOT.jar  \
org.aspress.prohadoop.c3.WordCountNewAPI \
<INPUT_PATH> <OUTPUT_PATH>
```

Because there is no need for the HADOOP_CLASSPATH, it is not included here. As described earlier, the JAR file is sent over to all the data nodes and the local classpath on those nodes are configured to include this JAR. The input and output paths refer to paths in the HDFS. The hadoop command is responsible for performing these preparatory steps and finally executing the jobs in a distributed manner on the cluster.

# Third-Party Libraries in Hadoop Jobs

So far, only standard Java and Hadoop libraries were used in the Mapper and Reducer classes. They include libraries found in the Hadoop distribution and standard Java classes (such as String.class).

However, you cannot develop complex Hadoop jobs using just the classes in the standard libraries; sometimes, third-party libraries are used. As discussed earlier, these libraries used by the Mapper and the Reducer classes need to be sent and configured on the nodes for the execution of the Mapper and Reducer instances.

First, you need to write programs a little differently to use third-party libraries. Listing 3-4 shows a version of the WordCount program that uses third-party libraries. It is a contrived example, but it serves to demonstrate the concept.

Assume that you have file with a list of words, and some words comprise a combination of Unicode characters and digits (0-9). Others are just Unicode characters. Words that are a combination of numbers and letters are often used in contract documents in which various identifiers are a combination of letters and words. For convenience, words that have other characters, such as "-"and "$", are rejected.

Assume that you are interested only in words with a combination of letters and numbers. To decide whether a word is such a combination, use StringUtils.class from the commons-lang library. The method call that allows you to make the appropriate determination is this:

```
if(StringUtils.isAlphanumeric(key.toString()) && !StringUtils.isAlphaSpace(key.toString()) )
```

The full code is shown in Listing 3-4.

***Listing 3-4.*** WordCountUsingToolRunner.java

```
package org.aspress.prohadoop.c3;

import java.io.IOException;
import org.apache.hadoop.conf.*;
import org.apache.hadoop.fs.Path;
import org.apache.hadoop.io.*;
import org.apache.hadoop.mapreduce.Job;
import org.apache.hadoop.mapreduce.Mapper;
import org.apache.hadoop.mapreduce.Reducer;
import org.apache.hadoop.mapreduce.lib.input.FileInputFormat;
import org.apache.hadoop.mapreduce.lib.input.TextInputFormat;
import org.apache.hadoop.mapreduce.lib.output.FileOutputFormat;
import org.apache.hadoop.mapreduce.lib.output.TextOutputFormat;
import org.apache.hadoop.util.Tool;
import org.apache.hadoop.util.ToolRunner;
import org.apache.hadoop.util.GenericOptionsParser;
import org.apache.commons.lang.StringUtils;;

public class WordCountUsingToolRunner extends Configured implements Tool{

 public static class MyMapper extends Mapper<LongWritable, Text, Text, IntWritable> {
    private final static IntWritable one = new IntWritable(1);

    public void map(LongWritable key, Text value, Context context) throws IOException,
    InterruptedException {
                String w = value.toString();
                context.write(new Text(w), one);
    } }

 public static class MyReducer extends Reducer<Text, IntWritable, Text, IntWritable> {
        public void reduce(Text key, Iterable<IntWritable> values, Context context)
      throws IOException, InterruptedException {
        if(StringUtils.isAlphanumeric(key.toString()) && !StringUtils.isAlphaSpace(key.toString()) )
{
                int sum = 0;
            for (IntWritable val : values) {
                sum += val.get();
            }
            context.write(key, new IntWritable(sum));
        }
    }
 }

public  int run(String[] allArgs) throws Exception {
    Job job = Job.getInstance(getConf());
    job.setJarByClass(WordCountUsingToolRunner.class);
    job.setOutputKeyClass(Text.class);
    job.setOutputValueClass(IntWritable.class);
```

```
    job.setMapperClass(MyMapper.class);
    job.setReducerClass(MyReducer.class);

    job.setInputFormatClass(TextInputFormat.class);
    job.setOutputFormatClass(TextOutputFormat.class);
    String[] args = new GenericOptionsParser(getConf(), allArgs).getRemainingArgs();
    FileInputFormat.setInputPaths(job, new Path(args[0]));
    FileOutputFormat.setOutputPath(job, new Path(args[1]));
    job.submit();
    return 0;
 }

 public static void main(String[] args) throws Exception {
    Configuration conf = new Configuration();
    ToolRunner.run(new WordCountUsingToolRunner (), args);
 }

}
```

Listing 3-4 will be discussed in a moment. First, you should understand some of the POM file changes needed to enable the use of third-party libraries. The relevant sections of the POM file are shown in bold in Listing 3-5.

***Listing 3-5.*** Relevant Changes to the POM File

```
<dependencies>
    <dependency>
        <groupId>org.apache.hadoop</groupId>
        <artifactId>hadoop-client</artifactId>
        <version>2.2.0</version>
        <scope>provided</scope>
    </dependency>
    <dependency>
        <groupId>commons-lang</groupId>
        <artifactId>commons-lang</artifactId>
        <version>2.3</version>
    </dependency>

</dependencies>
<build>
    <plugins>
        <plugin>
            <groupId>org.apache.maven.plugins</groupId>
            <artifactId>maven-assembly-plugin</artifactId>
            <version>2.4</version>
            <configuration>
                <descriptorRefs>
                    <descriptorRef>jar-with-dependencies</descriptorRef>
                </descriptorRefs>
            </configuration>
            <executions>
                <execution>
                    <id>assemble-all</id>
```

```
                        <phase>package</phase>
                        <goals>
                            <goal>single</goal>
                        </goals>
                    </execution>
                </executions>
            </plugin>
        </plugins>
</build>
```

In this POM file, a commons-lang library with version 2.3 was added. Note the <scope> tag provided for the hadoop-client library. For now, the provided value interacts with the <build> tag to ensure that two separate application JAR files are created by the build process:

- prohadoop-0.0.1-SNAPSHOT.jar includes only custom classes in the org.apress package and its subpackages.
- prohadoop-0.0.1-SNAPSHOT-jar-with-dependencies.jar includes not only the custom code but also the class files from the commons-lang-2.3.jar file. The class files in the Hadoop client library are not included in this jar file since we have provided the value of scope as provided in the POM file.This setting includes the library during compile time but does not bundle it in the resulting JAR file.

Why were the Hadoop client and its dependent libraries not included in the the prohadoop-0.0.1-SNAPSHOT-jar-with-dependencies.jar file? Remember that Maven not only resolves that library but also any other libraries that the declared library (with the correct version no) depends on. Including the hadoop-client library and its dependent libraries results in a bloated application JAR file.. The Hadoop client libraries are already configured on all the nodes, so the classes from the hadoop-client libraries do not need to be included in the application JAR. This is an important refinement. The size of the JAR file with dependencies is considerably smaller due to the exclusion of the Hadoop client libraries(for example, 200 KB vs. 22 MB). Remember that this file needs to be transferred from the client node to the remote DataNodes. The size of the file affects job startup times because the job cannot start until the file is fully transferred. Try this by removing the <scope> tag and recompiling.

The program execution is discussed before we describe Listing 3-4. Note that there is an additional set of libraries included in the program execution. That library is commons-lang-2.3.jar, which is needed in the remote Mapper and Reducer classes. The WordCountNewAPIV2.class now extends and implements new classes, which are required to distribute the library JAR files needed to execute the program.

Execute the program by invoking the following set of commands:

```
export HADOOP_CLASSPATH=$HADOOP_CLASSPATH:<DEPENDENT_JARS_USED_BY_ CLIENT_CLASS>

export LIBJARS=$MYLIB/commons-lang-2.3.jar,<OTHER_JARS_USED_BY_REMOTE_COMPONENTS>

hadoop jar prohadoop-0.0.1-SNAPSHOT.jar \
org.aspress.prohadoop.c3. WordCountUsingToolRunner
-libjars $LIBJARS \
<INPUT_PATH> <OUTPUT_PATH>
```

The preceding commands, their parameters, and their environment variable settings are explained here:

- The HADOOP_CLASSPATH update ensures that the hadoop command located in the $HADOOP_HOME/bin folder has access to dependent JAR files used by the client program launching the MapReduce jobs. In this case, this variable is not needed since the client portion of our MapReduce program does not utilize any third party libraries

- The $LIBJARS variable is a comma-separated list of paths to all the library files needed by the Mapper and the Reducer classes that execute on the DataNodes. Note the differences in the separator when compared with libraries specified in the HADOOP_CLASSPATH variable.
- The application JAR file and the JAR files included in the $LIBJARS variable are then sent across to all the nodes in which the Map/Reduce tasks will be executed. This is an example of moving the code to the data (refer to Chapter 1).These JAR files are used to configure the CLASSPATH variable on the remote nodes. The ToolRunner class used in the main() function is responsible for this.
- <INPUT_PATH> and <OUTPUT_PATH> refer to paths relative to the HDFS.
- Executing this program ensures that the Reducers run without throwing the ClassNotFoundException because of not finding the library JAR files that contain StringUtils.class.

There is an alternative way to run this job without using the -libjars option; use the following command:

```
hadoop jar prohadoop-0.0.1-SNAPSHOT-jar-with-dependencies.jar \
org.aspress.prohadoop.c3. WordCountUsingToolRunner
<INPUT_PATH> <OUTPUT_PATH>
```

Note the key difference. The dependent libraries are now included inside the application JAR file, so we do not need to provide the libjars option. Each time the job executes, the application JAR file is sent and configured on the remote nodes. In fact, this method can be used with both the old and new API, and it can be used without being forced to use the ToolRunner class described in Listing 3-4. The only downside is that when the number of libraries is very large, as is typically the case in a real-world application, the JAR file with dependent libraries bundled in it can be too large, which increases compile time during the build/test cycle. If your testing environment is remote from your development environment (for example, if you have a test cluster configured on the cloud), it can take longer to move the larger JAR file with dependencies as you are making constant changes to your application. But remember that this is the most general-purpose method to execute the application.

In the remainder of this book, the new MapReduce API and the ToolRunner class will be used to execute jobs.

Finally, it is time to discuss Listing 3-4. How is it different from Listings 3-2 and 3-3? The differences are mainly in the way the program is executed on the command line and the way classes are written to ensure that the new libraries are accessible in the Map and Reduce processes on the remote nodes:

- Listing 3-4 must use the GenericOptionsParser.class to extract the arguments needed by the program, such as the input path and the output path. The following lines perform the extraction:

  ```
  new GenericOptionsParser(getConf(),
                  allArgs).getRemainingArgs()
  ```

- These lines remove the -libjars and $LIBJARS_PATH arguments and return only the arguments required by the application program, which is the Hadoop job being executed.
- The most crucial aspect of Listing 3-4 is that the Configuration instance passed through the ToolRunner.run() invocation in the main() method must be the same Configuration instance that is used when configuring the job in the run() method. To ensure this, the run() method consistently uses the getConf() method that is defined in the Configurable interface and implemented in the Configured class that the class extends. If the same Configuration instance is not used, the job will not be correctly configured, and the third-party JAR files will not be available for the remote Mapper and Reducer tasks.

- The `ToolRunner.run()` method is responsible for parsing the `-libjars` argument. It delegates this task to `GenericOptionsParser.class`. The value of the `-libjars` argument is added to the `Configuration` object, which is the way remote tasks are configured.
- Listings 3-2 and 3-3 did not have the capability to pass third-party libraries. If you want to use third-party libraries with Listings 3-2 and 3-3, you need to package them into the application JAR file. This file is `prohadoop-0.0.1-SNAPSHOT-jar-with-dependencies.jar`, which was produced by the Maven build. (The process and purpose of building this JAR were discussed earlier in this chapter.)

The various components of the MapReduce program described in Listing 3-4 are listed in Table 3-3.

***Table 3-3.*** *Components of WordCount using the New API and ToolRunner*

| Component | Name |
|---|---|
| Client class | `WordCountUsingToolRunner.java` |
| Mapper class | `WordCountUsingToolRunner.MyMapper.java` |
| Reducer class | `WordCountUsingToolRunner.MyReducer.java` |
| Application JAR | `prohadoop-0.0.1-SNAPSHOT.jar` |
| Alternative application JAR | `prohadoop-0.0.1-SNAPSHOT-jar-with-dependencies.jar` |
| Client-side libraries | None |
| Remote libraries | `commons-lang-2.3.jar` (only when using `prohadoop-0.0.1-SNAPSHOT.jar` as the application JAR) |

# Summary

In this chapter, you got started with the Hadoop platform. You learned how to install the Hadoop development environment using a VM and you wrote your first few Hadoop programs. Admittedly, the programs were simple, but you were introduced to complex concepts in the Hadoop platform through these simple programs.

You learned about various configuration options in Hadoop as well as the way a Hadoop job is configured. You are now familiar with the array of classes in the Hadoop platform that enable job configuration as well as highly customizable I/O formats. Finally, you were led through an example to demonstrate the way a set of library files is transferred from the client to the various remote nodes in the Hadoop cluster.

CHAPTER 4

# Hadoop Administration

This chapter discusses Hadoop administration on a multinode cluster. You will explore the Hadoop configuration files in detail and learn how Hadoop can support multitenancy where multiple groups within an organization can utilize a single Hadoop cluster. Hadoop has various types of schedulers to support this functionality. The goal of this chapter is to enable you to appreciate the nuances of what happens on each node in the cluster when a Hadoop job executes.

## Hadoop Configuration Files

After you understand the various configuration files in Hadoop and their purposes, you can explore concepts such as the scheduler and Hadoop administration.

Every machine in the Hadoop Cluster has its own set of configuration files. Earlier versions of Hadoop had a single configuration file: `hadoop-site.xml`. Subsequent versions of Hadoop split this file into multiple files based on functionality. Additionally, there are two types of configuration files: `*-default.xml` and `*-site.xml`. The `*-site.xml` file configurations override the ones in the `*-default.xml` file.

The `*-default.xml` files are read-only and are read directly from the JAR files in the classpath. These files and their corresponding locations are listed in Table 4-1.

***Table 4-1.*** *Hadoop Default Configuration Files*

| Configuration File | Purpose |
| --- | --- |
| `core-default.xml` | Default core Hadoop properties. The file is located in the following JAR file: `hadoop-common-2.2.0.jar` (assuming version 2.2.0). |
| `hdfs-default.xml` | Default HDFS properties. The file is located in the following JAR file: `hadoop-hdfs-2.2.0.jar` (assuming version 2.2.0). |
| `mapred-default.xml` | Default MapReduce properties. The file is located in the following JAR file: `hadoop-mapreduce-client-core-2.2.0.jar` (assuming version 2.2.0). |
| `yarn-default.xml` | Default YARN properties. The file is located in the following JAR file: `hadoop-yarn-common-2.2.0.jar` (assuming version 2.2.0). |

The site-specific configuration files are typically located in the `$HADOOP_CONF_DIR` folder. These configuration files override the default ones mentioned in Table 4-1 and represent the unique properties of your cluster (site). Any property not mentioned in the site-specific file uses the property value from the default file. These files are listed in Table 4-2.

*Table 4-2. Hadoop Site-Specific Configuration Files*

| Configuration File | Purpose |
|---|---|
| core-site.xml | Site-specific common Hadoop properties. Properties configured in this file override the properties in the core-default.xml file. |
| hdfs-site.xml | Site-specific HDFS properties. Properties configured in this file override the properties in the hdfs-default.xml file. |
| mapred-site.xml | Site-specific MapReduce properties. Properties configured in this file override the properties in the mapred-default.xml file. |
| yarn.xml | Site-specific YARN properties. Properties configured in this file override the properties in the mapred-default.xml file. |

# Configuring Hadoop Daemons

In addition to the configuration files mentioned previously, Hadoop administrators can use the following script files to configure the Hadoop cluster:

- hadoop-env.sh
- yarn-env.sh
- mapred-env.sh

These script files are responsible for setting up properties such as the following:

- Java home directory
- Various log file locations
- JVM options for various daemons

Administrators can configure individual daemons using the configuration options shown in Table 4-3.

*Table 4-3. Daemon Configuration Variables*

| Daemon | Environment Variable |
|---|---|
| NameNode | HADOOP_NAMENODE_OPTS |
| DataNode | HADOOP_DATANODE_OPTS |
| Secondary NameNode | HADOOP_SECONDARYNAMENODE_OPTS |
| Resource Manager | YARN_RESOURCEMANAGER_OPTS |
| Node Manager | YARN_NODEMANAGER_OPTS |

For example, to configure NameNode to use parallelGC, the following statement should be added to hadoop-env.sh:

```
export HADOOP_NAMENODE_OPTS="-XX:+UseParallelGC ${HADOOP_NAMENODE_OPTS}"
```

Other useful properties that can be configured include these:

- YARN_LOG_DIR: Directories on the node in which the log files are stored. These directories are automatically created during startup if they do not exist.
- YARN_HEAPSIZE: Specify the amount of heap size to use in MB. Their default value is 1,000, indicating that by default Hadoop provides 1GB of heap size to each of its daemons. Although only one heap size is specified, YARN comprises several server services: Resource Manager and Node Manager. If a different heap size is desired for each service, values can be provided separately in variables such as these:
  - YARN_RESOURCEMANAGER_HEAPSIZE
  - YARN_NODEMANAGER_HEAPSIZE

# Precedence of Hadoop Configuration Files

Every node in the Hadoop cluster must have a copy of the configuration files, including the client nodes. The configurations in the file are applied as follows (in the order of highest to lowest):

1. Values specified in the JobConf or Job object when initializing the MapReduce job
2. *-site.xml on the client node
3. *-site.xml on the slave node
4. Default values from the *-default.xml, which are the same on all the nodes

However, if there are properties you do not want the client or the program to modify, mark them as final, as shown here:

```
<property>
  <name>{PROPERTY_NAME}</name>
  <value>{PROPERTY_VALUE}</value>
  <final>true</final>
</property>
```

If a property is marked as final on the slave node, the configuration on the client version of the *-site.xml file cannot modify it. If the property is marked as final on the client node, the job configuration instance cannot override it.

# Diving into Hadoop Configuration Files

This section explores some key parameters used to configure the Hadoop cluster through the configuration files mentioned earlier. This section shows you how the Hadoop cluster is set up at the individual node level. The various defaults for the configuration files are documented at the following links:

- http://hadoop.apache.org/docs/stable/hadoop-project-dist/hadoop-common/core-default.xml
- http://hadoop.apache.org/docs/stable/hadoop-project-dist/hadoop-hdfs/hdfs-default.xml
- http://hadoop.apache.org/docs/stable/hadoop-mapreduce-client/hadoop-mapreduce-client-core/mapred-default.xml
- http://hadoop.apache.org/docs/stable/hadoop-yarn/hadoop-yarn-common/yarn-default.xml

These links represent not only the default but also the current awareness of what constitutes best practices for Hadoop cluster setup. However, Hadoop is both a complex and a young system. As organizations continue to use Hadoop, the understanding of what constitutes "best practices" will change.

---

■ **Note** The properties are constantly being deprecated. In this book we will use the Hadoop version is 2.2.0, but the 2.4.1 was released in June 2014. This URL provides the current status: `http://hadoop.apache.org/docs/current/hadoop-project-dist/hadoop-common/DeprecatedProperties.html`

---

## core-site.xml

The `core-default.xml` and `core-site.xml` files configure the common properties for the Hadoop system.

This section explores the key properties of `core-site.xml`, including the following:

- `hadoop-tmp-dir`: Base for other temporary directories. The default value is `/tmp/hadoop-${user.name}`. We referenced this property on a few occasions as the root directory for several properties in the `hdfs-site.xml` file.
- `fs.defaultFs`: Name of the default path prefix used by HDFS clients when none is provided. In Chapter 3, "Getting Started with the Hadoop Framework," you configured it to be `hdfs://localhost:9000` for the pseudo-cluster mode. This property specifies the name and port for the NameNode (for example, `hdfs://<NAMENODE>:9000`). For local clusters, the value of this property is `file:///`. When the High Availability (HA) feature of HDFS is used, this property should be set to the logical HA URI. (See the Hadoop documentation for configuration of HDFS HA.)
- `io.file.buffer-size`: Chapter 2, "Hadoop Concepts," described the mechanics of the HDFS file create and HDFS file read processes. This property is relevant to those processes. It refers to the size of buffer to stream files. The size of this buffer should be a multiple of hardware page size (4096 on Intel x86), and it determines how much data is buffered during read and write operations. The default is 4096.
- `io.bytes.per.checksum`: Hadoop transparently applies checksums to all data written to it and verifies them at the time of reading. This parameter defines the number of bytes to which a checksum is applied. The default value is 512 bytes. The applied CRC-32 checksum is 4 bytes long. Thus, by default there is an overhead of approximately 1% per 512 bytes stored (with the default setting). Note that this parameter must not be a higher value than the `io.file.buffer-size` parameter because the checksum will be calculated on the data while in the memory where streaming during the HDFS read/write process. It needs to be calculated during the read process to verify the checksums stored during the write process.
- `io.compression.codecs`: A comma-separated list of the available compression codec classes that can be used for compression/decompression. The default settings that indicate the available compression codecs are `org.apache.hadoop.io.compress.DefaultCodec`, `org.apache.hadoop.io.compress.GzipCodec`, `org.apapache.hadoop.io.compress.BZip2Codec`, `org.apache.hadoop.io.compress.DeflateCodec`, and `org.apache.hadoop.io.compress.SnappyCodec`. (Compression schemes and how they apply to Hadoop are discussed in Chapter 7, "Hadoop Input/Output.")

## hdfs-*.xml

The hdfs-default.xml and hdfs-site.xml files configure the properties for the HDFS. Together with the core-site.xml file described next, the HDFS is configured for the cluster. As you learned in Chapter 2, NameNode and Secondary NameNode are responsible for managing the HDFS. The hdfs-*.xml files configure the NameNode and the Secondary NameNode components of the Hadoop system. The hdfs-*.xml set of files Iare also used for configuring the runtime properties of the HDFS as well the properties associated with the physical storage of files in HDFS on the individual data nodes. Although the list of properties covered in this section is not exhaustive, it provides a deeper understanding of the HDFS design at the physical and operational level.

This section explores the key properties of the hdfs-*.xml file. Some of the important properties of the hdfs-site.xml file include the following:

- dfs.namenode.name.dir: Directories on the local file system of the NameNode in which the metadata file table (the fsimage file) is stored. Recall that this file is used to store the HDFS metadata since the last snapshot. If this is a comma-delimited list of directories, the file is replicated to all the directories for redundancy. (Ensure that there is no space after the comma in the comma-delimited list of directories.) The default value for this property is file://${hadoop.tmp.dir}/dfs/name. The hadoop.tmp.dir property is specified in the core-site.xml (or core-default.xml if core-site.xml does not override it).
- dfs.namenode.edits.dir: Directories on the local file system of the NameNode in which the metadata transaction file (the edits file) is stored. This file contains changes to the HDFS metadata since the last snapshot. If this is a comma-delimited list of directories, the transaction file is replicated to all the directories for redundancy. The default value is the same as dfs.namenode.name.dir.
- dfs.namenode.checkpoint.dir: Determines where the Secondary NameNode should store the temporary images to merge on the local/network file system accessible to the Secondary NameNode. Recall from Chapter 2 that this is the location where the fsimage file from the NameNode is copied into for merging with the edits file from the NameNode. If this is a comma-delimited list of directories, the image is replicated in all the directories for redundancy. The default value is file://${hadoop.tmp.dir}/dfs/namesecondary.
- dfs.namenode.checkpoint.edits.dir: Determines where the Secondary NameNode should store the edits file copied from the NameNode to merge the fsimage file copied in the folder defined by the dfs.namenode.checkpoint.dir property on the local/network file system accessible to the Secondary NameNode. If it is a comma-delimited list of directories, the edits files are replicated in all the directories for redundancy. The default value is the same as dfs.namenode.checkpoint.dir.
- dfs.namenode.checkpoint.period: The number of seconds between two checkpoints. As an interval equal to this parameter elapses, the checkpoint process begins, which merges the edits file with the fsimage file from the NameNode. (This process is described in detail in Chapter 2.)
- dfs.blocksize: The default block size for new files, in bytes. The default is 128 MB. Note that block size is not a system-wide parameter; it can be specified on a per-file basis.
- dfs.replication: The default block replication. Although it can be specified per file, if not specified it is taken as the replication factor for the file. The default value is 3.
- dfs.namenode.handler.count: Represents the number of server threads the NameNode uses to communicate with the DataNodes. The default is 10, but the recommendation is about 10% of the number of nodes, with a minimum value of 10. If this value is too low, you might notice messages in the DataNode logs indicating that the connection was refused by the NameNode when the DataNode tried to communicate with the NameNode through heartbeat messages.

- `dfs.datanode.du.reserved`: Reserved space in bytes per volume that represents the amount of space to be reserved for non-HDFS use. The default value is 0, but it should be at least 10 GB or 25% of the total disk space, whichever is lower.
- `dfs.hosts`: This is a fully qualified path to a file name that contains a list of hosts that are permitted to connect with the NameNode. If the property is not set, all nodes are permitted to connect with the NameNode.

## mapred-site.xml

This section explores the key properties of the `mapred-site.xml` file. Note that although MR v1 has now been replaced with YARN, the original terminology of JobTracker and TaskTracker has carried forward into the `mapred-site.xml` configuration file. Before these properties are discussed, here is a quick recap of MR v2 and a review of how the MR v1 services map to MR v2 services.

In Hadoop 2, MapReduce has been split into two components. The cluster resource management responsibilities are now delegated to YARN, whereas the MapReduce-specific capabilities are now a separate Application Master for MapReduce jobs. (YARN architecture is discussed in Chapter 2.)

In MR v1, the Hadoop cluster was managed by the JobTracker service. The TaskTracker services that ran on the individual nodes were responsible for launching the Map and Reduce tasks and notifying JobTracker about their health and progress. JobTracker kept track of the progress of the overall cluster and the jobs running/completed on the cluster.

In MR v2, the function of the JobTracker has now been split into four services:

- Resource Manager, which is a YARN service, receives requests for and launches the Application Master instance. The Resource Manager arbitrates all the cluster resources enabling the management of the distributed applications running on the YARN system.
- A pluggable (in the Resource Manager) scheduler component. The role of the Scheduler is to allocate resources to running processes. It does not monitor or track resources. The Resource Manager has other services which perform the role of Monitoring and Tracking processes.
- MapReduce is an application with its own Application Master launched by the Resource Manager. The MapReduce Application Master manages its own MapReduce job and is shut down when the job completes. This is a departure from MR v1, in which JobTracker was a continuously running service and presented a single point of failure.
- The JobTracker responsibility of providing information about completed jobs has moved to the Job History Server.

The Task Manager functionality is now moved to the Node Manager, which is responsible for launching containers, each of which can house Map or Reduce tasks. These relationships are depicted in Figure 4-1.

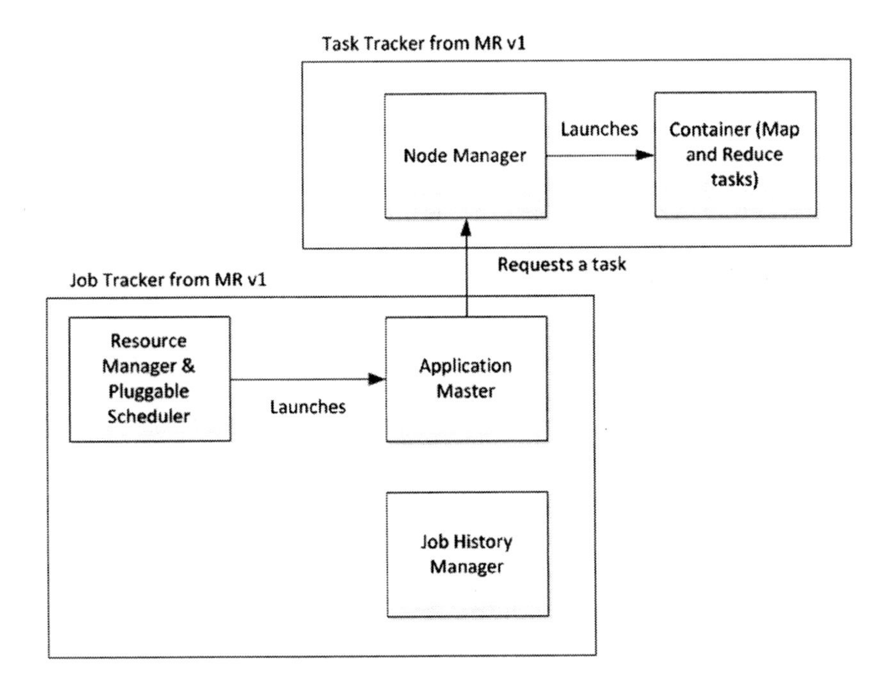


***Figure 4-1.*** *Mapping Between MR v1 and MR v2*

When reading the properties in the following list, bear in mind the relationships in the figure. The terminology has carried over from MR v1. Although the properties still imply that there are multiple TaskTrackers and one JobTracker, the actual service that the parameter refers to depends on the context provided by the previous discussion.

Some of the important properties defined in the `mapred-site.xml` file are these:

- `mapreduce.framework.name`: Determines whether the MapReduce job is submitted to a YARN cluster or runs locally using Local Job Runner. The valid values for this property are `yarn` and `local`.
- `mapred.child.java.opts`: The JVM heap size for Map or Reduce tasks. The default is `-Xmx200m` (200 MB of heap space). It is recommended that a higher value be specified such as 512 MB or 1 G, depending on the requirements. The value should be less than or equal to the value specified in the `mapreduce.map.memory.mb` and `mapreduce.reduce.memory.mb` properties as these properties are used by the Application Master to negotiate resources from the Resource Manager. The Node Manager is responsible for starting the containers using the JVM heap size specified by these properties. The actual Map and Reduce tasks are executed within the container using the JVM heap size specified by the `mapred.child.opts` property. If the value of `mapred.child.opts` is higher than the property of the `mapreduce.*.memory.mb` properties, the tasks fail (discussed in the section "Memory Allocations in YARN)."
- `mapreduce.map.memory.mb`: Specifies the amount of memory that needs to be allocated to the container that runs the Map Task. The default value is 1024 MB.
- `mapreduce.reduce.memory.mb`: Specifies the amount of memory that needs to be allocated to the container that runs the Reduce task. The default value is 1024 MB.

- mapreduce.cluster.local.dir: Local directory in which MapReduce stores intermediate data files. It can be a comma-separated list of directories on different devices to improve disk I/O. Examples of these files are the output of the Mapper processes and intermediate results of the sort during the shuffle-sort process. The default value is ${hadoop.tmp.dir}/mapred/local; this is where the value dfs.du.reserved (specified in the core-site.xml) matters. The general recommendation is to set this configuration to allow not more than 75% of the disk to be used by HDFS, enabling 25% of the disk to be used for intermediate data files.
- mapreduce.jobtracker.handler.count: Number of server threads used by the JobTracker (equivalent in YARN). It should be roughly 4% of the number of slave nodes, with a minimum value of 10.The default value is 10.
- mapreduce.job.reduce.slowstart.completedmaps: Percentage of Maps in the job that should be complete before Reducers are scheduled for the job. The default is 0.05, but it is recommended to use a value between 0.5 and 0.8.
- mapreduce.jobtracker.taskscheduler: Class responsible for scheduling the tasks. The default is the FIFO scheduler that is specified as org.apache.hadoop.mapred.JobQueueTaskScheduler. However, it is recommended to use the Fair Scheduler or Capacity Scheduler (discussed in the section called "Scheduler," later in this chapter).
- mapreduce.map.maxattempts: Maximum number of attempts per Map task. The framework attempts to execute a Map task a certain number of times before giving up and failing the job. The default is 4.
- mapreduce.reduce.maxattempts: Maximum number of attempts per Reduce task. The framework attempts to execute a Reduce task a certain number of times before giving up and failing the job. The default is 4.

## yarn-site.xml

MapReduce has undergone a complete overhaul from hadoop-0.23. It is now known as MapReduce 2.0 (MRv2) or YARN. The fundamental MR v2 is Resource Management, which has been separated from job execution. Each type of job is managed by a type-specific Application Master that requests resources from the generalized Resource Manager using a well-defined protocol. The yarn-site.xml file is used to configure the properties of generalized service daemons provided by the YARN framework such as Resource Manager and Node Manager.

The yarn-default.xml and yarn-site.xml files are both read by Hadoop YARN daemons. The yarn-default.xml file provides the defaults, and the yarn-site.xml file overrides the properties in the yarn-default.xml. Some of the key properties defined in yarn-*.xml include the following:

- yarn.resourcemanager.hostname: The host name of the Resource Manager.
- yarn.resourcemanager.address: The host name and port number on which the Resource Manager's server runs. The default value is http://${*yarn.resourcemanager.hostname*}:8032.
- yarn.nodemanager.local-dirs: A comma-separated list of local directories in which the containers initiated by the Node Manager store files during the course of their execution. Such files are typically additional configurations, libraries needed by the job of every master data files distributed via the distributed cache. (Distributed cache is discussed in Chapter 6, "Advanced MapReduce Development.") These files are deleted when the application terminates.
- yarn.nodemanager.aux-services: A comma-separated list of auxiliary services executed by the Node Manager. Its default value is blank. In Chapter 3, you configured the Hadoop installation to use the auxiliary service mapreduce_shuffle. In Hadoop 2.x, the Shuffle service is configured as an auxiliary service in yarn-site.xml. See Chapter 3 for additional details on how the mapreduce_shuffle needs to be configured.

- yarn.nodemanager.resource.memory-mb: The total amount of physical memory that can be allocated to the containers instantiated by the Node Manager executing on a node. The default value is 8192. Most nodes have significantly higher memory than 8GB and this value should be increased after making allowance for memory needed by OS and memory needed to run other Hadoop daemons.
- yarn.nodemanager.vmem-pmem-ratio: Ratio between virtual memory to physical memory when setting memory limits for containers. Container allocations are expressed in terms of physical memory, and virtual memory usage is allowed to exceed this allocation by this ratio. If the value yarn.scheduler.maximum-allocation-mb is left at its default value of 8192, a setting of 2.1 for the parameter allows the virtual memory upper limit for each container to be 8 GB x 2.1 = 16.2 GB. If this ratio is exceeded, the YARN framework might fail the containers.
- yarn.scheduler.minimum-allocation-mb: Minimum allocation for every container request at the Resource Manager, in megabytes. Memory requests lower than this are ignored, and the specified value gets allocated the minimum. The default value is 1024, which amounts to 1 GB.
- yarn.scheduler.maximum-allocation-mb: Maximum allocation for every container request at the Resource Manager, in megabytes. Memory requests higher than this are capped at this value. The default value is 8192, which amounts to 8 GB.
- yarn.scheduler.minimum-allocation-vcores: Minimum allocation for every container request at the Resource Manager, in terms of virtual CPU cores. Requests lower than this do not take effect, and the specified value gets allocated the minimum. The default value is 1.
- yarn.scheduler.maximum-allocation-vcores: Maximum allocation for every container request at the Resource Manager, in terms of virtual CPU cores. Requests higher than this do not take effect and get capped to this value. The default value is 32.

## Memory Allocations in YARN

As discussed in the section "Configuring Hadoop Daemons," Hadoop allocates 1,000 MB of memory by default for each of the Hadoop daemons, so the Node Manager and DataNode take up 1,000 MB each on each slave node. The property yarn.nodemanager.resource.memory-mb, whose default value is 8,192 MB, specifies the remaining memory used for the containers. Using the default setting ensures that there is at least 8,192 MB plus 2,000 MB available on the slave node to execute the Hadoop containers.

When the Application Master requests a Map or Reduce task, the Resource Manager requests the Node Manager to start a container with the size specified by the property mapreduce.map.memory.mb (or mapreduce.reduce.memory.mb). It reduces the amount of memory available to launch additional containers on the node in which the container is being launched. The total memory (in MB) allocated to start containers in YARN is defined by the yarn.nodemanager.resource.memory-mb property.

If the machine has 36 GB of RAM, and it is decided to utilize 24 GB for YARN containers and 8 GB for everything else, the value of this property is defined as 24576. If the value of the mapreduce.map.memory.mb is defined as 2 GB (2048), and a container for a Map task is launched, the amount of memory available to start additional containers drops from 24 GB to 22 GB.

After the container is launched by the Node Manager, it launches a child JVM Map or Reduce task with a heap size specified by the mapred.child.java.opts property. Note that the parent process (the launched container) should contain more memory than the child JVM (Map or Reduce task) because additional memory is needed to host entities such as native libraries. The requirement is that the container JVM memory size represents the maximum memory that can be used by the container or its launched child tasks. If this constraint is broken, the Node Manager terminates the container and marks it as failed.

Figure 4-2 depicts the relationship between the Container task and the Map/Reduce tasks launched within the container.

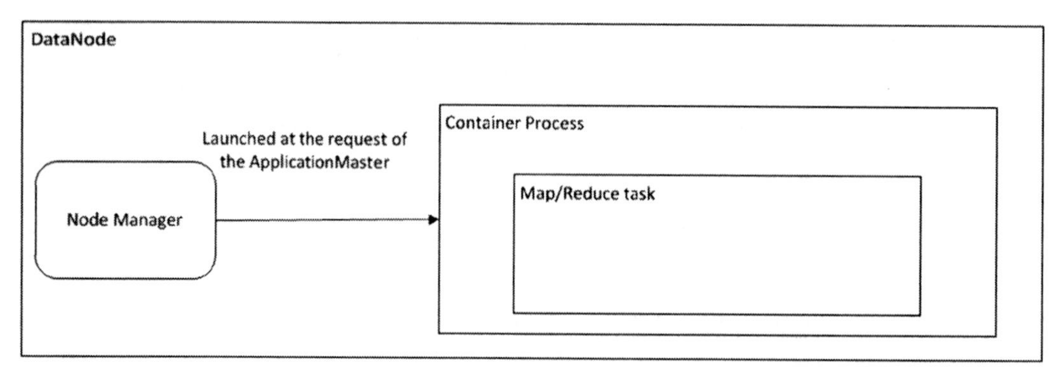


**Container** : Heap size defined by mapreduce.map.memory.mb or mapreduce.reduce.memory.mb property)
**Map/Reduce task** : Heap size defined by the mapred.child.java.opts property). Launched by the Container task to execute as a child process

mapreduce.map/reduce.memory.mb>mapred.child.java.opts

***Figure 4-2.*** *Relationship between a Container task and a Map/Reduce Task*

Additional constraint parameters can be imposed by the schedulers (as discussed in the following section).

# Scheduler

Early versions of Hadoop used only the FIFO scheduler. As the name suggests, the FIFO scheduler allocates resources based on the order in which the jobs are submitted to the cluster. It can cause a long-running job to monopolize the cluster resources and prevent smaller jobs from running. Subsequently, job priority was added to Hadoop. It can be configured by setting the `mapred.job.priority` property for each job to one of the following values:

- `VERY_HIGH`
- `HIGH`
- `NORMAL`
- `LOW`
- `VERY_LOW`

Although the ability to control Job Priority using the `mapred.job.priority` property alleviates the problem a bit, without support for job preemption, it is still possible for a lower-priority job to monopolize resources (as discussed earlier) to prevent a higher-priority job from running. This can happen if the lower-priority job is launched before the higher-priority job enters the queue. In this scenario, without a preemption feature, a long-running, lower-priority job prevents a higher-priority job from running until it completes. More-complex schedulers were added subsequently to handle this problem: the Fair Scheduler and Capacity Scheduler, which will be discussed in detail.

The scheduler, which is a component of the Resource Manager, is responsible for allocating resources to the running applications. Resources are subject to constraints such as system capacity and queues. The role of the scheduler is to support multitenancy in the Hadoop cluster. In multitenancy, multiple organization groups can execute their applications simultaneously on the Hadoop cluster with minimum resource guarantees.

In an Enterprise scenario, each organization typically has sufficient resources to meet its SLAs under peak conditions. Because each organization is not operating under peak conditions most of the time, and because different organizations within the Enterprise have peak conditions occurring at different times, there is poor overall utilization of the cluster. Multitenancy aims to utilize a common cluster for the Enterprise that is shared by various organizations to meet each organization's SLAs under peak conditions while achieving a high average utilization of the cluster. To fulfill this role, a scheduler needs to have the following goals:

- Ensure that users and applications get a guaranteed share of the cluster
- Maximize overall utilization of the cluster

A scheduler has a pluggable plug-in policy. Hadoop 2.2.0 supports two types of pluggable schedulers: Capacity Scheduler and Fair Scheduler. Ensuring a pluggable policy allows developers or third-party vendors to create specialized schedulers.

The next two sections discuss Capacity Scheduler and Fair Scheduler in detail.

## Capacity Scheduler

Capacity Scheduler currently allocates resources using a unit of allocation based on the amount of memory (it will include CPU in the future). This is a departure from Hadoop 1.x, in which the units of allocation were Map/Reduce slots.

The Capacity Scheduler supports multitenancy through the following configurations:

- Configure guaranteed capacities across organizational groups
- Configure resource limits to ensure against a single large application from monopolizing the cluster resource
- Configure resource and access control limits for various users within an organization group

The Capacity Scheduler can be configured by making the setting listed in Table 4-4 the property in the `$HADOOP_CONF_DIR/yarn-site.xml` file.

***Table 4-4.*** *Configuring the Capacity Scheduler*

| Property | Value |
| --- | --- |
| `yarn.resourcemanager.scheduler.class` | `org.apache.hadoop.yarn.server.resourcemanager.scheduler.capacity.CapacityScheduler` |

The Capacity Scheduler has its own configuration file, which is located in `$HADOOP_CONF_DIR/capacity-scheduler.xml`.

The Capacity Scheduler is based on the concept of hierarchical queues and has a predefined queue called `root`. All queues in the system are the children of the `root` queue, and the children can have their own children. Queue capacity and limits can be configured at a group level after which they can be reallocated for members of the group.

In the following sections, the preceding concepts will become clear. We assume the following scenario—the imaginary organization is divided into three projects:

- Nightly ETL of the online transaction processing (OLTP) databases. Transaction data is aggregated in the Hadoop cluster for data warehousing purposes.
- Tracking user behavior on the e-commerce web site.
- Resource planning application for the Manufacturing group.

Users of the three projects are divided into the following three user groups, respectively:

- Operations
- Marketing
- Manufacturing

## Configuring Capacity Guarantees Across Organizational Groups

The Capacity Scheduler enables minimum capacity to be guaranteed across a group of users and applications. It configures `root` level queues for each of the groups: Operations, Marketing, and Manufacturing: `yarn.scheduler.capacity.root.queues="operationsmarketing, manufacturing"`.

The actual syntax for configuring the preceding in the Capacity Scheduler configuration file `$HADOOP_CONF_DIR/capacity-scheduler.xml` is shown here. For the sake of brevity, the previous notation is used instead of the actual notation in the XML syntax:

```
<property>
  <name>yarn.scheduler.capacity.root.queues</name>
  <value>operations, marketing, manufacturing</value>
  <description>The queues below the root queue(root is the root queue).
  </description>
</property>
```

Next, the cluster capacity is divided among the preceding groups in terms of percentage of the overall cluster capacity (100%). At this point, minimum capacity guarantees are provided to each of the previous groups:

```
yarn.scheduler.capacity.root.operations.capacity=40
yarn.scheduler.capacity.root.marketing.capacity=20
yarn.scheduler.capacity.root.manufacturing.capacity=40
```

The previously mentioned limits are not for MapReduce jobs only; they define minimum guarantees for all processes running in the Hadoop (YARN) 2.x cluster. This is an important distinction. YARN was designed to allow pluggable distributed frameworks, so it supports frameworks other than MapReduce. Resources are allocated as containers by the Node Manager at the request of the scheduler (which runs in the Resource Manager). Thus, the Operations group is provided a minimum guarantee of 40% of the cluster, regardless of the nature of the jobs it runs. Similarly, the Marketing group is guaranteed 20%, and the Manufacturing group is guaranteed 40% of the cluster resources.

## Enforcing Capacity Limits

Sometimes the applications of one of the groups (queues) takes up considerably lower than the minimum allocated resources to it. At this time, it is important to allocate this unused capacity to other queues. For example, in this example, the Operations and the Manufacturing groups may run the bulk of their jobs as nightly processes. During the day, the cluster might be used more by the `marketing` queue. If the cluster has 50% unused capacity, it is appropriate to allocate this capacity to the Marketing group. However, we want to limit the maximum capacity that can be allocated to the Marketing group. We set this maximum limit as 50% by using the following configuration in the `yarn.xml` configuration file:

```
yarn.scheduler.capacity.root.Marketing.maximum-capacity=50
```

A typical Hadoop cluster has long-running jobs and short-running jobs. One of the limitations of the FIFO scheduler from Hadoop 1.x is that a long-running job can monopolize the cluster resources and prevent the short-running jobs from executing for a long time. This is a real risk, and the appropriate way to mitigate it is to create subgroups that are meant only for long-running jobs and limit the resources allocated to them. The following example is a configuration in which a separate queue is defined for a long-running job (typical in ETL applications) with a capacity of minimum 20% and maximum 20%:

```
yarn.scheduler.capacity.root.queues="operations,operations-long, marketing, manufacturing"
yarn.scheduler.capacity.root.operations-long.capacity=20
yarn.scheduler.capacity.root.operations-long.maximum-capacity=20
```

## Enforcing Access Control Limits

Finally, users in each group need to be limited. In a typical Hadoop system, power users have more cluster resource privileges than others due to the critical nature of the jobs they execute. A good system should give such users higher priority without keeping regular users from using the cluster. First, we configure the following to allow users to submit jobs to a queue:

```
yarn.scheduler.capacity.root.operations.acl_submit_applications="joe,james,ahmed,ricardo,maria"
```

Next, we limit their ability to request resources on the cluster:

```
yarn.scheduler.capacity.root.operations.minimum-user-limit-percent=20
```

The configuration limits any user running a job in the `operations` queue to 20% of the capacity of the `operations` queue.

## Updating Capacity Limit changes in the Cluster

The Capacity Scheduler queues, capacities, and user ACLs can be dynamically updated. To achieve this, we update the preceding configuration and run the following administrative command on the cluster to refresh the scheduler configuration without restarting any services:

```
$HADOOP_YARN_HOME/bin/yarn rmadmin -refreshQueues
```

## Capacity Scheduler Scenario

In the previous section, five users we assigned to the Operations group, and each of these users has been assigned 20% of the total queue capacity. Let us assume that user Joe submits a job, and Joe is the only user from the Operations group active in the cluster. In this scenario, the entire capacity of the Operations group is assigned to Joe.

If user Ahmed submits another job, the `operations` queue capacity is split evenly (50% to Joe and 50% to Ahmed) between Joe and Ahmed. If for any reason Ahmed needs only 30% capacity, Joe will be assigned the remaining 70%. Note that no tasks are killed to make way for Ahmed's jobs. A typical job in the Hadoop cluster comprises several tasks that are scheduled. As Joe's tasks complete, new tasks are not allocated to Joe unless the resource utilization for Joe falls below the prescribed limits.

The previous process continues until all five users submit their jobs, and each has 20% of the Operations group capacity. As each user's job completes, the additional capacity is allocated to the remaining users.

If a sixth user arrives, the cluster capacity is not further divided. The Capacity Scheduler first attempts to complete the jobs of the five users. Only after one of the jobs completes are the resources allocated to the sixth user's job. This feature prevents the cluster resources from being divided excessively by a large number of users.

The Capacity Scheduler offers several configuration knobs. Please refer to the Apache documentation for the various configuration parameters supported by the Capacity Scheduler.

# Fair Scheduler

The goals of the Fair Scheduler are similar to those of the Capacity Scheduler, but there are subtle differences. The Capacity Scheduler is designed so that a FIFO cluster is simulated for each user of each group, but the Fair Scheduler is designed to ensure that, on average, each job gets an equal share of resources. If there is a single job running, it takes up the entire cluster. As more jobs are submitted, the freed resources are divided between jobs so that each job gets approximately the same amount of resources.

The motivation for the Fair Scheduler began at Facebook, which initially had few jobs to analyze the content and log data accumulated on daily basis. However, other groups within Facebook started using Hadoop, and the number of production jobs increased. Eventually, users started submitting *ad hoc* queries through Hive (Facebook developed SQL as a query language for Hadoop). These short-running jobs often competed for resources with long-running jobs. A separate cluster for long-running jobs would have been very expensive, not just from a cost perspective but also from an operations perspective because the same data would have to be present on both clusters. Facebook developed the Fair Scheduler to allocate resources evenly between jobs while offering capacity guarantees.

The Fair Scheduler organizes applications into queues and shares resources fairly between these queues. The default queue is named `default`. If an application requests a specific queue in its container resource request, it is submitted to that queue. Queues can be assigned by default based on configuration. Application assignment to the queue in such cases depends on the user's principal information such as the username or the UNIX group to which the user is assigned.

The Fair Scheduler can be configured by making the setting in Table 4-5 the property in the `$HADOOP_CONF_DIR/yarn-site.xml` file.

***Table 4-5.*** *Configuring the Fair Scheduler*

| Property | Value |
| --- | --- |
| `yarn.resourcemanager.scheduler.class` | `org.apache.hadoop.yarn.server.resourcemanager.scheduler.fair.FairScheduler` |

The goals for the Fair Scheduler are the following:

- Allow multiple users to share the cluster resources
- Allow short *ad hoc* jobs to share the cluster with long-running jobs
- Allow cluster resources to be managed proportionately, ensuring the efficient utilization of the cluster resources

## Fair Scheduler Configuration

Similar to the Capacity Scheduler, the Fair Scheduler allows configuration of hierarchical queues. The name of the topmost queue is `root`, and all the queues descend from this queue. Resources are allocated to applications that are scheduled only on the leaf queues.

Similar to the Capacity Scheduler scenario, there are four queues for the Fair Scheduler:

- `root.operations`
- `root.operations-long`
- `root.marketing`
- `root.manufacturing`

Each queue can be configured with its own custom policy. A new custom policy can be created by extending the class as follows:

```
org.apache.hadoop.yarn.server.resourcemanager.scheduler.fair. SchedulingPolicy
```

The following policies are available:

- FIFO
- fair (default): Bases its scheduling decision only on the basis of memory.
- DominantResourceFairness (drf): Scheduling decisions based on both memory and CPU, which was developed by Ghodsi *et al.* The original paper is named "Dominant Resource Fairness: Fair Allocation of Multiple Resource Types."

Configuring the Fair-Scheduler involves updating two sets of files:

- **$HADOOP_CONF_DIR/yarn-site.xml:** Scheduler-wide options are set in this file.
- **Allocation File Listing:** The file location that contains this listing is configured in a yarn-site.xml property called yarn.scheduler.fair.allocation.file. The actual queues, their capacities and other properties are configured in this file. This file is reloaded every 10 seconds, which allows changes to the queues to be applied at runtime.

## yarn-site.xml Configurations

This section covers the scheduler-wide options set in the $HADOOP_CONF_DIR/yarn-site.xml file. Some of the configurations that can be made to the Fair Scheduler through this file are the following:

- yarn.scheduler.fair.allocation.file: Path to the allocation file. An allocation file is an XML manifest describing queues and their properties, in addition to certain policy defaults. This file must be in the XML format described in the next section. If a relative path is given, the file is searched for on the classpath (which typically includes $HADOOP_CONF_DIR). The default value is fair-scheduler.xml.
- yarn.scheduler.fair.preemption: Whether to use preemption if a queue is falling below minimum resource limits. Preemption causes containers in other queues to be prematurely terminated to make resources available to the underutilized queue. Remember that it does not cause a job to fail because both Map and Reduce processes have maximum attempts configured in the mapred-site.xml configuration file. The default value for these parameters is 4, which means that the job is not failed as long as the same Map or Reduce process does not fail more than four times. However, note that preemption is experimental in version 2.2.0. The default value is false and it is prudent to continue maintaining it as false in version 2.2.0.
- yarn.scheduler.fair.sizebasedweight: Determines whether the resource allocation shares to individual applications are based on their size rather than providing an equal share to all apps regardless of size. When set to true, applications are weighted by the natural logarithm of one plus the applications total requested memory, divided by the natural logarithm of 2. The default value is false.

For an exhaustive list of configurations, refer to the Hadoop documentation.

## Allocation File Format and Configurations

This is a custom file configured in the yarn-site.xml file. It is used to configure the queues and their runtime attributes such as minimum and maximum resources, weight scheduling policy, subqueues, and user-level constraints within the queues. A sample allocation file listing is shown here:

```
<?xml version="1.0"?>
<allocations>
  <queue name="sample_queue">
    <minResources>10000 mb,0vcores</minResources>
    <maxResources>90000 mb,0vcores</maxResources>
    <maxRunningApps>50</maxRunningApps>
    <weight>2.0</weight>
    <schedulingPolicy>fair</schedulingPolicy>
    <queue name="sample_sub_queue">
      <minResources>5000 mb,0vcores</minResources>
    </queue>
  </queue>
  <user name="sample_user">
    <maxRunningApps>30</maxRunningApps>
  </user>
  <userMaxAppsDefault>5</userMaxAppsDefault>
</allocations>
```

The preceding file shows that the Fair Scheduler supports hierarchical queues. A child queue inherits properties from the parent queue. A child queue can modify any property from the parent queue. In the file, the root tag is allocations, which contains the queue tags that are used to configure the individual queues. The configuration elements are described here.

Queue represents the property of queues. It contains the following elements:

- minResources: The minimum resources the queue is entitled to. It is defined as "X mb, Y cores". When the schedulingPolicy attribute value is fair, the cores specification is ignored, and the shares are based only on memory. If a queue's minimum share is not satisfied, it is offered resources before any other queue under the same parent.
- maxResources: The maximum resources a queue is allowed. It is defined as "X mb, Y cores". Similar to minResources, when the schedulingPolicy attribute value is fair, the cores specification is ignored and the shares are only based on memory. A queue is never assigned a container so that its aggregate usage exceeds this limit.
- maxRunningApps: Applies a maximum limit on the number of applications that can be executed from the queue.
- weight: Allows cluster resources to be shared nonproportionately. The default value is 1. If it is set to 2, the queue receives twice as many resources as a queue with the default weight.
- schedulingPolicy: Defines the scheduling policy of jobs for a specific queue, as discussed previously. The allowable values are FIFO, fair, and drf. The default value is fair. Alternatively, a custom schedulingPolicy can be defined, which is a class name that extends the org.apache.hadoop.yarn.server.resourcemanager.scheduler.fair.SchedulingPolicy class. This property interacts with the minResources property. For example, under fair configuration, a queue is considered unsatisfied if its memory usage falls below its minimum memory share. Under drf configuration, a queue is considered unsatisfied if the usage of its dominant resource with respect to cluster capacity falls below the minimum share for that resource. (Dominant resource sharing is discussed in the next section.)

`User` represents settings governing the behavior of individual users. It contains the following properties:

- `maxRunningApps`: Limits the number of applications that can be simultaneously run by a user.
- `userMaxAppsDefault`: Applies to users whose `maxRunningApps` setting is not explicitly specified.
- `defaultQueueSchedulingPolicy`: Applies to queues whose `schedulingPolicy` element is not explicitly specified. The default value is `fair`.

---

■ **Note** In earlier versions of Hadoop, the term used for "queue" was "pool." For the purposes of backward compatibility, the preceding queue elements can also be called pool elements.

---

## Determine Dominant Resource Share in drf Policy

Under the `drf` policy, two types of resources can be specified:

- Memory in MB
- CPU in cores

Queues can utilize different amounts of each resource subject to its `minResouces` and `maxResources` limitations. Consider a cluster in which the total amount of cluster resources is the following:

- 1,000 GB
- 300 CPU core

Assume two queues: Q1 and Q2. Assume that jobs running in these queues utilize cluster resources as follows and shown in Figure 4-3.

- Q1: 600 GB, 100 CPU cores
- Q2: 200 GB, 200 CPU cores

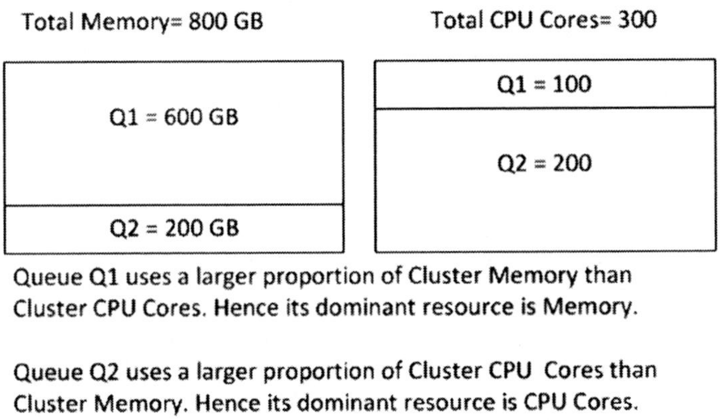

Total Memory= 800 GB

Q1 = 600 GB

Q2 = 200 GB

Total CPU Cores= 300

Q1 = 100

Q2 = 200

Queue Q1 uses a larger proportion of Cluster Memory than Cluster CPU Cores. Hence its dominant resource is Memory.

Queue Q2 uses a larger proportion of Cluster CPU Cores than Cluster Memory. Hence its dominant resource is CPU Cores.

***Figure 4-3.*** *Depiction of Queue Utilization of Server Resources*

Based on the distribution of Memory and CPU resources between queues, you can infer that Q1 is running I/O-intensive jobs, and Q2 is running CPU-intensive jobs. Q1 accounts for 60% of the overall cluster memory and 33% of the CPU cores. Q2 accounts for 20% of the cluster memory and 66% of CPU cores, so the dominant resource for Q1 is memory, and the dominant resource for Q2 is the CPU cores.

Now assume that we change the `minResources` property for Q1 to 700,000 MB, 50 vcores and that for Q2 as 100,000 MB, 250 vcores. Due to this change, under the `drf` scheduling policy, both of the queues are not satisfying their min share. The Fair Scheduler attempts to move some of the CPU core resources from Q1 to Q2 while transferring the memory allocation from Q2 to Q1.

# Slaves File

In a typical setup, one machine in the cluster is designated as the NameNode, one machine is designated as a Secondary NameNode, and one machine is designated as the Resource Manager. Each of the remaining machines is in the role of the Node Manager as well as the DataNode. They are referred to as *slaves*. The host names or IP addresses of these slaves are listed in the `$HADOOP_CONF_DIR/slaves` file as one slave per line.

The rack awareness feature discussed next pertains to making the Hadoop cluster rack aware with respect to the Hadoop and YARN components executing on the slave nodes.

# Rack Awareness

Hadoop is designed to run on a large cluster of commodity machines that are often set up as spanning multiple physical racks. Each physical rack could be a single point of failure due to using a low-cost single switch, so Hadoop attempts to save replicas in separate racks. The replication behavior in a rack-aware cluster is to store one replica in one rack and the remaining replicas (usually two) in another rack.

## Providing Hadoop with Network Topology

By default, Hadoop does not know anything about node locations with respect to racks. It assumes that all nodes belong to one rack: `/default-rack`.

To make the Hadoop cluster rack aware, the following parameters in the `$HADOOP_CONF_DIR/core-site.xml` file have to be updated:

- `topology.node.switch.mapping.impl`: Specifies a Java class implementation. You can plug in your own class here, but the default setting is `org.apache.hadoop.net.ScriptBasedMapping`. This implementation takes a user-provided script. The script is specified in the property discussed in the next bullet.

- `net.topology.script.file.name`: Specifies the name and path of the file that is executed by the Java class specified in the default setting of the parameter in the previous bullet. The role of this script is to take DNS names and resolve them to network topology names. For example, the script would take `10.1.1.1` and return `/rack1` as the output. If the value is not set for this property, the default value of `/default-rack` is returned for all node names.

- `net.topology.script.number.args`: The maximum number of arguments with which the script configured with `net.topology.script.file.name` should be run. Each argument is a DNS name or IP address. The default value is 100.

A sample listing for the setting `net.topology.script.file.name` is shown in Listing 4-1. The listing was borrowed from the following link and modified slightly: `http://wiki.apache.org/hadoop/topology_rack_awareness_scripts`).

***Listing 4-1.*** Sample Listing for net.topology.script.file.name

```
HADOOP_CONF=$HADOOP_HOME/etc/hadoop/

while [ $# -gt 0 ] ; do
  nodeArg=$1
  exec< ${HADOOP_CONF}/topology.data
  result=""
  while read line ; do
    ar=( $line )
    if [ "${ar[0]}" = "$nodeArg" ] ; then
      result="${ar[1]}"
    fi
  done
  shift
  if [ -z "$result" ] ; then
    echo -n "/default/rack "
  else
    echo -n "$result "
  fi
done
```

This script parses a text file from `$HADOOP_CONF_DIR/topology.data` that contains the mapping from the host name or IP address to the rack name. A sample listing is shown in Listing 4-2.

***Listing 4-2.*** Listing for $HADOOP_CONF_DIR/topology.data

```
Hadoop1      /dc1/rack1
Hadoop2      /dc1/rack1
Hadoop3      /dc1/rack1
10.1.1.1     /dc1/rack1
10.1.1.2     /dc1/rack1
10.1.1.3     /dc1/rack2
```

When the cluster starts up, the `dfs.hosts` property in the `hdfs-site.xml` file is read. If it is set to the full path to the `$HADOOP_CONF_DIR/slaves` file mentioned in the earlier section, Listing 4-1 is called with all the host names from the `slaves` files in one call, with the limit for the number of slaves that can be invoked in each call specified by the property `net.topology.script.number.args`. If the property `dfs.hosts` is empty, Listing 4-1 is called once per DataNode, as reported to the NameNode during startup.

# Cluster Administration Utilities

This section reviews the cluster administration utilities that do the following:

- Check the HDFS for corrupt blocks
- Perform command-line Hadoop administration
- Add/remove nodes from the cluster
- Rebalance HDFS data

This list is not exhaustive. Access this page (note the version number in the URL) to get the exhaustive list of commands:

```
http://hadoop.apache.org/docs/r2.2.0/hadoop-project-dist/hadoop-common/CommandsManual.html
```

All commands are invoked by calling the $HADOOP_HOME/bin/hdfs command. The general syntax for invoking the command is as follows:

```
hdfs [--config confdir] [COMMAND] [GENERIC_OPTIONS][COMMAND_OPTIONS]
```

- -config confdir: Overrides the default configuration directory, which is $HADOOP_CONF_DIR. We will not override this property in our examples.
- COMMAND: The name of the command (for example, fsck and dfsadmin).
- GENERIC_OPTIONS: The list of generic parameters that can be passed to the job. The only ones that apply to this section are -config <configuration file>, which specifies an application configuration file, and -D <property>=<value>, which allows passing property name/value pairs to the command. Although this is similar to passing system properties to a Java command line, note the difference in the syntax: there is a space between -D and <property>.
- COMMAND_OPTIONS: Additional command-specific parameters.

## Check the HDFS

The File System Check (fsck) utility is used to check the HDFS for corrupt or missing data blocks. The syntax for this command is the following:

```
hdfs fsck [GENERIC_OPTIONS] <path> [-move | -delete | -openforwrite] [-files [-blocks [-locations |
-racks]]]
```

Table 4-6 shows the various command options of the fsck command.

***Table 4-6.*** *fsck Command Options*

| Property | Value |
| --- | --- |
| Path | HDFS path being checked. To check all files, provide the value as "/". |
| -move | Moves corrupted files to /lost+found folder |
| -delete | Deletes corrupted files. |
| -openForWrite | Prints out files that are opened for write. |
| -files | Prints out files being checked. |
| -blocks | Prints out block reports. |
| -locations | Prints out locations for every block. |
| -racks | Prints out network topology for data node locations. It is an alternative to -locations. |

Sample fsck commands are listed here:

- `bin/hdfs fsck/`: Gets a report for the entire file system. A sample output follows:

```
Connecting to namenode via http://ip-10-139-18-58.ec2.internal:9101
FSCK started by hadoop (auth:SIMPLE) from /10.139.18.58 for path / at Sun Jan 12 16:52:17 UTC 2014
..Status: HEALTHY
 Total size:    288324 B
 Total dirs:    14
 Total files:   2
 Total symlinks:                0
 Total blocks (validated):      2 (avg. block size 144162 B)
 Minimally replicated blocks:   2 (100.0 %)
 Over-replicated blocks:        0 (0.0 %)
 Under-replicated blocks:       0 (0.0 %)
 Mis-replicated blocks:         0 (0.0 %)
 Default replication factor:    1
 Average block replication:     1.0
 Corrupt blocks:                0
 Missing replicas:              0 (0.0 %)
 Number of data-nodes:          2
 Number of racks:               1
FSCK ended at Sun Jan 12 16:52:17 UTC 2014 in 6 milliseconds
```

- `bin/hdfs fsck / -files`: Gets a report for all files in the HDFS. A sample output is as follows:

```
Connecting to namenode via http://ip-10-139-18-58.ec2.internal:9101
FSCK started by hadoop (auth:SIMPLE) from /10.139.18.58 for path / at Sun Jan 12 16:53:54 UTC 2014
/ <dir>
/tmp <dir>
/tmp/hadoop-yarn <dir>
/tmp/hadoop-yarn/staging <dir>
/tmp/hadoop-yarn/staging/hadoop <dir>
/tmp/hadoop-yarn/staging/hadoop/.staging <dir>
/tmp/hadoop-yarn/staging/history <dir>
/tmp/hadoop-yarn/staging/history/done <dir>
/tmp/hadoop-yarn/staging/history/done/2014 <dir>
/tmp/hadoop-yarn/staging/history/done/2014/01 <dir>
/tmp/hadoop-yarn/staging/history/done/2014/01/12 <dir>
/tmp/hadoop-yarn/staging/history/done/2014/01/12/000000 <dir>
/tmp/hadoop-yarn/staging/history/done/2014/01/12/000000/job_1389544292752_0001-
1389544406547-hadoop-streamjob4494792172397765215.jar-1389544533181-19-5-SUCCEEDED-
default.jhist 205681 bytes, 1 block(s):  OK
/tmp/hadoop-yarn/staging/history/done/2014/01/12/000000/job_1389544292752_0001_conf.xml
82643 bytes, 1 block(s):  OK
/tmp/hadoop-yarn/staging/history/done_intermediate <dir>
/tmp/hadoop-yarn/staging/history/done_intermediate/hadoop <dir>
Status: HEALTHY
 Total size:    288324 B
 Total dirs:    14
 Total files:   2
 Total symlinks:                0
```

```
  Total blocks (validated):      2 (avg. block size 144162 B)
  Minimally replicated blocks:   2 (100.0 %)
  Over-replicated blocks:        0 (0.0 %)
  Under-replicated blocks:       0 (0.0 %)
  Mis-replicated blocks:         0 (0.0 %)
  Default replication factor:    1
  Average block replication:     1.0
  Corrupt blocks:                0
  Missing replicas:              0 (0.0 %)
  Number of data-nodes:          2
  Number of racks:               1
 FSCK ended at Sun Jan 12 16:53:54 UTC 2014 in 8 milliseconds
```

- `bin/hdfs fsck / -files -block`: Gets a block report.
- `bin/hdfs fsck / -files -block -locations`: Gets a block report with the locations of each block reported.
- `bin/hdfs fsck / -files -block -racks`: Gets a block report with the racks hosting each block reported.

---

■ **Note** It is a good practice to run the `fsck` command during low usage times as a scheduled job to proactively identify problems in the HDFS.

---

## Command-Line HDFS Administration

Hadoop includes the `dfsadmin` utility for HDFS administration. This section explores various features of the `dfsadmin` command.

### HDFS Cluster Health Report

Execute the following command to obtain basic statistics about the HDFS cluster heath. It provides the following feedback about the cluster: NameNode status, DataNode status, available disk capacity, and so on.

```
hdfs dfsadmin -report
```

A sample output for the preceding command is as follows:

```
Configured Capacity: 1800711258112 (1.64 TB)
Present Capacity: 1800711258112 (1.64 TB)
DFS Remaining: 1800710930432 (1.64 TB)
DFS Used: 327680 (320 KB)
DFS Used%: 0.00%
Under replicated blocks: 0
Blocks with corrupt replicas: 0
Missing blocks: 0
```

```
-------------------------------------------------
Datanodes available: 2 (2 total, 0 dead)

Live datanodes:
Name: 10.240.106.239:9200 (domU-12-31-39-04-65-01.compute-1.internal)
Hostname: domU-12-31-39-04-65-01.compute-1.internal
Decommission Status : Normal
Configured Capacity: 900355629056 (838.52 GB)
DFS Used: 12288 (12 KB)
Non DFS Used: 0 (0 B)
DFS Remaining: 900355616768 (838.52 GB)
DFS Used%: 0.00%
DFS Remaining%: 100.00%
Last contact: Sun Jan 12 17:01:33 UTC 2014

Name: 10.96.157.76:9200 (domU-12-31-39-16-9E-A2.compute-1.internal)
Hostname: domU-12-31-39-16-9E-A2.compute-1.internal
Decommission Status : Normal
Configured Capacity: 900355629056 (838.52 GB)
DFS Used: 315392 (308 KB)
Non DFS Used: 0 (0 B)
DFS Remaining: 900355313664 (838.52 GB)
DFS Used%: 0.00%
DFS Remaining%: 100.00%
Last contact: Sun Jan 12 17:01:35 UTC 2014
```

To get a more-detailed status, execute the following command:

```
hdfs dfsadmin -metasave filename
```

A sample listing of the preceding `filename` is the following:

```
16 files and directories, 2 blocks = 18 total
Live Datanodes: 2
Dead Datanodes: 0
Metasave: Blocks waiting for replication: 0
Mis-replicated blocks that have been postponed:
Metasave: Blocks being replicated: 0
Metasave: Blocks 0 waiting deletion from 0 datanodes.
Metasave: Number of datanodes: 2
10.240.106.239:9200 IN 900355629056(838.52 GB) 12288(12 KB) 0.00% 900355616768(838.52 GB) Sun Jan 12
17:21:15 UTC 2014
10.96.157.76:9200 IN 900355629056(838.52 GB) 315392(308 KB) 0.00% 900355313664(838.52 GB) Sun Jan 12
17:21:14 UTC 2014
```

## Add/Remove Nodes

Hadoop allows new nodes to be added to the cluster. These nodes can be configured using two properties:

- `dfs.hosts`: As discussed in earlier sections, it is a path to a file that lists nodes that are allowed to communicate with the NameNode.
- `dfs.hosts.exclude`: Fully qualified path of the file that contains a list of hosts that are not permitted to connect to the NameNode.

To add or remove nodes from the cluster, update the previous files and run the following command:

```
hdfs dfsadmin -refreshNodes
```

## Placing the HDFS in Safemode

During startup, the NameNode loads the file system state from the `fsimage` and the `edits` log file. The NameNode waits for DataNodes to report their blocks to avoid prematurely replicating blocks that are adequately replicated. During this operation, the NameNode is in Safemode. During this state, the HDFS can be read, but no updates can be made to the HDFS or blocks. The NameNode leaves this mode when the DataNodes have reported a sufficient amount of block information to the NameNode. It is specified by the `dfs.safemode.threshold.pct` parameter in the `hdfs-site.xml` configuration file.

The HDFS can be manually placed in Safemode, which prevents modifications to the HDFS. No replication is performed in this mode, and no file creation or deletion is allowed. The command syntax to work in the Safemode is the following:

```
hdfs dfsadmin -safemode action
```

The `action` can take the following values:

- `enter`: The HDFS enters the Safemode.
- `leave`: The HDFS is forced to exit the Safemode.
- `get`: Returns a string indicating the current Safemode status as ON or OFF.
- `wait`: Waits until Safemode has exited and returns. It is appropriate for shell scripts that need to perform operations on the HDFS only when the Safemode is OFF.

## Rebalancing HDFS Data

To add new nodes to the Hadoop cluster, follow these steps:

1. Install the same version of Hadoop and apply the same configuration as the rest of the cluster.
2. Add the new node to the slaves file. This file is usually pointed to by the `dfs.hosts` property.
3. Start the DataNode, and it should join the cluster.

The new DataNode does not initially have any data. As new files are added, they are stored in the new DataNode as well as the existing DataNodes. This cluster is currently unbalanced. The `balancer` utility allows the cluster to get balanced.

The balancer utility is implemented by the class

```
org.apache.hadoop.hdfs.server.balancer.Balancer
```

The role of this class is to balance the blocks across the nodes to achieve an even distribution of blocks within a given threshold, expressed as a percentage. The default value is 10%.

The syntax for the balancer script is the following:

```
bin/start-balancer.sh -threshold utilization-threshold%
```

An example invocation is `bin/start-balancer.sh -threshold 5`, which indicates to the `balancer` that the nodes of the HDFS cluster must achieve even distribution within a 5% utilization threshold of each other. Achieving a lower threshold value takes longer, and achieving 0% is impossible. Node utilization in this context is the amount of data storage utilized on a node. A node is underutilized if its utilization is less than the average utilization calculated across all nodes. Conversely, it is overutilized if its utilization is above the average utilization. The overall disk utilization on the nodes is measured. The `balancer` utility does not consider block placements. The number of blocks on each node could be very different, especially if files are using their own block sizes instead of the default block size specified in `dfs.blocksize` property in `hdfs-site.xml`.

The `balancer` automatically terminates when the specified utilization threshold goal is achieved, when an error occurs, or when it cannot find more candidate blocks to move to achieve a better balance. The `balancer` can always be terminated safely by the administrator by running `bin/stop-balancer.sh`.

The balancing script is typically run during low cluster utilization. However, when it needs to be run along with the rest of the jobs in the cluster, there could be concerns about it using large amounts of bandwidth, thereby negatively impacting other jobs. The property of the `dfs.datanode.balance.bandwidthPerSec` configuration parameter can be used to limit the number of bytes per second that each DataNode can use to rebalance its own storage. The default value is 1,048,576 (1 MB/sec). The general recommendation is to use 10% of the network speed. For example, in a 1 GB/sec network, use 10 MB/second.

## Copying Large Amounts of Data from the HDFS

It is common to require that the data from the HDFS be moved to another cluster (for backups) or to move the data from another distributed storage such as Amazon S3 to the HDFS. Hadoop installation comes with a distributed copy (`distcp`) utility. Following is the syntax for the `discp` command:

```
bin/hadoop distcp srcdir destdir
```

Following are the parameters:

- `srcdir`: A fully qualified path to the source directory. The path can even include the NameNode host and port when data is being moved from one cluster to another (for example, `/user/mydir`, `hdfs://namenode1:9000/user/mydir`).
- `destdir`: A fully qualified path to the destination directory. The path can even include the NameNode host and port when data is being moved from one cluster to another (for example, `/user/mydestdir`, `hdfs://namenode2:9000/user/mydir`).
- S3 paths can be specified as `s3://bucket-name/key`.

The preceding paths are assumed to be directories and are recursively copied from source to destination. The `distcp` utility starts a MapReduce job to copy the data from source to destination in parallel.

## Summary

This chapter explored the underlying details of the Hadoop system in general and HDFS in particular. The various configuration files and multitenancy in Hadoop using various schedulers were described. Starting from the very basic FIFO scheduler, you learned about the more complex Capacity and Fair Schedulers and how they can be used to maximize cluster utilization for an Enterprise. Finally, several HDFS administration utilities were discussed. The goal of this chapter was to help you appreciate the physical and operational design of Hadoop.

Armed with this knowledge, you can perform basic administration tasks on the Hadoop system. Furthermore, understanding the Hadoop internals also enables you to debug tricky programming issues as well as design Hadoop-based applications.

CHAPTER 5

# Basics of MapReduce Development

In earlier chapters, you were introduced to Hadoop as a platform. You learned the concepts behind Hadoop architecture, saw the basics of Hadoop administration, and even wrote basic MapReduce programs.

In this chapter, you learn the basic fundamentals of MapReduce programming and common design patterns of MapReduce programs with sample use-cases. These basic fundamentals will help you dive into the advanced concepts of MapReduce in the chapters that follow.

## Hadoop and Data Processing

The fundamentals of MapReduce are described here by using practice examples. RDBMS and SQL are ubiquitous in data-processing environments; we use language elements from SQL to explain the basic concepts of MapReduce. Discussing the basic MapReduce concepts in the context of SQL language elements not only creates a familiar context for you to understand MapReduce but you will also appreciate the suitability of MapReduce for solving familiar data-processing problems when the datasets are very large.

To achieve the goals of this chapter, we use an airline dataset that comprises arrival and departure flight details of all commercial flights from 1987 through 2008. We introduce this dataset first; you can then start tackling familiar data-processing problems that we mentioned in terms of SQL language elements by using MapReduce. In the process, you will gain familiarity with various components of MapReduce, including these:

- Mapper
- Reducer
- Combiner
- Partitioner

## Reviewing the Airline Dataset

This chapter uses an airline dataset that consists of flight arrival and departure details of commercial domestic flights in the United States from 1987 to 2008. This dataset holds approximately 120 million records in comma-separated value (CSV) format. The total disk size of this dataset in uncompressed format is 120 GB. Although it is a large dataset that is suitable for Hadoop processing, it is not too large, so it is suitable for the purpose of this book. The other reason for selecting this dataset is to discuss MapReduce in the context of structured data processing. Although it is true that Hadoop is used for unstructured data processing, its most common use is to perform large-scale ETL on structured datasets and use it as a data warehouse for large datasets. The airline dataset is highly structured and suitable for explaining MapReduce concepts in terms of SQL language elements. The dataset can be downloaded from `http://stat-computing.org/dataexpo/2009/the-data.html`.

The various dataset fields are shown in Table 5-1.

***Table 5-1.** Data Dictionary for Airline Dataset*

| Field | Documentation |
|---|---|
| Year | Year of the scheduled flight (1987-2008). |
| Month | Month of the scheduled flight (1-12). |
| DayofMonth | Day of the month (1-31). |
| DayOfWeek | Identifier for the day of the week; for example, 1=Monday, 7=Sunday. |
| DepTime | Actual departure time of the flight in its local time zone, expressed in *HH/MM* format (*HH* = hours of 24-hour clock; *MM* = minutes from 00 to 59). |
| CRSDepTime | Scheduled departure time in *HH/MM* format (see `DepTime`). |
| ArrTime | Actual arrival time in *HH/MM* format. |
| CRSArrTime | Scheduled arrival time in *HH/MM* format. See also `ArrTime`. |
| UniqueCarrier | Carrier company code. |
| FlightNum | Flight number used to uniquely identify the flight. |
| TailNum | Aircraft registration number used to uniquely identify the aircraft. It is similar to the license plate number on a car. |
| ActualElapsedTime | Actual flight time (in minutes). |
| CRSElapsedTime | Scheduled flight time (in minutes). See also `ActualElapsedTime`. |
| AirTime | Total airtime (in minutes). |
| ArrDelay | Arrival delay (in minutes. |
| DepDelay | Departure delay (in minutes). |
| Origin | Origin code of the airport. |
| Dest | Destination code of the airport. |
| Distance | Total flight distance (in miles). |
| TaxiIn | Taxi in time during arrival (in minutes). |
| TaxiOut | Taxi out time during departure (in minutes). |
| Cancelled | Flag indicating whether the flight was cancelled (1 = yes, 0 = no). |
| CancellationCode | Reason for cancellation (A = carrier, B = weather, C = NAS, D = security). See causes of delay following. |
| Diverted | Flag indicating whether the flight was diverted (1 = yes, 0 = no). |
| CarrierDelay | Delay (in minutes) caused by factors within control of the carrier. See this link for details: `http://aspmhelp.faa.gov/index.php/Types_of_Delay`. |
| WeatherDelay | Delay (in minutes) caused by extreme weather conditions that may be forecast or occur during departure, en route, or at arrival. |
| NASDelay | Delay (in minutes) that is within the control of the National Airspace System (NAS). See this link for details: `http://aspmhelp.faa.gov/index.php/Types_of_Delay`. |
| SecurityDelay | Security delay (in minutes) caused by security reasons. See the following link for details: `http://aspmhelp.faa.gov/index.php/Types_of_Delay`. |
| LateAircraftDelay | Delay (in minutes) due to the same aircraft arriving late at a previous airport. It is caused by the propagation of delays. |

We also use master data to interpret codes in the dataset:

- `carriers.csv`: Unique carrier code
- `airports.csv`: Airport codes

These master files are in CSV format and can be downloaded from `http://stat-computing.org/dataexpo/2009/supplemental-data.html`.

The airline data is in BZIP2 compressed format. Compression schemes are discussed in more detail in Chapter 7 but you need to know now that Hadoop handles certain compressed formats transparently, and BZIP2 is one of those formats.

## Preparing the Development Environment

To simplify the development effort, a small subset of input data has been added to the project folder `src/main/resources/input/devairlinedataset/txt/`. This folder has two uncompressed files:

- `1987`: A small subset of the records from 1987
- `1988`: A small subset of the records from 1988

You should run the development programs against these files to ensure that the jobs run quickly in the development environment.

The master data is placed in the `src/main/resources/input/masterdata` folder. This folder has two files:

- `airports.csv`: Master data for all airports referenced by the airport code in the airline dataset
- `carriers.csv`: Master data for all carriers referenced in the airline dataset

## Preparing the Hadoop System

We need to transfer our airline data into the HDFS. Assume that the data is located in a local directory of the machine: `/user/local/airlinedata/`. This directory contains files with the following file names: `1987.csv.bz2`, `1988.csv.bz2` ... `2008.csv.bz2`.

All the master data (`carriers.csv` and `airports.csv`) is assumed to be present in the folder`/user/local/masterdata/`

Next, we create three directories by executing the following command. These folders will be created in the HDFS home folder, which is assumed to be `/user/hdfs/`.

```
hdfs  dfs -mkdir airlinedata
hdfs  dfs -mkdir sampledata
hdfs  dfs -mkdir masterdata
```

The first folder contains the entire airline dataset. The second folder is created to have data only for the years 1987 and 1988. This folder should be used to execute the jobs on the pseudo-cluster for testing because the entire dataset will take a long time to complete on a pseudo-cluster. The final folder contains the master data files.

Now you can copy local files into their respective directories:

```
hdfs dfs -copyFromLocal /user/local/airlinedata/* /user/hdfs/airlinedata/
hdfs dfs -copyFromLocal /user/local/airlinedata/1987.bz2 /user/hdfs/sampledata/
hdfs dfs -copyFromLocal /user/local/airlinedata/1988.bz2 /user/hdfs/sampledata/
hdfs dfs -copyFromLocal /user/local/masterdata/* /user/hdfs/masterdata/
```

The Hadoop system is now ready to execute the Hadoop programs in this chapter.

# MapReduce Programming Patterns

As discussed earlier, this chapter explores typical MapReduce design patterns in the context of the following SQL language features:

- `SELECT`: Selects a subset of columns from a large number of columns in a table.
- `WHERE`: Filters out rows of a table based on Boolean criteria applied on various column values.
- `AGGREGATION`: Computes aggregation values—typically values such as `MIN`, `MAX`, and `SUM`—based on the grouping of certain attributes. In SQL aggregation, queries are usually based on the `GROUP BY` and `HAVING` clauses.

Chapters 6 and 7 explore advanced MapReduce concepts in the context of the following SQL features:

- `SORTING`: Required when the output has to be sorted by certain criteria
- `JOIN`: Combines separate tables based on similar column values of separate tables

## Map-Only Jobs (SELECT and WHERE Queries)

We first explore Map-only jobs (basic MapReduce programs were discussed in Chapter 3). A typical MapReduce programs comprises two phases: the Map phase and the Reduce phase. Each phase is implemented by a custom class extending the Mapper and Reducer base class, respectively, from the Hadoop Framework:

- **Map**: Implementation of a custom class extending the `org.apache.hadoop.mapreduce.Mapper` base class from the Hadoop Framework. Its responsibility is to process input records received in key-value format and produce zero or more key-value pairs as output.
- **Reduce**: Implementation of a custom class extending the `org.apache.hadoop.mapreduce.Reducer` base class from the Hadoop Framework. It takes the Map output of each key and the key's respective values collectively. The Reducer can iterate through each of the values for a given key and produce zero or more key-value pairs of its own. The Reduce phase is optional. Certain use-cases do not require a Reduce phase. Examples of such use-cases are the `SELECT` and `WHERE` clauses in SQL when applied on a single table. For such functions, you can use a Map-only job.

## Problem Definition: SELECT Clause

From the airline dataset, we want to produce the output for the entire dataset, which comprises only the following attributes:

- Flight date in *MM/DD/YYYY* format (for example, 01/13/1987, 03/28/1988)
- Day of the week
- Departure time
- Arrival time
- Origin airport code
- Destination airport code
- Total flight distance in miles
- Actual flight time

- Scheduled flight time
- Departure delay
- Arrival delay

This is a simple SELECT clause on the table with some basic computations:

- Flight date is a combination of the Month, DayOfMonth, and Year columns formatted in specified manner.
- The rest of the fields are all simple as-is projections of the corresponding fields.

The program that performs the SELECT clause is org.apress.prohadoop.c5.SelectClauseMRJob.

We also use another utility class that contains utility functions that are described as we go along. The name of the class is
org.apress.prohadoop.utils.AirlineDataUtils.

Listing 5-1 shows the listing for SelectClauseMRJob.

***Listing 5-1.*** SelectClauseMRJob.java

```
package org.apress.prohadoop.c5;

//import statements, skipped for brevity
public class SelectClauseMRJob extends Configured implements Tool{
 public static class SelectClauseMapper
                 extends Mapper<LongWritable, Text, NullWritable, Text> {

    public void map(LongWritable key, Text value, Context context)
                             throws IOException, InterruptedException {
       if(!AirlineDataUtils.isHeader(value)){
            StringBuilder output = AirlineDataUtils.mergeStringArray(
                      AirlineDataUtils.getSelectResultsPerRow(value),
                      ",");
            context.write(NullWritable.get(),new Text(output.toString()));
       }
    }
}

public int run(String[] allArgs) throws Exception {
   Job job = Job.getInstance(getConf());
    job.setJarByClass(SelectClauseMRJob.class);
    job.setInputFormatClass(TextInputFormat.class);
    job.setOutputFormatClass(TextOutputFormat.class);

    job.setOutputKeyClass(NullWritable.class);
    job.setOutputValueClass(Text.class);

    job.setMapperClass(SelectClauseMapper.class);
    job.setNumReduceTasks(0);
```

```
        String[] args = new GenericOptionsParser(getConf(), allArgs)
                                    .getRemainingArgs();
        FileInputFormat.setInputPaths(job, new Path(args[0]));
        FileOutputFormat.setOutputPath(job, new Path(args[1]));
        boolean status = job.waitForCompletion(true);
        if(status){
            return 0;
        }
        else{
            return 1;
        }
    }

    public static void main(String[] args) throws Exception {
        Configuration conf = new Configuration();
        ToolRunner.run(new SelectClauseMRJob(), args);
    }
}
```

We use the new MapReduce API and ToolRunner classes (see Chapter 3) consistently through the book because they provide maximum flexibility through the use of the GenericOptionsParser class, which you will encounter frequently in this chapter.

The main characteristic of this program is that it is a Map-only program. The decision about whether you need only the Map phase or the Map and Reduce phase is the key to designing high-performance MapReduce programs. Map-only programs run significantly faster because they do not have the Sort/Shuffle phase which we discuss later in this chapter. This phase is responsible for moving the Mapper output from the Mapper nodes to the Reducer nodes over the network. In a Map-only job the Mapper simply writes out its results, usually on the same machine on which it is executed, which reduces network communication overhead, so the job completes more quickly.

## run() method

Key aspects of the `run( )` method are the following:

- `InputFormatClass`: This is the `TextInputFormat`, which reads the input data line by line. Remember that in the example, the input files are compressed. The Hadoop Framework uncompresses these files transparently and serves the individual lines to the Mapper of this program.
- `OutputFormatClass`: This is the `TextOutputFormat`, and the output is plain uncompressed text files. Compression is discussed in Chapter 6, in which output can be declared to be compressed using user-defined compression schemes.
- Because the output is a CSV file, the key is declared as `NullWritable`, and the value is declared as `Text` (both implement of the `Writable` interface). The importance of the `Writable` interface is discussed in more detail in Chapter 6, in which you learn how custom `Writable` classes can be implemented. If the key was not declared as NullWritable but Text instead, our output file would have the key followed by the tab (`\t`) character and then the value. Making the key as NullWritable ensures that each line of the output only contains a value which is comma separated string.

- The output key and value classes need to be declared under certain conditions. In Chapter 7 we will explore how the the implementations of the `InputFormat` interface define the Mapper input key and value types. But the Hadoop Framework needs to know the output data types from the Mapper and Reducer to create instances of those types at runtime to deserialize values between the Mapper and Reducer as well as during serializing instances from the Reducer to the output files. It is not possible to infer the types at runtime from the class definitions of the Mapper and Reducer since Java Generics uses Type Erasure (see note below). The `InputFormat` classes define the Mapper Input key and value types. The `TextInputFormat` defines these types as `LongWritable` and `Text` respectively. By default the Mapper output key and value types are assumed to be the same as the Mapper input key and value types. If they are different we need to specify to this to the Hadoop Framework. In our case we have a Map only job. Hence we just define `setOutputKeyClass` and `setOutputValueClass`. The latter is optional since our Mapper Output Value class is same as the default (`Text`). In fact both are optional in this case due to how the TextOutputFormat is defined internally. The TextOutputFormat simply writes the records line by line. Each output line comprises of the key and value seperated by a tab. It first converts them to `String` instances by calling the `toString` method on the key and value instances. It also handles NullWritable keys as a special case, where only the value portion is written out to the output file. However, this exception is a side effect of how `TextOutputFormat` is defined. In Chapter 7 we will encounter `SequenceFileOutputFormat` which stores the metadata about the file contents. This includes the key and value types of its contents. The `SequenceFileOutputFormat` depends on calls to `getOutputKeyClass()` and `getOutputValueClass()` on the `Job` instance for this information. Later we will encounter two more methods namely, `setMapOutputKeyClass` and `setMapOutputValueClass`. These will be used when the output key and value types of the Mapper and Reducer are not the same. In this section you have been provided the general rules behind why these methods need to be invoked by the client on the `Job` instance. As discussed there are subtle exceptions to these rules based on how specific OutputFormat classes are implemented. However, they are not relevant to writing good MapReduce programs. If you follow the rules above your MapReduce programs will run without throwing any runtime exceptions.
- Our job is the Map-only job, so we specify the Mapper class but do not specify a Reducer class. We explicitly indicate that we do not need a Reducer by invoking the `job.setNumReduceTasks(0)` method, which is required for a Map-only job. By default, the Hadoop Framework configures one Reducer and utilizes a class called an `IdentityReducer`. The Mapper output is sent to this `IdentityReducer`, which then writes all its input to the output. This process is highly inefficient due to the additional network overhead incurred from moving the Mapper output unnecessarily from the Mapper nodes to the Reducer node.
- We then run the job using the `job.waitForCompletion(true)` method. The parameter name is `verbose`, which indicates whether the job should indicate the progress to the user.

---

■ **Note** There are various input and output formats, but this example uses `TextInputFormat` and `TextOutputFormat`, which are also the default input and output formats, respectively. You learn about various I/O formats in Chapter 7.

---

---

■ **Note** The Java Programming language implemented the Generics Feature using Type Erasure. This was done to ensure binary compatability with legacy classes since Java Genrerics appeared late in the adoption life cycle of Java. Due to Type Erasure there is no distinction between a class such as `java.util.ArrayList<Integer>` and `java.util.ArrayList <String>`. A call to `getClass()` will return `java.util.ArrayList` at runtime. Thus at runtime the type information is lost. Type Erasure works as follows, if you defined the following code snippet in your source code

```
List<String> l=new ArrayList<String>();l.put("t");String s=l.get("t");
```

The compiler will compile it the following set of lines

```
List l=new ArrayList();l.put("t");String s=(String)l.get("t");
```

Thus the type information is lost at runtime. Hence it is often said that Java Generics is simply syntactic sugar.

---

## SelectClauseMapper Class

All the action happens in the Mapper class. First, we describe the configuration of the Mapper class:

```
SelectClauseMapper extends Mapper<LongWritable, Text, NullWritable, Text>
```

- The input key class is `LongWritable`, which is the byte offset of the line in the file. It is the task of the `TextInputFormat` to provide the appropriate key and value instances to the Mapper.
- The input value class is `Text`, which is the actual line of text in the input file.
- The output key class is `NullWritable`, which is a singleton object with no data.
- The output value class is the `Text` class. We utilize the `TextOutputFormat` which defines the output specification of the Job. Based on this selection the output file has lines of text.

Next, we consume each line of input. We skip the header lines; for nonheader lines, the following method is invoked:

```
AirlineDataUtils.getSelectResultsPerRow(value)
```

Listing 5-2 shows the implementation of this method.

***Listing 5-2.*** AirlineDataUtils.getSelectResultsPerRow

```
public static String[] getSelectResultsPerRow(Text row){
     String[] contents = row.toString().split(",");
     String[] outputArray = new String[10];
     outputArray[0]=AirlineDataUtils.getDate(contents);
     outputArray[1]=AirlineDataUtils.getDepartureTime(contents);
     outputArray[2]=AirlineDataUtils.getArrivalTime(contents);
     outputArray[3]=AirlineDataUtils.getOrigin(contents);
     outputArray[4]=AirlineDataUtils.getDestination(contents);
     outputArray[5]=AirlineDataUtils.getDistance(contents);
     outputArray[6]=AirlineDataUtils.getElapsedTime(contents);
```

```
    outputArray[7]=AirlineDataUtils
                        .getScheduledElapsedTime(contents);
    outputArray[8]=AirlineDataUtils
                        .getDepartureDelay(contents);
    outputArray[9]=AirlineDataUtils.getArrivalDelay(contents);
    return outputArray;
}
```

The method splits each line of text using a " , " separator, extracts various components of the SELECT clause, and returns an output array in the proper order. The AirlineDataUtils.mergeStringArray call merges the array returned in the previous call and returns a line whose components are comma-separated.

```
StringBuilder output = AirlineDataUtils.mergeStringArray(
    AirlineDataUtils.getSelectResultsPerRow(value), ",");
```

Finally, the next call works with the TextOutputFormat instance configured for the job to send the line to the output file in the HDFS:

```
context.write(NullWritable.get(),new Text(output.toString()));
```

## Running the SELECT Clause Job in the Development Environment

The job can be executed in the development environment in local mode. If you are using the Eclipse IDE, simply execute the preceding class using the following set of parameters:

- ./src/main/resources/input/devairlinedataset/txt: Input folder that contains tiny versions of the 1987 and 1988 flight records
- ./output/c5/select/: Output folder relative to the Eclipse project home folder

The job executes in local mode. It is useful to execute all jobs in local mode prior to running them in cluster mode. All programs in this book can be executed in this mode in a similar manner. Job executions in the development environment in subsequent sections are skipped because they are very similar to the current example.

## Running the SELECT Clause Job on the Cluster

Finally, we execute a Maven build. The result is a JAR file called prohadoop-0.0.1-SNAPSHOT.jar.

Execute the job in the Hadoop environment as follows:

```
hadoop jar prohadoop-0.0.1-SNAPSHOT.jar org.apress.prohadoop.c5.SelectClauseMRJob /user/hdfs/
sampledata \
/user/hdfs/output/c5/select
```

The various components of the above command are the following:

- Hadoop JAR file prohadoop-0.0.1-SNAPSHOT.jar indicates the JAR file to execute.
- org.apress.prohadoop.c5.SelectClauseMRJob is the fully qualified class name.
- /user/hdfs/sampledata is the name of the input folder in HDFS containing the input files/ user/hdfs/output/c5/select is the name of the folder in HDFS to which the job writes the output files.

When the job is executed, you should see an output on your screen similar to the one shown in Listing 5-3. The key characteristics of the output are shown in bold.

- The first bold line indicates the number of files being processed. We process files for 1987 and 1988 from the sampledata folder, so there are two files.
- The next bold section (abbreviated) indicates the job progress from 0–100 percent.
- The third bold section indicates that two map tasks were started, and both of them were local to the data. In other words, each Map task read the input files from its local disk and did not have to fetch it from the network from another node's disk. The MapReduce framework strives to make this possible as it minimizes the network overhead. (You will revisit how the InputFormat class achieves this when the InputFormat and RecordReader classes are discussed in Chapter 6.)

The last two bold lines are interesting because they tell you how many lines of text were processed. Recall that each line of text is one record. Each file also has a header line that is ignored. Quite correctly, the input records exceed the output records by two because we ignore the header lines in the Mapper.

***Listing 5-3.*** Job Execution Screen Log

```
14/01/26 06:39:30 WARN mapred.JobClient: Use GenericOptionsParser for parsing the arguments.
Applications should implement Tool for the same.
14/01/26 06:39:32 INFO input.FileInputFormat: Total input paths to process : 2
14/01/26 06:39:32 INFO mapred.JobClient: Running job: job_201401251457_0002
14/01/26 06:39:33 INFO mapred.JobClient:  map 0% reduce 0%
14/01/26 06:40:47 INFO mapred.JobClient:  map 1% reduce 0%
........
14/01/26 06:45:12 INFO mapred.JobClient:  map 98% reduce 0%
14/01/26 06:45:15 INFO mapred.JobClient:  map 99% reduce 0%
14/01/26 06:45:19 INFO mapred.JobClient:  map 100% reduce 0%
14/01/26 06:45:30 INFO mapred.JobClient: Job complete: job_201401251457_0002
14/01/26 06:45:30 INFO mapred.JobClient: Counters: 24
14/01/26 06:45:30 INFO mapred.JobClient:   File System Counters
14/01/26 06:45:30 INFO mapred.JobClient:     FILE: Number of bytes read=0
14/01/26 06:45:30 INFO mapred.JobClient:     FILE: Number of bytes written=319500
14/01/26 06:45:30 INFO mapred.JobClient:     FILE: Number of read operations=0
14/01/26 06:45:30 INFO mapred.JobClient:     FILE: Number of large read operations=0
14/01/26 06:45:30 INFO mapred.JobClient:     FILE: Number of write operations=0
14/01/26 06:45:30 INFO mapred.JobClient:     HDFS: Number of bytes read=62170169
14/01/26 06:45:30 INFO mapred.JobClient:     HDFS: Number of bytes written=290559727
14/01/26 06:45:30 INFO mapred.JobClient:     HDFS: Number of read operations=4
14/01/26 06:45:30 INFO mapred.JobClient:     HDFS: Number of large read operations=0
14/01/26 06:45:30 INFO mapred.JobClient:     HDFS: Number of write operations=2
14/01/26 06:45:31 INFO mapred.JobClient:   Job Counters
14/01/26 06:45:31 INFO mapred.JobClient:    Launched map tasks=2
14/01/26 06:45:31 INFO mapred.JobClient:    Data-local map tasks=2
14/01/26 06:45:31 INFO mapred.JobClient:   Total time spent by all maps in occupied slots (ms)=539064
14/01/26 06:45:31 INFO mapred.JobClient:   Total time spent by all reduces in occupied slots (ms)=0
14/01/26 06:45:31 INFO mapred.JobClient:   Total time spent by all maps waiting after reserving
slots (ms)=0
14/01/26 06:45:31 INFO mapred.JobClient:    Total time spent by all reduces waiting after reserving
slots (ms)=0
14/01/26 06:45:31 INFO mapred.JobClient:    Map-Reduce Framework
```

```
14/01/26 06:45:31 INFO mapred.JobClient:     Map input records=6513924
14/01/26 06:45:31 INFO mapred.JobClient:     Map output records=6513922
14/01/26 06:45:31 INFO mapred.JobClient:     Input split bytes=270
14/01/26 06:45:31 INFO mapred.JobClient:     Spilled Records=0
14/01/26 06:45:31 INFO mapred.JobClient:     CPU time spent (ms)=217780
14/01/26 06:45:31 INFO mapred.JobClient:     Physical memory (bytes) snapshot=225677312
14/01/26 06:45:31 INFO mapred.JobClient:     Virtual memory (bytes) snapshot=1383579648
14/01/26 06:45:31 INFO mapred.JobClient:     Total committed heap usage (bytes)=121503744
```

## View SELECT Clause Job Results

Having executed the job, you need to process the results of the job. The output is a set of files stored in comma-separated format in the HDFS. You might want to use this output for various purposes such as visualization or downstream analytics. But first, you should view these files to verify the output. Each of these files is very large. This section reviews the steps required to view small portions of these files for verification purposes.

First, get a listing of the output directory, as follows (assuming your current hdfs folder is `/user/hdfs/`):

```
hdfs dfs -ls output/c5/select
```

This code should produce the following listing:

```
-rw-r--r--   3 hadoop hadoop 0 2014-01-26 06:45 output/c5/select/_SUCCESS
drwxr-xr-x   - hadoop hadoop 0 2014-01-26 06:39 output/c5/select/_logs
-rw-r--r--   3 hadoop hadoop 232132738 2014-01-26 06:45 output/c5/select/part-m-00000
-rw-r--r--   3 hadoop hadoop 58426989 2014-01-26 06:42 output/c5/select/part-m-00001
```

The files that contain the output of the program are `part-m-00000` and `part-m-00001.` These files can be viewed by executing the `hdfs` version of the `cat` command.

```
hdfs dfs -cat output/c5/select/part-m-00000
hdfs dfs -cat output/c5/select/part-m-00001
```

This produces a lot of output, however. The last few lines of these files can be viewed by executing the `hdfs` version of the `tail` command.

```
hdfs dfs -tail output/c5/select/part-m-00000
hdfs dfs -tail output/c5/select/part-m-00001
```

Note that each input file was processed by its own Mapper, so each output file was also produced by the respective Mapper. Thus, each output file contains listings for only 1 year, similar to the input file. There is no order to this, however. On our system, `part-m-00000` contains output for 1988 records, and `part-m-00001` contains the output for 1987 records. Each separate run can be different,

Listings 5-4 and 5-5 show the last few lines of each output file.

***Listing 5-4.*** Last Five Lines of the part-m-00000 File

```
...
12/12/1988,1325,2043,HNL,LAX,2556,318,309,0,9
12/13/1988,1325,2038,HNL,LAX,2556,313,309,0,4
12/14/1988,1325,2045,HNL,LAX,2556,320,309,0,11
12/01/1988,2027,2152,ATL,MCO,403,85,78,0,7
12/02/1988,2106,2229,ATL,MCO,403,83,78,39,44
```

***Listing 5-5.*** Last Five Lines of the part-m-00001 File

```
...
12/11/1987,1530,1825,ORD,EWR,719,115,113,0,2
12/13/1987,1530,1815,ORD,EWR,719,105,113,0,-8
12/14/1987,1530,1807,ORD,EWR,719,97,113,0,-16
12/01/1987,1525,1643,BOS,EWR,200,78,73,0,5
12/02/1987,1540,1706,BOS,EWR,200,86,73,15,28
```

## Recapping Key Hadoop Features Explored with SELECT

As you implemented the SELECT clause, you explored the following key features of Hadoop and learned the following:

- How to create folders and copy local files into the HDFS
- How to write a Map-only job
- How to interpret the console log output of the MapReduce job
- How to obtain a directory listing of the MapReduce job output directory
- How to view the actual output of the MapReduce job

# Problem Definition: WHERE Clause

Now you can refine the previous result to add a `WHERE` clause. You will filter the results to include flights that had a delay in arrival or departure by more than 10 minutes. You can also add an additional attribute that indicates whether the delay was during arrival, departure, or both. This attribute is called Point Of Delay, and it takes one of the following values:

- `O` for delays at origin
- `D` for delays at destination
- `B` for delays at both origin and destination

Note that we used values `O`, `D`, and `B` instead of `Origin`, `Destination` and `Both`. This substitution is not just for brevity. When running a typical MapReduce program, considerable output can be produced, and the main bottleneck for most MapReduce programs is disk or network I/O. When large amounts of data are being written out to disk or moved over the network, restricting the size of a field can result in a significant improvement in the runtime performance of MapReduce jobs.

The program that performs the `WHERE` clause is `org.apress.prohadoop.c5.WhereClauseMRJob`.

Listing 5-6 shows the relevant sections of the source code listing for `WhereClauseMRJob`.

***Listing 5-6.*** WhereClauseMRJob.java

```
package org.apress.prohadoop.c5;

//import statements

public class WhereClauseMRJob extends Configured implements Tool{
   public static class WhereClauseMapper
            extends Mapper<LongWritable, Text, NullWritable, Text> {
```

```
    private int delayInMinutes = 0;

    @Override
    public void setup(Context context){
        this.delayInMinutes = context.getConfiguration()
                                    .getInt("map.where.delay",1);
    }

  public void map(LongWritable key, Text value, Context context)
                       throws IOException, InterruptedException {
    if (AirlineDataUtils.isHeader(value)) {
       return;//Only process non header rows
    }
    String[] arr = AirlineDataUtils.getSelectResultsPerRow(value);
    String depDel = arr[8];
    String arrDel = arr[9];
        int iDepDel = AirlineDataUtils.parseMinutes(depDel, 0);
        int iArrDel = AirlineDataUtils.parseMinutes(arrDel, 0);
    StringBuilder out = AirlineDataUtils.mergeStringArray(arr, ",");
        if (iDepDel >=this.delayInMinutes
             && iArrDel >=this.delayInMinutes) {
      out.append(",").append("B");
      context.write(NullWritable.get(), new Text(out.toString()));
        }
        else if (iDepDel >=this.delayInMinutes) {
      out.append(",").append("O");
      context.write(NullWritable.get(), new Text(out.toString()));
        }
        else if (iArrDel >=this.delayInMinutes) {
      out.append(",").append("D");
      context.write(NullWritable.get(), new Text(out.toString()));
        }
   }
  }

  public  int run(String[] allArgs) throws Exception {
    Job job = Job.getInstance(getConf());
    ...
    job.setMapperClass(WhereClauseMapper.class);
    String[] args = new GenericOptionsParser(getConf(),
                          allArgs).getRemainingArgs();
    FileInputFormat.setInputPaths(job, new Path(args[0]));
    FileOutputFormat.setOutputPath(job, new Path(args[1]));
    Boolean status = job.waitForCompletion(true);
    ....
  }

  public static void main(String[] args) throws Exception {
    Configuration conf = new Configuration();
    int res = ToolRunner.run(new WhereClauseMRJob(), args);
  }
}
```

This example builds on the SELECT clause job developed in the prior section. It applies the WHERE clause on top of the SELECT results. Also we do not want to hard-code the delay in minutes; we want it to be a runtime parameter supplied by the user. In the preceding listing, this parameter is called `map.where.delay` and read in the setup method of the Mapper, which is described in the next section.

## WhereClauseMapper Class

Every Mapper (or Reducer) has a setup method that is invoked once when the Mapper (or Reducer) starts. The setup method is used to initialize Mapper (or Reducer) level attributes or to initialize task-level resources.

There is also a cleanup method in the Mapper (or Reducer) that is invoked once at the end of the job. It is used to perform end-of-task actions such as releasing resources back to their resource pools.

You will encounter these methods again when `DistributedCache` is discussed in Chapter 6.

---

■ **Note** A common use of the setup method is to open network resources, and the cleanup method is used to close network resources. For example, if you use MapReduce to bulk index documents to a search engine such as Apache Solr, the setup method is used to open a connection to the Solr and retain the connection as an attribute in the Mapper-or Reducer-level instance. A cleanup method is used to close the connection to the search engine.

---

The delay in minutes is read from the `Context` instance as follows:

```
context.getConfiguration().getInt("map.where.delay",1);
```

A default of 1 minute is provided to ensure that if the user forgets to provide the value, only flights with delays are returned.

The `Configuration` instance used to configure the job is accessible in all the remote tasks (Mapper and Reducer). It is accessed through the `Context` instance, which is different for the Mapper and Reducer:

- Mapper: `org.apache.hadoop.mapreduce.Mapper.Context`
- Reducer: `org.apache.hadoop.mapreduce.Reducer.Context`

The `WhereClauseMapper` is very similar in configuration to the `SelectClauseMapper` in the previous section. The following sequence of actions is performed in it:

1. `AirlineDataUtils.getSelectResultsPerRow(value)` gets the required columns out of the larger input line.
2. `depDel = arr[8]` and `arrDel = arr[9]` extract the departure delay and arrival delay, respectively,
3. `AirlineDataUtils.parseMinutes(String minutes, int defaultValue)` converts the preceding delays in String format to an integer. The dataset uses `NA` to indicate null. The function returns the value of `defaultValue` if the string to integer parsing fails. A default value of `0` is used in the example, which indicates that there is no delay when the attribute value is `NA`.
4. The next few lines write the line (comprising only the fields retrieved in the first bullet) to output after appropriately augmenting it with an indicator whose value indicates where the delay occurred.

## Running the WHERE Clause Job on the Cluster

Finally, we rebuild using Maven and execute the job in the Hadoop environment as follows:

```
hadoop jar prohadoop-0.0.1-SNAPSHOT.jar org.apress.prohadoop.c5.WhereClauseMRJob -D map.where.delay=10 \
/user/hdfs/sampledata /user/hdfs/output/c5/where
```

The key difference in the preceding command is how the parameter is passed to the job. It is very similar to the way `System` parameters are passed to Java programs through the command line using the `-D` option. However, there is a key difference: unlike Java programs, there is a space character between `-D` and the name/value pair in MapReduce programs.

The `-D` options passed to the job are passed through the `Configuration` object. In the example, `-D map.where.delay=10` is equivalent to making the following direct call in the `run()` function:

```
job.getConfiguration().set("map.where.delay", "10");
```

Functions such as `getInt()` in the `Configuration` class are helper methods on top of the `get(String paramName, String paramValue)` method.

## GenericOptionsParser Revisited

`GenericOptionsParser` was discussed in Chapter 3, when its utility was explored in using third-party libraries with Hadoop jobs. In the previous section, `GenericOptionsParser` was used to pass named parameters to the job.

To summarize, `GenericOptionsParser` can be used to do the following:

- Third-party libraries are passed using `-libjars <comma separated jar paths>`.
- Custom named parameters are passed using `-D <name>=<value>`.

The call used to extract the user-defined parameters is the following:

```
new GenericOptionsParser(getConf(), allArgs).getRemainingArgs();
```

In the example, an array is returned that contains the input path and output path for the job.

## Key Hadoop Features Explored with the WHERE Clause

As we implemented the `WHERE` clause, we discussed the following new features of Hadoop:

- Role of the `GenericOptionsParser` to pass named custom parameters to the job
- Roles of the setup and cleanup methods

# Map and Reduce Jobs (Aggregation Queries)

The next common language element of SQL is aggregation queries. With aggregation functions such as `SUM`, `COUNT`, `MAX`, and `MIN`—in combination with the `GROUP BY` and `HAVING` clauses—SQL is commonly used to compute data aggregations.

This section explores how to use MapReduce jobs to compute aggregations. For the first time in this chapter, you will encounter a job that uses both Map and Reduce phases.

## Problem Definition: GROUP BY and SUM Clauses

Let us determine whether there is any pattern in delays based on the month of the scheduled flight. You can determine the proportion of on-time flights by month and get the proportion of cancelled/diverted flights by month. In this section, you explore the following MapReduce features:

- Use of the Mapper to perform partial computations to reduce I/O between Mapper and Reducer. The decision of whether a flight arrived/departed on time is made in the Mapper. A simple indicator variable is passed on the Mapper to the Reducer to indicate it. This is significantly cheaper than passing the entire input to the Reducer and performing the computations there. Such decisions are critical to good Map Reduce development.
- Use of the Reducer to compute aggregations.
- The output dataset contains the following attributes:
    - Month of departure
    - Total number of flight records for the month
    - Proportion of flights with on-time arrival
    - Proportion of flights with arrival delays
    - Proportion of flights with on-time departure
    - Proportion of flights with departure delays
    - Proportion of cancelled flights
    - Proportion of diverted flights
- For the sake of this example, if the fields related to delayed departure or delayed arrival are NA, we treat the value to be 0, which indicates that there was no delay for that activity. For the `Cancellation` attribute, we assume NA to imply that the flight was not cancelled; for the `Diverted` attribute, we assume NA to imply that the flight was not diverted.

The program that performs the aggregations just described is the following:

```
org.apress.prohadoop.c5.AggregationMRJob
```

## run() method

The key lines in the `run()` method that are different for this job are the following:

```
job.setMapOutputKeyClass(Text.class);
job.setMapOutputValueClass(IntWritable.class);
job.setOutputKeyClass(NullWritable.class);
job.setOutputValueClass(Text.class);
job.setMapperClass(AggregationMapper.class);
job.setReducerClass(AggregationReducer.class);
job.setNumReduceTasks(1);
```

Note the first four lines. This is an example where the Mapper output key and value classes are different from the Reducer output key and value classes. It is required to explicitly make the calls to `setMapOutputKeyClass`, `setMapOutputValueClass` since their actual values are different from their default values (`LongWritable` and `Text` repectively for reasons described earlier). The `setOutputKeyClass` and `setOutputValueClass` declare the output key and value classes as we had defined in the section about MapReduce implementation of the Select Clause. If the Mapper output key and value were the same the Reducer output key and value, then it would be enough to just invoke the `setOutputKeyClass` and `setOutputValueClass`. The framework would assume the Mapper output key Class and Mapper output value class by invoking the `getOutputKeyClass` and `getOutputValueClass` methods. This example, clearly illustrates when it is required to set both sets of functions. Revisit this section when you have read the sections on the `AggregationMapper` and `AggregationReducer` classes.

We set the Mapper and Reducer class and indicate that we will use one Reducer. The last line is optional because the MapReduce framework uses one Reducer for a MapReduce job by default. Chapter 6 explores situations in which this default should be changed.

## AggregationMapper Class

We define `AggregationMRJob` class-level variables that are instances of `IntWritable` as follows:

```
public static final IntWritable RECORD=new IntWritable(0);
public static final IntWritable ARRIVAL_DELAY=new IntWritable(1);
public static final IntWritable ARRIVAL_ON_TIME=new IntWritable(2);
public static final IntWritable DEPARTURE_DELAY=new IntWritable(3);
public static final IntWritable DEPARTURE_ON_TIME=new IntWritable(4);
public static final IntWritable IS_CANCELLED=new IntWritable(5);
public static final IntWritable IS_DIVERTED=new IntWritable(6);
```

Each of these lines is a marker variable that the Mapper passes to the Reducer. The key in each Mapper output is the Month value coded as a two-digit, "0" left-padded string.

Listing 5-7 shows the `map()` method for the `AggregationMapper` inner class.

***Listing 5-7.*** map() method of AggregationMapper

```
public void map(LongWritable key, Text value, Context context) throws IOException,
InterruptedException {
   if(!AirlineDataUtils.isHeader(value)){
    String[] contents = value.toString().split(",");
    String month = AirlineDataUtils.getMonth(contents);
    int arrivalDelay = AirlineDataUtils
                       .parseMinutes(AirlineDataUtils
                                     .getArrivalDelay(contents),0);
    int departureDelay =AirlineDataUtils
                        .parseMinutes(AirlineDataUtils
                                     .getDepartureDelay(contents),0);
    boolean isCancelled =  AirlineDataUtils
                            .parseBoolean(AirlineDataUtils
                                     .getCancelled(contents),false);
    boolean isDiverted = AirlineDataUtils
                            .parseBoolean(AirlineDataUtils
                                     .getDiverted(contents),false);
         context.write(new Text(month), RECORD);
    if(arrivalDelay>0){
            context.write(new Text(month), ARRIVAL_DELAY);
    }
```

```
        else{
                context.write(new Text(month), ARRIVAL_ON_TIME);
        }
        if(departureDelay>0){
                context.write(new Text(month), DEPARTURE_DELAY);
        }
        else{
                context.write(new Text(month), DEPARTURE_ON_TIME);
        }
        if(isCancelled){
                context.write(new Text(month), IS_CANCELLED);
        }
        if(isDiverted){
                context.write(new Text(month), IS_DIVERTED);
        }
    }
```

The method extracts the key fields from the airline data, such as the following:

- Arrival delay in minutes
- Departure delay in minutes
- Whether the flight was cancelled
- Whether the flight was delayed

The `context.write()` calls are made based on various conditions:

- The `context.write(new Text(month), RECORD)` call is made for every record. We are calculating proportions. This call enables the calculation of the total number of records for a given month in the Reducer.
- The `context.write(new Text(month), ARRIVAL_DELAY)` call indicates a record with an arrival delay.
- The `context.write(new Text(month), ARRIVAL_ON_TIME)` call indicates a record with an on-time arrival.
- The `context.write(new Text(month), DEPARTURE_DELAY)` call indicates a record with a departure delay.
- The `context.write(new Text(month), DEPARTURE_ON_TIME)` call indicates a record with an on-time departure.
- The `context.write(new Text(month), IS_CANCELLED)` call indicates a record with a cancelled flight.
- The `context.write(new Text(month), IS_DIVERTED)` call indicates a record with a diverted flight.

## Reduce Phase

The Reducer has three important phases:

- **Sort**: The output of the Mappers is sorted by the key (the exact mechanism of sorting is discussed in Chapter 6). For now, accept that Mapper output is ordered by the Month key lexicographically from 01-12.
- **Shuffle**: The MapReduce framework copies the output from Mappers to Reducers. The Shuffle phase is discussed in more detail in Chapter 6.
- **Reduce**: In this step, all associated records must be processed together by a single entity that is the instance of the custom Reducer class. Listing 5-8 shows the `AggregationReducer` definition.

***Listing 5-8.*** AggregationReducer Definition

```
public static class AggregationReducer extends Reducer<Text, IntWritable, NullWritable, Text> {
     public void reduce(Text key, Iterable<IntWritable> values, Context context)
   ...
  }
}
```

The definition of the `AggregationReducer` makes the following assertions:

- The input Key class for the Reducer is `Text`. This is the same as the Key class of the Mapper output. The value of this Key is the month formatted as a two character string (Example- 01,02..,12)
- The input Value class for the Reducer is `IntWritable`. This is same as the Value class of the Mapper output
- Next note the definition of the `reduce` method. The first parameter is an instance of the Key class (`Text`). The second parameter is an instance of `Iterable<IntWritable>`. Each instance emerging from the `Iterable` instance indicates whether the record is a simple counter (`RECORD`), whether the flight was on-time with respect to arrival or departure, whether the flight was delayed on arrival or departure, or whether the flight was cancelled or diverted.
- Similar to earlier examples, since the output is a CSV file, the reducer output key class is declared as `NullWritable`, and the reducer output value class is declared as `Text`.

Listing 5-9 shows how the actual counters are incremented in the for loop. After the for loop is exited, all counters for that month are guaranteed to be processed completely, and the final output is prepared and written.

***Listing 5-9.*** Reduce() method of the AggregationReducer

```
public void reduce(Text key, Iterable<IntWritable> values, Context context) {
   double totalRecords = 0;
   double arrivalOnTime = 0;
   double arrivalDelays = 0;
   double departureOnTime = 0;
   double departureDelays = 0;
   double cancellations = 0;
   double diversions = 0;
   for(IntWritable v:values){
      if(v.equals(RECORD)){
         totalRecords++;
      }
```

```
        if(v.equals(ARRIVAL_ON_TIME)){
           arrivalOnTime++;
        }
        if(v.equals(ARRIVAL_DELAY)){
           arrivalDelays++;
        }
        if(v.equals(DEPARTURE_ON_TIME)){
           departureOnTime++;
        }
        if(v.equals(DEPARTURE_DELAY)){
           departureDelays++;
        }
        if(v.equals(IS_CANCELLED)){
           cancellations++;
        }
        if(v.equals(IS_DIVERTED)){
           diversions++;
        }
     }
     DecimalFormat df = new DecimalFormat( "0.0000" );
        //Prepare and produce output
     StringBuilder output = new StringBuilder(key.toString());
     output.append(",").append(totalRecords);
     output.append(",").append(df.format(arrivalOnTime/totalRecords));
     output.append(",").append(df.format(arrivalDelays/totalRecords));
     output.append(",").append(df.format(departureOnTime/totalRecords));
     output.append(",").append(df.format(departureDelays/totalRecords));
     output.append(",").append(df.format(cancellations/totalRecords));
     output.append(",").append(df.format(diversions/totalRecords));
        context.write(NullWritable.get(), new Text(output.toString()));
}
```

Note the following guarantees of the MapReduce framework:

- All output associated with a specific key that is part of the Mapper output is processed by the same Reducer in a single reduce invocation. This is true even when the same key is produced by multiple Mappers. In this example, we chose to have a single Reducer, but this rule applies even if the number of Reducers is greater than one.
- Only the keys processed by a Reducer are sorted. The values associated with a key received by the Reducer in an instance of `Iterable` and processed in a `reduce` call are not sorted by default, and their order might change between runs. In Chapter 6 we will discuss Secondary Sort, a method for sorting reduce-side values (as opposed to the reduce-side input keys) based on user defined criteria.
- Although all keys are sorted from the perspective of the Reducer, they are not sorted across Reducers. Because only a single Reducer was used, the output is sorted correctly by month, with January output preceding the February output and so on until December. However, if we had used two Reducers, it would be possible for the first Reducer to get keys 01,03,05,07,09,11; and the second Reducer to get keys 02,04,06,08,10,12. Notice that the keys are ordered for each Reducer, but are not ordered across Reducers. You learn more about total ordering across Reducers when sorting is discussed in Chapter 6.

---

**Caution** A common mistake made by new MapReduce programmers is to reuse the instance of the value across iterations in the for loop inside the Reducer. This process is done when there are calculations that need to be performed across iterations using values from previous iterations, and it leads to incorrect results. The MapReduce framework does not provide a new instance of `IntWritable` (in the example) during each iteration. Instead, the `Iterator` in the Hadoop Reducer uses a single instance whose value changes during each iteration. Reusing these references across iterations leads to errors that are very hard to debug.

---

## Running the GROUP BY and SUM Clause Job in the Cluster

Finally, we rebuild by using Maven and execute the job in the Hadoop environment as follows:

```
hadoop jar prohadoop-0.0.1-SNAPSHOT.jar org.apress.prohadoop.c5.AggregationMRJob /user/hdfs/
sampledata \
/user/hdfs/output/c5/aggregation
```

## Results for the Entire Dataset

Running the job for the entire dataset from the years 1987–2008 and graphing the output with the month on the X-axis produces the graph shown in Figure 5-1.

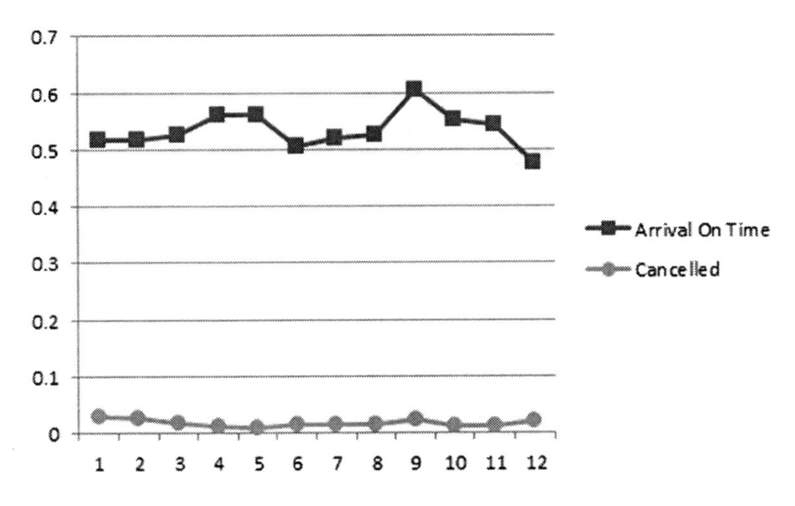


***Figure 5-1.** Delay/Cancelations/Diverted Patterns by Flight Month*

Figure 5-1 shows the following:

- Delays increase during summertime from June through August. Schools are closed in the United States during these months, so June through August is a popular vacation season.
- Delays increase during the winter months from December through March, probably due to bad weather resulting from snowfall.
- Cancellations also tend to go up in the months of December through March, indicating bad weather. Interestingly, there is an uptick in the cancellations for the month of September. Could this be the result of massive cancellations on September 11, 2001? More detailed analysis is needed to test this hypothesis.

The entire listing for the output using the full dataset is shown in Listing 5-10.

***Listing 5-10.*** Output of AggregationMRJob Executed on Complete Dataset

```
01,8473442.0,0.5176,0.4824,0.5851,0.4149,0.0295,0.0029
02,7772317.0,0.5166,0.4834,0.5853,0.4147,0.0258,0.0025
03,8606145.0,0.5258,0.4742,0.5845,0.4155,0.0189,0.0021
04,8305404.0,0.5603,0.4397,0.6288,0.3712,0.0114,0.0018
05,8510026.0,0.5596,0.4404,0.6330,0.3670,0.0103,0.0019
06,8424353.0,0.5061,0.4939,0.5639,0.4361,0.0141,0.0028
07,8715855.0,0.5204,0.4796,0.5712,0.4288,0.0143,0.0028
08,8768265.0,0.5270,0.4730,0.5827,0.4173,0.0140,0.0025
09,8192348.0,0.6037,0.3963,0.6780,0.3220,0.0249,0.0017
10,8908173.0,0.5524,0.4476,0.6336,0.3664,0.0117,0.0017
11,8451547.0,0.5442,0.4558,0.6124,0.3876,0.0113,0.0019
12,8740941.0,0.4751,0.5249,0.5269,0.4731,0.0220,0.0029
```

## Recapping the Key Hadoop Features Explored with GROUP BY and SUM

As the aggregation feature was implemented, we explored the following new features of Hadoop:

- Developing custom Reducers
- Techniques to perform custom aggregations

# Improving Aggregation Performance Using the Combiner

This section discusses an important optimization that can be added to MapReduce programs to considerably improve their performance. In the aggregation example discussed in the prior section, every record was transferred from the Mapper to the Reducer. In the Reducer, we performed aggregations and wrote only 12 records: one for each month. Aggregations in several cases can be accumulated. A partial set of records can be aggregated and then the partial aggregations can be combined into a final aggregation. MapReduce provides a method of performing partial aggregations in the Mapper node. This section elaborates on this feature of the MapReduce API.

# Problem Definition: Optimized Aggregators

In the `AggregationMRJob` for each record, `AggregationMapper` outputs a minimum of three and a maximum of four records.

- Every record has a record marker.
- Every record has an arrival on-time marker or arrival delay marker.
- Every record has a departure on-time marker or a departure delay marker.
- A record can have a cancellation marker, but a record marked as cancelled cannot have a diverted marker.
- A record can have a diverted marker, but a record marked as diverted cannot have a cancellation marker.

All these records need to be moved to the Reducer. Reducer tasks run on a separate machine from most Mappers. Given that there are more than 100 million records, 300 to 400 million records are moving from the Mapper machines to the Reducer machines. This is an expensive operation.

Considerable I/O savings can be achieved with a Combiner class, which is a mini-Reducer that operates on the Mapper machines on the Mapper output. In fact, the Combiner is implemented as a subclass of the Reducer class. The final output contains only 12 rows. For simplicity, assume that one Mapper is started for each input file, so there are 22 Mappers. If an intermediate output is produced once (the `SUM` of appropriate values for each year) per Mapper, each Mapper sends only 12 (number of month keys) x 4 = 48 lines to the Reducer. For 22 Mappers, 1056 lines are sent from Mapper machines to Reducer machines. Compare this with the 400 million lines we were transferring across the network. Again, this is an approximation; these intermediate lines are generated in memory. It is possible that due to memory constraints, the Mapper invokes the Combiner several times after a certain threshold of memory is reached as input lines are being processed. (You will revisit this feature when Sort/Shuffle is discussed in more detail at the end of this chaper).

The program that performs the aggregation described in the previous section is this:

```
org.apress.prohadoop.c5.AggregationWithCombinerMRJob
```

In the run for the full dataset, the following numbers were obtained:

- Launched map tasks = 22
- Mapper input records = 101,868,834
- Mapper output records = 307,596,518
- Combiner input records = 307,596,518
- Combiner output records = 3,395
- Reducer input records = 3,395

The Combiner output records count indicates that the Combiner is invoked approximately 4 times per Mapper.

## AggregationMapper Class from the AggregationWithCombinerMRJob Class

When you use the Combiner, you have to follow the constraints that its use requires, which usually leads to modifying the output key and value classes of the Mapper, Combiner, and Reducer.

The key constraint enforced by the Combiner is that the Combiner class's input and output key and value classes must match the corresponding output key and value classes of the Mapper. Recall that a Combiner is a mini-Reducer that works on the Mapper node. The Combiner output is also sent to the Reducer. The Reducer does not know whether the input it is receiving is coming directly from the Mapper or processed by a Combiner en route, so the output key and value classes of the Mapper should match the input key and value classes of the Combiner. Similarly, the Output key and value classes of the Combiner must match the input key andvalue classes of the Reducer. This naturally implies that the Combiner output key andvalue class must be identical to the output key and value classes of the Mapper, so we have to rewrite the Mapper and Reducer a bit. Earlier, the `AggregationMapper` simply had to send an `IntWritable` marker along with the key, which is month. In the Reducer, we checked for the existence of the marker and incremented the relevant variable by 1 of the respective marker (for example, `ARRIVAL_ON_TIME`).

But with the user of the Combiner, it is not enough to send this marker. We need to send the aggregation (`SUM`) value along with the marker type, which is the partial sum computed at the end of each Mapper. The output of `AggregationMapper` is `MapWritable`. It has two key value pairs:

- `TYPE`, which represents the marker type (for example, `ARRIVAL_ON_TIME`)
- `VALUE`, which is `IntWritable(1)`

The `AggregationCombiner` acts as an intermediate Reducer on the Mapper machine and produces an output that is a `MapWritable`. It has two key value pairs:

- `TYPE`, which represents the marker type (for example, `ARRIVAL_ON_TIME`)
- `VALUE`, which is an intermediate sum of all the `IntWritable` instances received from the Mapper for the type

The `AggregationReducer` receives the output from the Mapper machine after the Combiner has operated on it and performs the final computations, similar to the `AggregationCombiner`.

Listing 5-11 shows the key changes in the `AggregationMapper` class source code.

***Listing 5-11.*** AggregationMapper in the AggregationWithCombinerMRJob

```
public static class AggregationMapper
                    extends Mapper<LongWritable, Text, Text, MapWritable> {
    public void map(LongWritable key, Text value, Context context)
                                 throws IOException, InterruptedException {

        if(!AirlineDataUtils.isHeader(value)){
        ......
           context.write(new Text(month),
                       getMapWritable(RECORD,new IntWritable(1)));
           if(arrivalDelay>0){
              context.write(new Text(month),
                       getMapWritable(ARRIVAL_DELAY,new IntWritable(1)));
           }
           else{
              context.write(new Text(month),
                       getMapWritable(ARRIVAL_ON_TIME,new IntWritable(1)));
           }
           if(departureDelay>0){
              context.write(new Text(month),
                       getMapWritable(DEPARTURE_DELAY,new IntWritable(1)));
           }
           else{
              context.write(new Text(month),
                       getMapWritable(DEPARTURE_ON_TIME,new IntWritable(1)));
           }
```

```
            if(isCancelled){
               context.write(new Text(month),
                     getMapWritable(IS_CANCELLED,new IntWritable(1)));
            }
            if(isDiverted){
               context.write(new Text(month),
                     getMapWritable(IS_DIVERTED,new IntWritable(1)));
            }
         }
      }

      private MapWritable getMapWritable(IntWritable type,IntWritable value){
         MapWritable map = new MapWritable();
         map.put(TYPE, type);
         map.put(VALUE, value);
         return map;
      }
}
```

Listing 5-12 shows the AggregationCombiner class source code. Note that it extends the Reducer class just like the AggregationReducer. Also notice that it performs intermediate aggregations (SUM) and invokes context.write.

***Listing 5-12.*** AggregationCombiner in the AggregationWithCombinerMRJob

```
public static class AggregationCombiner extends Reducer<Text, MapWritable, Text, MapWritable> {
   public void reduce(Text key,
                      Iterable<MapWritable> values,
                      Context context) throws IOException,
                                              InterruptedException {
      int totalRecords = 0;
      int arrivalOnTime = 0;
      int arrivalDelays = 0;
      int departureOnTime = 0;
      int departureDelays = 0;
      int cancellations = 0;
      int diversions = 0;
      for(MapWritable v:values){
               IntWritable type = (IntWritable)v.get(TYPE);
               IntWritable value = (IntWritable)v.get(VALUE);
               if(type.equals(RECORD)){
                  totalRecords=totalRecords+value.get();
               }
               if(type.equals(ARRIVAL_ON_TIME)){
                  arrivalOnTime=arrivalOnTime+value.get();
               }
               if(type.equals(ARRIVAL_DELAY)){
                  arrivalDelays= arrivalDelays+value.get();
               }
               if(type.equals(DEPARTURE_ON_TIME)){
                  departureOnTime=departureOnTime+value.get();
               }
```

```
                if(type.equals(DEPARTURE_DELAY)){
                    departureDelays=departureDelays+value.get();
                }
                if(type.equals(IS_CANCELLED)){
                    cancellations=cancellations+value.get();
                }
                if(type.equals(IS_DIVERTED)){
                    diversions=diversions+value.get();
                }
        }
        context.write(key, getMapWritable(RECORD,new IntWritable(totalRecords)));
        context.write(key, getMapWritable(ARRIVAL_ON_TIME,
                                    new IntWritable(arrivalOnTime)));
        context.write(key, getMapWritable(ARRIVAL_DELAY,
                                    new IntWritable(arrivalDelays)));
        context.write(key, getMapWritable(DEPARTURE_ON_TIME,
                                    new IntWritable(departureOnTime)));
        context.write(key, getMapWritable(DEPARTURE_DELAY,
                                    new IntWritable(departureDelays)));
        context.write(key, getMapWritable(IS_CANCELLED,
                                    new IntWritable(cancellations)));
        context.write(key, getMapWritable(IS_DIVERTED,
                                    new IntWritable(diversions)));
    }
    private MapWritable getMapWritable(IntWritable type,IntWritable value){
    //
    }
}
```

Listing 5-13 shows the source code's new version of the AggregationReducer class, which is very similar to the earlier version from the last section. Notice that it performs the final aggregations using the values of the intermediate aggregations received from the Combiner.

***Listing 5-13.*** *AggregationReducer in the AggregationWithCombinerMRJob*

```
public static class AggregationReducer
                    extends Reducer<Text, MapWritable, NullWritable, Text> {
     public void reduce(Text key,
                         Iterable<MapWritable> values,
                         Context context)
                                   throws IOException, InterruptedException {
        double totalRecords = 0;
        double arrivalOnTime = 0;
        double arrivalDelays = 0;
        double departureOnTime = 0;
        double departureDelays = 0;
        double cancellations = 0;
        double diversions = 0;
        for(MapWritable v:values){
           IntWritable type = (IntWritable)v.get(TYPE);
           IntWritable value = (IntWritable)v.get(VALUE);
```

```
            if(type.equals(RECORD)){
                        totalRecords=totalRecords+value.get();
            }
            if(type.equals(ARRIVAL_ON_TIME)){
                        arrivalOnTime=arrivalOnTime+value.get();
            }
            if(type.equals(ARRIVAL_DELAY)){
                        arrivalDelays=arrivalOnTime+value.get();
            }
            if(type.equals(DEPARTURE_ON_TIME)){
                        departureOnTime=departureOnTime+value.get();
            }
            if(type.equals(DEPARTURE_DELAY)){
                        departureDelays=departureDelays+value.get();
            }
            if(type.equals(IS_CANCELLED)){
                        cancellations=cancellations+value.get();
            }
            if(type.equals(IS_DIVERTED)){
                        diversions=diversions+value.get();
            }
        }
        DecimalFormat df = new DecimalFormat( "0.0000" );
            //Prepare and produce output StringBuilder output = new StringBuilder(key.toString());
            output.append(",").append(totalRecords);
            output.append(",").append(df.format(arrivalOnTime/totalRecords));
            output.append(",").append(df.format(arrivalDelays/totalRecords));
            output.append(",").append(df.format(departureOnTime/totalRecords));
            output.append(",").append(df.format(departureDelays/totalRecords));
            output.append(",").append(df.format(cancellations/totalRecords));
            output.append(",").append(df.format(diversions/totalRecords));
            context.write(NullWritable.get(), new Text(output.toString()));
    }
}
```

## run() Method

The only changes to the run() method are the following lines:

```
job.setMapOutputValueClass(MapWritable.class);
job.setCombinerClass(AggregationCombiner.class);
```

---

▒ **Note** Using a Combiner does not automatically produce better performance. The use of the Combiner was appropriate in this instance because the Combiner produced a significantly lower number of lines (3,000 versus 300 million), which were transferred from the Mapper nodes to the Reducer. This justified the additional cost of running the Combiner. However, if this requirement is not met, using the Combiner might slow execution due to the additional processing required.

---

## Running the Combiner-Based Aggregation Job in the Cluster

Finally, we rebuild using Maven and execute the job in the Hadoop environment as follows:

```
hadoop jar prohadoop-0.0.1-SNAPSHOT.jar org.apress.prohadoop.c5.AggregationWithCombinerMRJob /user/
hdfs/sampledata \
/user/hdfs/output/c5/combineraggregation
```

## Key Aggregation Hadoop Features Explored with Aggregation

As the aggregation features were implemented, we covered the following new feature of Hadoop: Reducing I/O between Mappers and Reducers by using a custom Combiner.

# Role of the Partitioner

The previous section discussed the use of Reducers, which receive keys and the list of values for those keys. The keys are sorted from the perspective of the Reducer. Although for simplicity we assume a single Reducer, it is not a good choice in practice for performance reasons. In practice, it is common to use tens or even hundreds of Reducers. The Hadoop API guarantees that the Reducer processes the key-value pairs received by it, sorted by the key. This does not imply that Reducer-1 will have keys which are lower in sort order to keys received by Reducer-0. By default, the keys are randomly allocated to reach the Reducer from the Mapper. However, this random allocation is consistent. The same key is always routed to the same Reducer, regardless of the Mapper it comes from. Within the Reducer, the key and value pairs received from the Mapper are processed in a sorted (by Reducer input key or Mapper output key) order.

This section discusses the component that decides which key goes to which Reducer: the Partitioner class. This class and how to create custom implementations of this class are discussed in this section. First, let us create a use-case context for this discussion.

# Problem Definition: Split Airline Data by Month

Suppose that we need to split the airline dataset into separate files, one for each month, and we want to analyze data for specific months only. For example, we might want to analyze the delay patterns for days of the week in the month of January (a typical bad weather season) and then compare the same analysis for the month of June (a vacation season).

The listing for the preceding program is in `org.apress.prohadoop.c5.SplitByMonthMRJob`.

## run() Method

The key lines in the `run()` method that are different for this job are the following:

```
job.setMapperClass(SplitByMonthMapper.class);
job.setReducerClass(SplitByMonthReducer.class);
  job.setPartitionerClass(MonthPartioner.class);
  job.setNumReduceTasks(12);
```

The key difference is the inclusion of the Partitioner class, which is responsible for ensuring that the keys go to the appropriate Reducer.

Although we chose to hard-code the number of Reducers, we could pass number of Reducers as a parameter on the command line using the following option:

```
-D mapred.reduce.tasks=12
```

## Partitioner

Let us now examine how the keys are distributed between the Reducers. The class used to distribute keys across the Reducer is an implementation of org.apache.hadoop.mapreduce.Partitioner, An instance of Partitioner executes in the same JVM as the Mapper instance. Its getPartition method is called for each invocation of the context.write in the Mapper instance. The method getPartition has this signature:

```
int getPartition(K key, V value, int numReduceTasks)
```

It takes an instance of each key and value emitted by the Mapper as well as the number of Reduce tasks. The number of Reduce tasks is known through the job configuration and passed by the framework in each invocation of getPartition. The return value is the index of the Reducer. For 10 Reducers, it takes values between 0 and 9.

When a custom Partitioner is not defined, the Hadoop framework uses a default defined by the class org.apache.hadoop.mapreduce.lib.partition.HashPartitioner.

The HashPartitioner calls the hashCode() method on the key instance and performs a Modulo operation on it using numReduceTasks as the denominator. It explains how each Reducer gets a random sampling of keys. The sorting by key is applied *only after* the Partitioner separates the keys for each Reducer. Thus, each Reducer sees input sorted by keys, but the keys are not sorted across Reducers. The exact formula utilized by the HashPartitioner is

```
key.hashCode() & Integer.MAX_VALUE) % numReduceTasks.
```

### AVOID PARTITION ERRORS

A common error when computing the partitions is to use the Math.abs(key.hashCode())%NO_OF_REDUCERS call, which can lead to a negative value returned. If one of the key.hashCode() returns a negative value equal to -2147483648, the Math.abs() call on the value returns the same negative value because its absolute value is 1 greater than Integer.MAX_VALUE. Due to the manner of representing negative numbers internally in the computer, the absolute value of an int variable with value -2147483648 is the same number and negative. The workaround to the problem is the way the HashPartitioner implements this function. The key line of code in the HashPartitioner is

```
key.hashCode() & Integer.MAX_VALUE) % numReduceTasks.
```

& Integer.MAX_VALUE operates at the bit level, and the previous computation, key.hashCode() & Integer.MAX_VALUE, returns 0 for a hashCode() value of -2147483648 since all the bits for -2147483648 are different from the bit representation of Integer.MAX_VALUE. Another way to address this problem is, type cast the key.hashCode() to long and apply the Math.abs() on the long variable. Now if the return value of the hashCode() is -2147483648, once it is type cast to a long, the absolute value of a long instance of this value is 2147483648 because there are more bits in a long variable to store this value. Alternatively, Math.abs(key.hashCode()%NO_OF_REDUCERS) can be used, in which the absolute function is applied after the Modulo function is applied on the hashCode function. Note that the previous methods all return different values for the partition ID.

## SplitByMonthMapper Class

Listing 5-14 shows the source code for the SplitByMonthMapper class. Notice how the key was formulated for the Mapper output. The Mapper output key is a IntWritable instance representing the month of the flight record. It takes values from 1 through 12.

***Listing 5-14.*** SplitByMonthMapper

```
public static class SplitByMonthMapper extends
                    Mapper<LongWritable, Text, IntWritable, Text> {
    public void map(LongWritable key, Text value, Context context)
                            throws IOException, InterruptedException {
       if (!AirlineDataUtils.isHeader(value)) {
          int month = Integer.parseInt(AirlineDataUtils.getMonth(
                                                value.toString()
                                                .split(","));
          context.write(new IntWritable(month), value);
       }
    }
}
```

## SortByMonthAndDayOfWeekReducer Class

Listing 5-15 shows the source code for the `SplitByMonthReducer` class, which iterates through the values for each month (which is the key) and sends them to the output file.

***Listing 5-15.*** SortByMonthAndDayOfWeekReducer

```
public static class SplitByMonthReducer extends
                    Reducer<IntWritable, Text, NullWritable, Text> {
    public void reduce(IntWritable key, Iterable<Text> values, Context context)
                              throws IOException, InterruptedException {
       for (Text output : values) {
          context.write(NullWritable.get(), new Text(output.toString()));
       }
    }
}
```

## MonthPartitioner Class

Listing 5-16 shows the source code for the `MonthPartitoner` class. Since we use twelve reducers by design each reducer index is simply (`month -1`) where month takes values from 1 through 12. This is the key component that ensures that each month goes to a separate Reducer in the proper order. That is, records for January go to Reducer 0, records for February go to Reducer 1, and so on until records for December go to Reducer 11.

***Listing 5-16.*** MonthPartitioner

```
public static class MonthPartioner extends Partitioner<IntWritable, Text> {
 @Override
    public int getPartition(IntWritable month, Text value, int numPartitions) {
            return (month.get() - 1);
    }
}
```

## Run the Partitioner Job in the Cluster

Finally, we rebuild using Maven and execute the job in the Hadoop environment as follows:

```
hadoop jar prohadoop-0.0.1-SNAPSHOT.jar org.apress.prohadoop.c5.SplitByMonthMRJob  /user/hdfs/
sampledata \
/user/hdfs/output/c5/partitioner
```

When the job completes, there are 12 files in the /user/hdfs/output/c5/partitioner HDFS directory. The file called `part-r-00000` has records for the month of January, `part-r-00001` has records for the month of February, and so on until `part-r-00011` has records for the month of December.

---

■ **Note** Multiple Reducers are used only when you run the job in psuedo-cluster or full cluster mode. Running this job locally in your single node cluster causes only a single Reducer to be used, which results in a single output file with its data sorted by month. In the local mode the Partitioner is bypassed and all the Mapper output records go to the single reducer.

---

## Key Hadoop Features Explored with Partitioner

This section covered the following new features of Hadoop:

- How Partitioners split keys between Reducers.
- How to develop a custom Partitioner. We will develop more-complex Partitioners in Chapter 6 when we discuss the sorting use-case.
- How to configure the number of Reducers to an MR program as a command-line parameter; for example: `-D mapred.reduce.tasks=12`. Note that this property has been deprecated in Hadoop 2 and replaced with `mapreduce.job.reduces`. However the older property is still valid and widely used. Hence we will continue to use the older property names in the book since you are more likely to encounter them in practice. In fact any property which begins with `mapred` instead of `mapreduce` is likely to be deprecated in Hadoop 2. Check the documentation for its replacement if you want to use the newer version. For Hadoop 2.2.0, the link is : `http://hadoop.apache.org/docs/r2.2.0/hadoop-project-dist/hadoop-common/DeprecatedProperties.html`.

# Bringing it All Together

The last few sections discussed key features of MapReduce. We examined the Mapper, Reducer, Partitioner, and Combiner components. Figure 5-2 brings them all together in one graphic to show how they work together. Pay attention to where these components execute, which is the key to truly understanding MapReduce.

Figure 5-2 does not describe how the input values are fed to the Mapper nodes; it is described in Chapter 7 when the`InputFormat` interface is discussed. For now, assume that the Mapper somehow gets the correct set of key and value pairs.

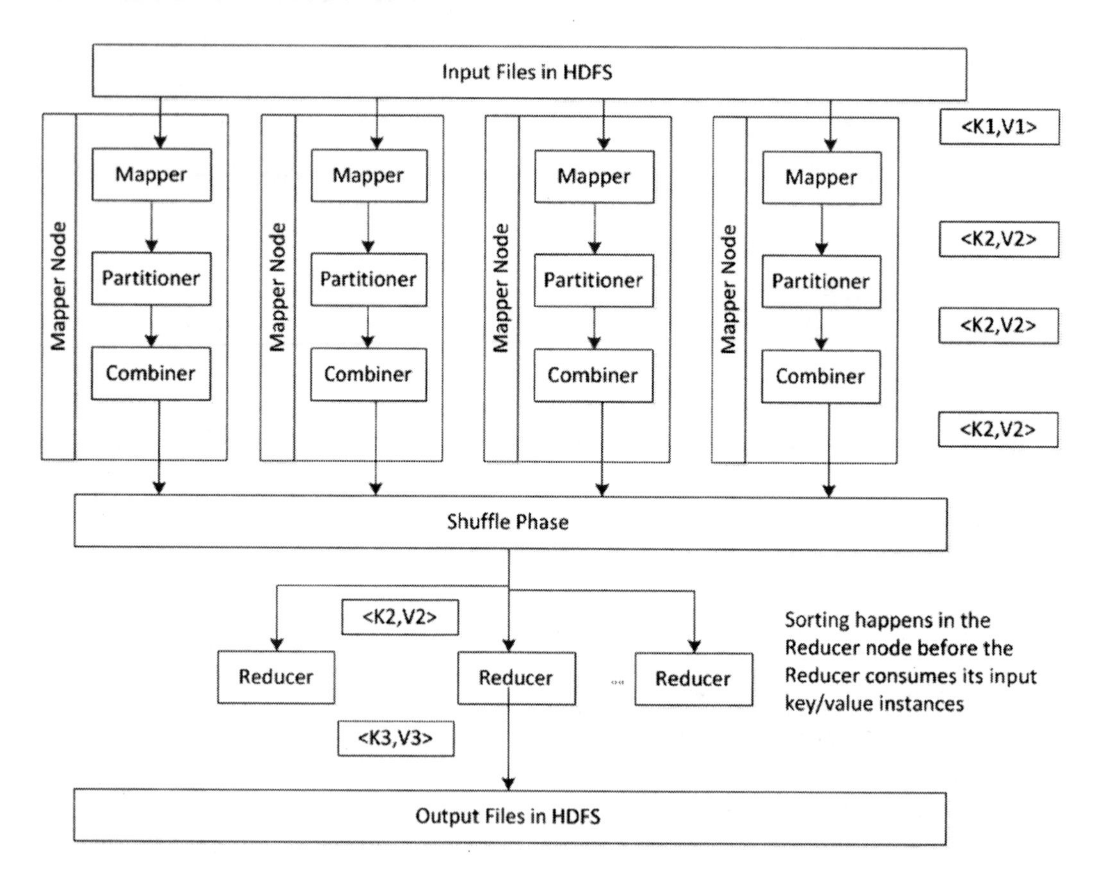


***Figure 5-2.*** *Bringing it all together*

The goal of the Mapper is to produce a partitioned file sorted by the Mapper output keys. The partitioning is with respect to the reducers the keys are meant to be processed by. After the Mapper emits its key andvalue pair, they are fed to a Partitioner instance that runs in the same JVM as the Mapper. The Partitioner partitions the Mapper output, based on the number of Reducers and any custom partitioning logic. The results are then sorted by the Mapper output key. At this point, the Combiner is invoked (if a Combiner is configured for the job) on the sorted output. Note that the Combiner is invoked after the Partitioner in the same JVM as the Mapper. Finally this partitioned, sorted and combined output is spilled to the disk. Optionally the Mapper intermediate outputs can be compressed. We will discuss compression in chapter 7. Compression reduces the I/O as these Mapper output files are written the disk on the Mapper node. Compression also reduces the network I/O as these compressed files are transferred to the reducer nodes.

On the Mapper node the following steps occur:

1. The records collected on the Mapper via the `context.write` invocation fill up a buffer maintained on the Mapper. The size of this buffer is defined by the property `mapreduce.task.io.sort.mb`. Its default value is `512` MB. The size of this buffer controls the efficiency of the sort process on the Mapper node. Higher the size more efficient the sort. However, memory is limited resource and increasing this size will limit the number of mappers you can start on a node. Thus there is a tradeoff between efficient sorting and the degree of parallelism that can be achieved.

2. When the buffer fills up to the extent defined by the property `mapreduce.map.sort.spill.percent`, a background thread will start spilling the contents to the disk. The default value for this property is `0.8` (80%). The Mapper side invocation of `context.write` will not block even if this threshold is exceeded if the spill is in progress. Thus the amount of data written to the disk in one spill might exceed this threshold. The spills are made to a file in the directories defined by the property mapreduce.cluster.local.dir in a round robin fashion.

3. Before data is written to the disk, the Partitioner is invoked and partitions are created per Reducer. Within each partition, an in-memory sort is performed by key (remember that the Reducer receives its keys in a sorted order. We will learn how to define the sorting order in Chapter 6). If a Combiner is defined it will be executed on the output of the sort.

4. The `mapreduce.map.sort.spill.percent` might be exceeded multiple times during the Mapper execution. A new spill file is created each time this threshold is exceeded. When the Mapper finally completes there may be several partitioned spill files on the local disk of the Mapper. They are merged into a single partitioned, sorted file. The number of spills to be merged in one pass is defined by the property `mapreduce.task.io.sort.factor`. Its default value is `10`. If there are more files to merge than defined by the property `mapreduce.map.combine.minspills`, the Combiner is invoked again before the final file is created. Remember, that Combiner can be invoked multiple times without affecting the final result.

The Shuffle phase ensures that the partitions reach the appropriate Reducers. During the Shuffle phase, each Reducer uses the HTTP protocol to retrieve its own partition from the Mapper nodes. Each Reducer uses five threads by default to pull its own partitions from the Mapper node as defined by the property `mapreduce.reduce.shuffle.parallelcopies`. But how do the Reducer's know which nodes to query to get their partitions? This happens through the Application Master. As each Mapper instance completes, it notifies the Application Master about the partitions it produced during its run. Each Reducer periodically queries the Application Master for Mapper hosts until it has received a final list of nodes hosting its partitions. The reduce phase begins when the fraction of the total mapper instances completed exceeds the value set in the property `mapreduce.job.reduce.slowstart.completedmaps`.

On the Reducer node, these presorted Mapper outputs are merged (the merge phase of the sort). Finally the sorted key and value pairs are processed by the reducer by invoking the `reduce` method call. Before the first `reduce` call is invoked on the Reducer instance, all the Mapper output key and value pairs meant for the Reducer are sorted by the Mapper output key (or Reduce input key). This is what ensures that the Reducer processes the Mapper output sorted by its input keys. Also the Iterable interface that the `reduce` call supports is just that, an interface. The entire value set for a key is not loaded in memory before invoking the `reduce` call for a given key. It served one by one for a given key as the iterations are performed at runtime on the Iterable interface (with appropriate in-memory buffering to ensure efficiency) inside the `reduce` invocation. The sorted file is read sequentially with each next () call on the Iterable instance inside the Reduce. As the key changes a new `reduce` invocation with the new key is made on the Reducer instance, and a new Iterable instance is created which will serve values for the new key by continuing to read the key and value pairs from the sorted file. This is the primary reason for the sort phase in MapReduce. It is the sort phase which allows millions of values to be processed efficiently (in one pass while reading a file whose records are sorted by the Reducer input key) for a given key inside the `reduce` invocation through the familiar `Iterable` interface without running out of memory or having to make multiple passes on a reducer input file. Some of the parameters which can be utilized to optimize the sort behavior on the Reducer node are

- `mapreduce.reduce.shuffle.input.buffer.percent` - This is the proportion of heap memory dedicated to storing map outputs retrieved from the mapper during the shuffle phase. If a specific mapper output is small enough it is retrieved into this memory or else it is persisted to the local disk of the reducer
- `mapreduce.reduce.shuffle.merge.percent` - This is the usage threshold of the heap memory defined by the `mapreduce.reduce.shuffle.input.buffer.percent` at which an in-memory merge of the mapper outputs maintained in memory will be initiated.
- `mapreduce.reduce.merge.inmem.threshold` - This is the threshold in terms of the number of mapper outputs accumulated in memory after which the in-memory merge process initiates. Note that one of the thresholds defined by, `mapreduce.reduce.merge.inmem.threshold` or `mapreduce.reduce.shuffle.merge.percent` needs to be reached for the in-memory merge process to be initiated.

# Summary

This chapter explored all the major components of MapReduce using a use-case-driven approach. We took the familiar SQL paradigm and developed common data-processing functions such as `SELECT`, `WHERE`, and `GROUP BY` aggregations; so you learned the basics of MapReduce development as it applies to real-world situations and large-scale datasets. Lastly, we discussed the internals of the MapReduce process. This discussion delved deep into how the files are moves across various nodes and components in the execution of a MapReduce program.

# CHAPTER 6

# Advanced MapReduce Development

Chapter 5 discussed the basics of MapReduce from the perspective of familiar SQL concepts. You learned how MapReduce can be used to solve familiar problems. You also learned how data is read from input files, processed in the Mappers, routed to the Reducers using a Partitioner, and finally processed in the Reducer and written to output files in the HDFS.

This chapter tackles the Sort and Join features of SQL, which require an introduction to more-complex concepts underlying MapReduce programs. You learn about how multiple output files can be written to from a single MapReduce program. Finally, you learn about counters, which can be used to collect statistics in MapReduce programs.

## MapReduce Programming Patterns

This chapter explores the following MapReduce design patterns in the context of SQL features:

- **Sort:** Required when output has to be sorted by certain criteria
- **Join:** Combines separate tables based on similar column values of separate tables

Before you look these use-cases, let us briefly explore the underlying I/O classes of the Hadoop Framework. This understanding will enable you to truly understand how Hadoop works under the covers.

### Introduction to Hadoop I/O

This chapter explores two common RDBMS-based use-cases: Sort and Join. But first you need to understand the Hadoop I/O system. Chapter 5 discussed the following classes:

- `Text`
- `IntWritable`

These classes, which are provided by the Hadoop Framework, have very specific implementations that enable them to be used as keys and values for Mappers and Reducers. We first discuss the characteristics of these classes that provide this capability. In the process, you also learn how to develop custom classes that can be used as key and value types within the Hadoop Framework.

All classes used for keys and values in Mappers and Reducers need to have the capability to be serialized to a raw byte stream and deserialized back into class instances. These class instances are deserialized when being read into the Mapper and then serialized to a byte stream and sent over the network for the Reducer to consume. Inside the Reducer nodes, the raw bytes are deserialized back to Reducer-side key/value class instances. The records are sorted by a Reducer-side input key (as discussed in Chapter 5, this process is partially done on the Mapper and completed on the Reducer) and then presented to the Reducer's `reduce` method. Finally, the Reducer output needs to be serialized into raw bytes and written to the disk.

The key characteristics of the key and value classes are the following:

- Classes used as keys and values should be capable of being serialized and deserialized.
- Classes used as Reducer-side keys should be capable of supporting user defined sorting.

Figure 6-1 demonstrates the flow from an input file to an output file in a MapReduce program.

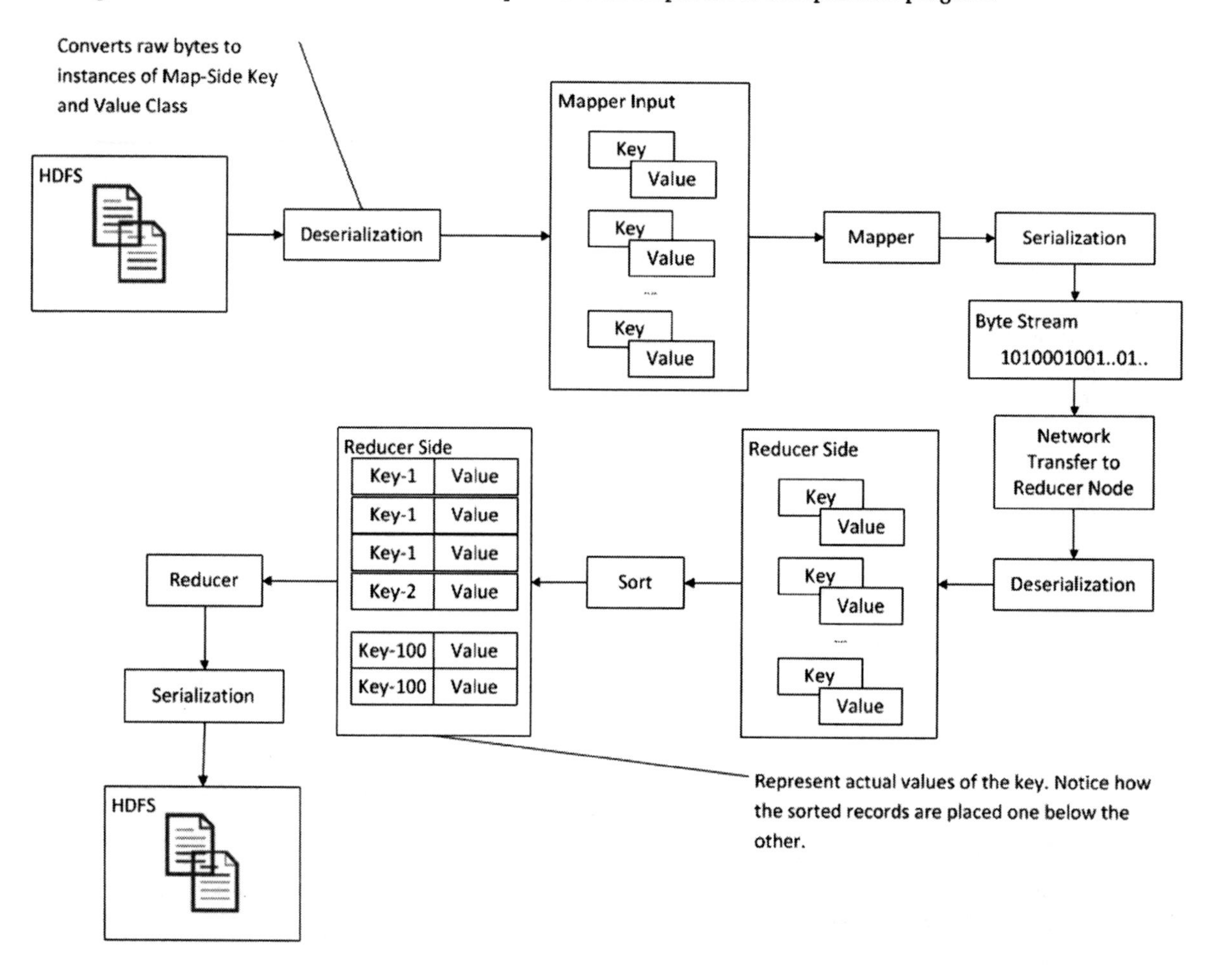


*Figure 6-1. Serialization/deserialization and sorting workflow in Hadoop*

Figure 6-1 demonstrates how the Mappers and Reducers communicate over a network. Key andvalue instances need to be serialized to write to files or transfer over the network. Byte streams need to be deserialized to key andvalue instances before they can be used in Mappers or Reducers.

## Writable and WritableComparable Interfaces

All objects that participate in Mapper and Reducer processes have to implement a specific interface: the `Writable` interface. Alternatively, the Reducer-side key class can implement the `WritableComparable` instances. Although the latter is not a requirement (alternative ways to support sorting are discussed later in the chapter), it is the simplest way to support custom sorting in a key class you will define. All the major Hadoop `Writable` instances, such as `Text` and `IntWritable`, implement this interface to support the default sorting criteria for instances of these classes.

Figure 6-2 shows the inheritance hierarchy of `Writable` and `WritableComparable` interfaces. The `Writable` interface has the methods that support object serialization and deserialization.

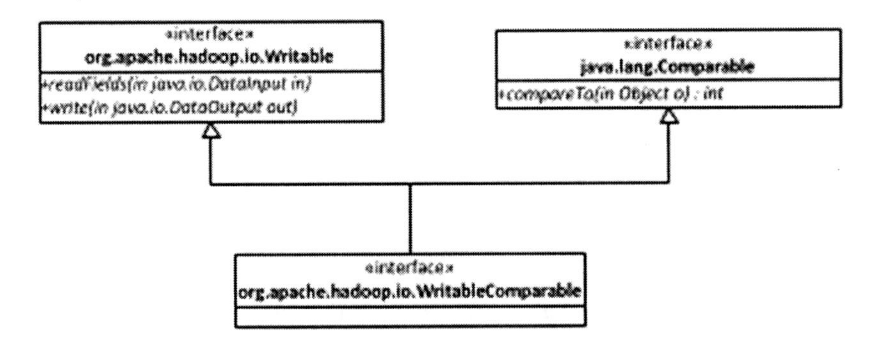


***Figure 6-2.*** *Hadoop I/O classes hierarchy*

The `Writable` interface has two methods:

- `write(java.io.DataOutput out)`: Writes all the primitive attributes of the instance to the output stream represented by the `java.io.DataOutput` instance. The `DataOutput` class has the methods to serialize basic Java types.
- `readFields(java.io.DataInput in)`: Re-creates the `Writable` instance by utilizing the data fetched from the input stream represented by the `java.io.DataInput` instance. The `DataInput` class has the methods to deserialize basic Java types.

The only contract between these methods is that the order of the fields should be identical in the `write()` and `readFields()` methods. The data is read in the same order as it was written out to the stream.

Hadoop also provides certain guarantees. The Reducer input is sorted by key. The `WritableComparable` interface can be used to customize the sorting behavior. The `WritableComparable` interface inherits its capability to serialize and deserialize records from the `Writable` class and its sorting capability from the standard `java.lang.Comparable` class.

As a good rule of thumb, Mapper output (or Reducer input) key class should implement the `WritableComparable` interface. All other types (such as key and value class used for the Mapper input, value class of Mapper output, and the key class of the Reducer output) have to implement only the `Writable` interface. Because the `WritableComparable` interface is a subtype of the `Writable` class, `WritableComparable` can always be used where `Writable` is used, but the converse is not true. All the classes you have encountered so far (e.g., `Text` and `IntWritable`) implement the `WritableComparable` interface, which enables them to be used as keys as well as values in Mappers and Reducers.

It is not a requirement for the Mapper output key class (or the Reducer input key class) to implement `WritableComparable`. If it implements only `Writable`, you have to create a class that implements the `RawComparator` interface. You will see examples of this later in the chapter.

## Problem Definition: Sorting

This section explores the Sort use-case using Hadoop. Sorting is very common in practice and is implemented using the `ORDER BY` keyword when using SQL.

By exploring the Sort functionality, you learn how to develop the custom `Writable` and `WritableComparable` classes discussed earlier. But we must first create a use-case context for this exercise. The Sort method implements the following features:

- The airline dataset will be sorted based on month and day of the week. To make it a bit more interesting, we will also add another requirement: the output will be sorted by month in an ascending order and day of the week in a descending order.
- The example program described below will support any number of reducers. If we use 12 reducers, the output will be split into 12 files so that each month's data is in its own file. For example, January data will be in the file `part-r-00000`, February data will be in the `part-r-00001` file, and so on until the December data is in the `part-r-00011` file. Within each file, the data will be sorted in descending order by week. Alternatively, we could use a single Reducer; the output would be sorted in ascending order by month and descending order by the day of the week. If we use number of reducers greater than 1 and not equal to 12, the records will be divided between files such that the output keys (Month and Day of Week combination) are evenly divided between files. They will also be sorted across files (called as Total Order Sorting). This implies that if we appended the files after sorting them in ascending order by name, the ordering by key will be the same as if we had used a single reducer.
- The output of the MapReduce program will be the records implemented as a line of text with only the following fields in a comma-separated format:
  - Month
  - Day of the week
  - Year
  - Date
  - Arrival delay
  - Departure delay
  - Origin airport code
  - Destination airport code
  - Carrier code

Some of the analysis we could do with this output is to determine whether there is any relationship between flight delays and the location of the delays or the flight carrier.

## Primary Challenge: Total Order Sorting

One of the guarantees of MapReduce is that the input to the Reducer is always sorted by the Reducer input key. If we use a single Reducer, sorting is straightforward: we simply choose the key to sort by and make it Reducer input key. In the Reducer the records arrive sorted by the key (as defined by the Sort Comparator implementation which we will examine soon).

Using a single Reducer is not the most efficient way to run jobs, however. In fact, it quickly becomes a bottleneck with large numbers of records. We want to use multiple Reducers, but although the keys to a given Reducer are sorted, there is no guarantee that they are sorted between Reducers. So if we use `IntWritable` instances as the key to the Reducers, it is very possible that Reducer 1 will get keys 1,2,5,8; and Reducer 2 will get keys 3,4,6,7, in that order. Even if they are sorted for a given Reducer, they are not sorted across Reducers. There will be two output files that are sorted individually, but the `part-00002` output file does not have all keys higher (lower in case of a descending sort) than all the keys in the `part-00001` output file, which is a requirement for Total Order Sorting across files.

The trick of Total Order Sorting lies in the `Partitioner` implementation. Let us utilize the use-case considered in this section: sorting by month in ascending order and day of the week in descending order. Because we have 12 months per year and 7 days per week, there are a total of 84 keys. The key order based on our criteria is as follows, where the first element is the month and the second element following the comma is day of the week:

1. 1,7
2. 1,6
3. ...
4. 1,1
5. 2,7
6. ...
7. 2,1
8. ...
9. 12,7
10. ...
11. 12,1

Next, we convert the preceding key range into an index number using the following formula `(Month_Id-1)*7 + (7-Day_of_Week Id)`, where `Month_Id` ranges from `1` through `12`, and `Day_of_Week` ranges from `1` through `7`.

Index 0 corresponds to January and Sunday, index 1 corresponds to January and Saturday, index 6 corresponds to January and Monday and so on until index 83 corresponds to December and Monday.

Finally, we partition the key index number range 0–83 by Reducer. If there are 12 Reducers, this index is partitioned evenly: Reducer-0 gets range 0–6, Reducer-1 gets range 7–13, and so on until Reducer-11 is reached, which gets range 77–83. In this case, each Reducer handles records for a specific month.

But what if we use a different Reducer count, say 20? In this scenario, we need more-complex partition logic. One way to handle it is to perform the calculation `84/20 = 4` (ignoring the remainder). Each Reducer handles the indexes as follows: Reducer-0 handles the range 0–3, Reducer-1 handles the range 4–7, and Reducer 18 handles the range 72–75, leaving Reducer-19 to handle the remaining 8 key indexes 76–83. Although this method of partitioning is simple, it leads to heavily unbalanced partitioning. Reducer-19 will take twice as long to complete as the other Reducers, which negatively affects the overall job performance. This problem gets worse as the calculation `Index_Range%#Reducers` (or remainder of the calculation `Index_Range/#Reducers`) grows. (In the simplified Partitioner implemented in the book, we used this method, but you should consider changing it in real-world applications based on the discussion that follows.)

A better partitioning scheme is to take the remainder of the calculation `84%20 = 4` and have the first 4 Reducers handle an additional key. For example, Reducer-0 handles key indexes 0–4; Reducer-1 handles key indexes 5–9; and Reducer-2 and Reducer-3 handle indexes 10–14 and 15–19, respectively. Starting with Reducer-4, each Reducer handles 4 consecutive index key ranges. For example, Reducer-5 handles range 20–23; Reducer-6 handles range 24–27; and so on until Reducer-20, which handles range 80–83.

The key steps in Total Order Sorting are as follows:

1. Implement the sorting logic in your WritableComparable keys or write a custom Comparator. This ensures that keys arrive at the Reducer sorted in a custom manner. In this example we will implement the sorting logic in the `compareTo` method of the `WritableComparable` key class.

2. Define the function to convert a Reducer key instance into an index number and define the range of these index numbers.

3. Implement a custom Partitioner in this way:
    a. It is aware of the entire Reducer key index range.
    b. It utilizes the key, the number of Reducers passed to the partition call, and the knowledge of the Reducer key index range to implement the partitioning logic described previously.

There is an important catch here. Our key range is fixed but if we had to configure the Partitioner at run time by dynamically passing the key range how can we do so? By default, the `Partitioner` interface does not support configuration. Passing this information with each key is very expensive because keys are transferred over the network. The trick is to have the custom Partitioner implement the `org.apache.hadoop.conf.Configurable` interface, which provides two important methods for configuration

- `getConf()`
- `setConf(Configuration conf)`

The previous method is used to pass the full range in the `Configuration` object. During the initialization phase, the Hadoop Framework creates an instance of the custom `Partitioner`. When the Hadoop reflection utilities instantiate this class, they check to see whether the class is the `Configurable` type. If it is, `setConf` is called, so the `Configuration` instance for the job is available for the `Partitioner` before the first call to `getPartition` is made.

---

■ **Note** A sample template of configurable custom Partitioners is available in the book source code in the following class (customize this template to create your own configurable Partitioner): `org.apress.prohadoop.c6.ConfigurablePartitioner`.

---

Hadoop supports Total Order Sorting through classes such as `TotalOrderPartitioner` and `InputSampler`. The idea is to have a separate MapReduce program execute prior to your own. This purpose of this program is to efficiently determinethe key range since in most cases, you are not aware of the key range. The Hadoop library provides the `InputSampler` class which was designed to identify the key range efficiently by sampling the records. Using the key range returned by the `InputSampler` class a partition file is created where each partition range is explicitly defined and it has as many partitions as the number of Reducers configured for your main MapReduce program. The partitions created using the `InputSampler` are based on the assumption that the keys are evenly distributed, which might not be valid for every use-case. This partition file is then used by the `TotalOrderPartitioner` class to pick the Reducer based on the key (remember, the partition file has as many partitions as there are reducers. Each partition has a minimum and maximum key value configured. Hence based on the key it is simple to pick the right reducer index). Note that the number of Reducers is fixed to the number of partitions in the partition file. To change the number of Reducers, first regenerate the partition file. You can also write your own custom class to create the Partition file instead of using the `InputSampler`. You should do this if the distribution of keys is not uniform and you want to exploit this knowledge to create partitions which are evenly sized.

## Implementing a Custom Key Class: MonthDoWWritable

We implement the Sort MapReduce routine using custom `Writable` classes instead of the default Hadoop-provided classes such as `Text` and `IntWritable`. The benefits of doing so will become apparent as the design evolves. First let us define the key class, which we call `MonthDoWWritable`. Listing 6-1 shows the implementation of this class.

***Listing 6-1.*** MonthDoWWritable implementation

```
package org.apress.prohadoop.c6;

import java.io.DataInput;
import java.io.DataOutput;
import java.io.IOException;

import org.apache.hadoop.io.IntWritable;
import org.apache.hadoop.io.WritableComparable;

public class MonthDoWWritable implements WritableComparable<MonthDoWWritable>{
   public IntWritable month=new IntWritable();
   public IntWritable dayOfWeek = new IntWritable();

      public MonthDoWWritable(){
      }

      @Override
      public void write(DataOutput out) throws IOException {
          this.month.write(out);
          this.dayOfWeek.write(out);
      }

      @Override
      public void readFields(DataInput in) throws IOException {
          this.month.readFields(in);
          this.dayOfWeek.readFields(in);
      }

      //Sort ascending my month and descending my day of the week
      @Override
      public int compareTo(MonthDoWWritable second) {
          if(this.month.get()==second.month.get()){
              return -1*this.dayOfWeek
                        .compareTo(second.dayOfWeek);
          }
          else{
              return 1*this.month
                        .compareTo(second.month);
          }
     }

      @Override
      public boolean equals(Object o) {
           if (!(o instanceof MonthDoWWritable)) {
               return false;
           }
           MonthDoWWritable other = (MonthDoWWritable)o;
           return (this.month.get() == other.month.get() &&
                   this.dayOfWeek.get() == other.dayOfWeek.get());
       }
```

```
    @Override
    public int hashCode() {
        return (this.month.get()-1)*7;
    }
}
```

The key features of this design include the following:

- Always implement WritableComparable for the Mapper output or Reducer input key classes.
- Provide a no-argument constructor. Although it is provided by default, if you provide another constructor that takes arguments, remember to add the no-argument constructor. The Hadoop Framework creates instances by invoking the no-argument constructor, and if it is not defined you will get an exception at runtime and your jobs will fail.
- The write() method is a serialization method that writes the fields to the output stream. Note that only the required information is being serialized out, which was a primary motive behind writing a new serialization framework for Hadoop. The Internet forums are rife with heated discussions on how the original Java serialization framework was adequate as well. For now, be aware that the Hadoop mechanism is very efficient and performs well.
- The readFields() method is responsible for deserialization of streams back to a Writable instance. Notice that the order of the fields is identical to the write() method.

  Recall that the Writable interface has two methods: write() and readFields(). Because WritableComparable extends Writable, these methods are inherited from the Writable interface. Because we are writing a custom class, we have to provide implementation for these two methods.
- The compareTo method is used to compare two instances of Writable. It is part of the java.lang.Comparable interface which the WritableComparable interface extends. The compareTo operation implementation ensures that the output is sorted in ascending order by month and descending order by day of the week. The multiplication factors of 1 and -1 ensure it.
- The role of hashCode is to simply provide a hash for an instance of MonthDoWWritable. We define it such that it has the value from 0 to 11 based on the month field of MonthDoWWritable. Note that it is not unique based on the attribute values of MonthDoWWritable. Hashcode's by definition do not need to be unique. Hence the equals method should be implemented along with hashCode to determine if two instances returning the same hashCode are indeed equal. Note that the hashCode return value is used by the HashPartitioner, the default Partitioner class in Hadoop. Hence if we do not define a custom Partitioner, a MapReduce program utilizing the MonthDoWWritable as its Mapper output key class would partition only on the basis of the month attribute of the MonthDoWWritable instance. In this example, we will define a custom Partitioner as our partitioning logic is based on month as well as the day of the week attributes of MonthDoWWritable.

## Implementing a Custom Value Class: DelaysWritable

Now let us define the value class, which we call DelaysWritable. Listing 6-2 shows the implementation of this class.

Why should there be a custom Writable class? Why not just extract the fields we need in the Reducer and pass the entire line of record to the Reducer using an instance of the key class we defined in the earlier section? Although that is possible, it increases the I/O between the Mapper and Reducer considerably. It is a good idea to extract only the fields needed in the Mapper before sending it to the Reducer.

In that case, why not just extract the fields we need and create a Text instance, which is a comma-separated string of only the required fields? Although this is an acceptable solution, Text instances contain only strings, and strings require more space than primitive variables. Most of the output is composed of numbers. So using a custom Writable class will enhance the I/O performance of the program. (You will learn about storing results in binary formats such as Sequence Files in Chapter 7.)

***Listing 6-2.*** DelaysWritable implementation

```
package org.apress.prohadoop.c6;
//Import statements

public class DelaysWritable implements Writable{
    public IntWritable year=new IntWritable();
    public IntWritable month = new IntWritable();
    public IntWritable date = new IntWritable();
    private IntWritable dayOfWeek = new IntWritable();
    public IntWritable arrDelay = new IntWritable();
    public IntWritable depDelay = new IntWritable();
    public Text originAirportCode = new Text();
    public Text destAirportCode = new Text();
    public Text carrierCode = new Text();

    public DelaysWritable(){
    }

    public void setDelaysWritable(DelaysWritable dw){
        this.year = dw.year;
        this.month = dw.month;
        this.date = dw.date;
        this.dayOfWeek = dw.dayOfWeek;
        this.arrDelay = dw.arrDelay;
        this.depDelay = dw.depDelay;
        this.originAirportCode = dw.originAirportCode;
        this.destAirportCode = dw.destAirportCode;
        this.carrierCode = dw.carrierCode;
    }

    @Override
    public void write(DataOutput out) throws IOException {
        this.year.write(out);
        this.month.write(out);
        this.date.write(out);
        this.dayOfWeek.write(out);
        this.arrDelay.write(out);
        this.depDelay.write(out);
        this.originAirportCode.write(out);
        this.destAirportCode.write(out);
        this.carrierCode.write(out);
    }
```

```
    @Override
    public void readFields(DataInput in) throws IOException {
        this.year.readFields(in);
        this.month.readFields(in);
        this.date.readFields(in);
        this.dayOfWeek.readFields(in);
        this.arrDelay.readFields(in);
        this.depDelay.readFields(in);
        this.originAirportCode.readFields(in);
        this.destAirportCode.readFields(in);
        this.carrierCode.readFields(in);
    }
}
```

The key features of this design are the following:

- For the value classes, it is enough to implement the `Writable` interface. Because `DelaysWritable` is never used as a key, implementing just the `Writable` interface is adequate.
- For reasons discussed in the previous section, provide a no-argument constructor.
- The `write()` and `readFields()` methods are responsible for serializing and deserializing.

## Sorting MapReduce Program

The client program used for implementing the sorting use-case is listed in the `org.apress.prohadoop.c6.SortAscMonthDescWeekMRJob` file.

Some of the features of this program are the following:

- It can support a variable number of Reducers and sorts records by Reducer input key across Reducer's (Total Order Sorting).
- It uses a smart Partitioner that selects a Reducer for a key to ensure total order sorting.
- Although the code works with any number of Reducers. However bear in mind the following reducer count characteristics:
    - Ideally the number of Reducers should be such that the full range of keys (84 keys corresponding to 12 months, and 7 days in week) is a multiple of the number of Reducers. So if 21 Reducers are selected, there are 4 keys per Reducer. If 20 Reducers are selected, the first 19 Reducers get 4 keys each, and the last Reducer gets 8 keys.
    - If the number of Reducers is selected to be greater than the range (84), the first 84 Reducers get 1 key each, and the remaining Reducers get no keys.

### run() method

The key lines in the `run()` method that differ for this job are as follows:

```
job.setMapOutputKeyClass(MonthDoWWritable.class);
job.setMapOutputValueClass(DelaysWritable.class);
job.setOutputKeyClass(NullWritable.class);
job.setOutputValueClass(Text.class);
```

```
job.setMapperClass(SortAscMonthDescWeekMapper.class);
job.setReducerClass(SortAscMonthDescWeekReducer.class);
job.setPartitionerClass(MonthDoWPartitioner.class);
```

For the first time in this book, key andvalue classes that are custom types are used. In this example, the output key andvalue pairs for the Mapper are both custom types.

The services of the Partitioner are also used. This class is responsible for ensuring that the keys go to the appropriate Reducer so the output is sorted in the correct way across all Reducers.

The number of Reducers has to be passed as a parameter (see Chapter 5):

```
-D mapred.reduce.tasks=12
```

To learn more, you should run the program using various numbers of Reducers.

## SortAscMonthDescWeekMapper Class

Listing 6-3 shows the source code for the `SortByMonthAndDayOfWeekMapper class`. Notice how the key for the output of the Mapper is formulated.

***Listing 6-3.*** SortByMonthAndDayOfWeekMapper

```
public static class SortAscMonthDescWeekMapper extends Mapper<LongWritable, Text, MonthDoWWritable,
DelaysWritable> {

    public void map(LongWritable key, Text value, Context context)
                        throws IOException, InterruptedException {
        if(!AirlineDataUtils.isHeader(value)){
            String[] contents = value.toString().split(",");
            String month = AirlineDataUtils.getMonth(contents);
            String dow = AirlineDataUtils.getDayOfTheWeek(contents);

                        MonthDoWWritable mw = new MonthDoWWritable();
                        mw.month=new IntWritable(Integer.parseInt(month));
                        mw.dayOfWeek=new IntWritable(Integer.parseInt(dow));

                        DelaysWritable dw = AirlineDataUtils
                                                      .parseDelaysWritable(value.toString());
            context.write(mw,dw);
        }
    }
 }
```

The code listing for the `parseDelaysWritable()` method is shown in Listing 6-4. It takes the line of airline data and converts it to an instance of `DelaysWritable()`.

***Listing 6-4.*** AirlineDataUtils.parseDelaysWritable()

```
public static DelaysWritable parseDelaysWritable(String line) {

   String[] contents = line.split(",");
   DelaysWritable dw = new DelaysWritable();
   dw.year = new IntWritable(Integer.parseInt(AirlineDataUtils
                                                .getYear(contents)));
   dw.month = new IntWritable(Integer.parseInt(AirlineDataUtils
                                                .getMonth(contents)));
   dw.date = new IntWritable(Integer.parseInt(AirlineDataUtils
                                                .getDateOfMonth(contents)));
   dw.dayOfWeek = new IntWritable(Integer.parseInt(AirlineDataUtils
                                                .getDayOfTheWeek(contents)));
   dw.arrDelay = new IntWritable(AirlineDataUtils.parseMinutes(
                         AirlineDataUtils.getArrivalDelay(contents), 0));
   dw.depDelay = new IntWritable(AirlineDataUtils.parseMinutes(
                         AirlineDataUtils.getDepartureDelay(contents), 0));
   dw.destAirportCode = new Text(AirlineDataUtils
                                               .getDestination(contents));
   dw.originAirportCode = new Text(AirlineDataUtils.getOrigin(contents));
   dw.carrierCode = new Text(AirlineDataUtils.getUniqueCarrier(contents));
   return dw;
}
```

## MonthDoWPartitioner Class

The core of the sorting is in the MonthDoWPartitioner class. Without this component, results are sorted within a Reducer, but not sorted across Reducers. Listing 6-5 shows the implementation of this class.

***Listing 6-5.*** MonthDoWPartitioner

```
public static class MonthDoWPartitioner extends
    Partitioner<MonthDoWWritable, Text> implements Configurable {
    private Configuration conf = null;
    private int indexRange = 0;

    private int getDefaultRange(){
        int minIndex = 0;
        int maxIndex = 11 * 7 + 6;
        int range = (maxIndex - minIndex) + 1;
        return range;
    }

    @Override
    public void setConf(Configuration conf) {
        this.conf = conf;
        this.indexRange = conf.getInt("key.range", getDefaultRange());
    }
```

```
    @Override
    public Configuration getConf() {
        return this.conf;
    }

    public int getPartition(MonthDoWWritable key,
                            Text value,int numReduceTasks) {
        return AirlineDataUtils.getCustomPartition(key,
                                                   indexRange,
                                                   numReduceTasks);

    }
}
```

As discussed earlier we use a configurable Partitioner. We need to implement the Configuragle interface for our custom Partitioner to make this possible. We configure our range in the setConf() method. Although we can pass the index range as a configuration parameter "`key.range`", we do not do so in our client program. This is because our range is fixed and `getDefaultRange()` method computes it. The line below is just an illustration of how a Partitioner can be configured from the Configuration instance.

```
this.indexRange = conf.getInt("key.range", getDefaultRange())
```

The actual logic for computing the reducer index is in the `getCustomPartition` method of the `AirlineDataUtils` class. Keeping it in a separate utility class enables us to test this method without having to deal with the various Hadoop-specific classes and the Hadoop environment. (This pattern is discussed in the chapter on Testing Hadoop Programs) Listing 6-6 shows the code for the `getCustomPartition()` method.

***Listing 6-6.*** AirlineDataUtils.getCustomPartition

```
public static int getCustomPartition(MonthDoWWritable key,
                                     int indexRange, int noOfReducers) {
   int indicesPerReducer = (int) Math.floor( indexRange/ noOfReducers);
   int index = (key.month.get()-1) * 7 + (7- key.dayOfWeek.get());
   /*
    *If the noOfPartitions is greater than the range just return the index
    *All Partitions above the (range-1) will receive no records.
    */
   if (indexRange < noOfReducers) {
           return index;
   }
   for (int i = 0; i < noOfReducers; i++) {
      int minValForPartitionInclusive = (i) * indicesPerReducer;
      int maxValForParitionExclusive = (i + 1) * indicesPerReducer;
      if (index >=minValForPartitionInclusive
             && index < maxValForParitionExclusive) {
           return i;
      }
   }
   /*
    * If the index>=indicesPerReducer*(noOfReducers) the last partition
    * gets the remainder of records.
    */
    return (noOfReducers - 1);
}
```

The features of the Partitioner logic are as follows:

- The range of the keys is identified. We will define each key as a number that is calculated as index = Month * 7 + (7-Day of Week) as described earlier.
- The total number of partitions handled are calculated as the floor of the calculation range/ noOfReducers where range = 84 because there are 84 distinct keys. Thus, if the number of Reducers is 19, this value will be floor (4.42) = 4.
- The range for each Reducer is calculated as follows
  - minValue = index of Reducer * keysPerReducer
  - maxValue = (index of Reducer+1) * keysPerReducer

  The minValue is inclusive, and the maxValue is exclusive. When using 19 reducers, these values are 0 and 4 respectively for Reducer 0. Reducer 0 handles index range 0-3; Reducer 1 handles index range 4–7 and so on.
- When the range is not a multiple of the number of Reducers, the last Reducer gets the remaining indexes. In our example of 19 reducers, the last Reducer index is 18, and its key index range is 72–75. However based on the previous discussion, it also handles keys with index values 76–83. To ensure a more even distribution of keys, the range should be a multiple number of Reducers. If we had used 21 as the number of Reducers, all Reducers would have evenly handled the key index range (four indexes per Reducer).
- If the number of Reducers is greater than the range, the Reducer index is simply the value of the index calculated previously. If the number of Reducers is 90, the Reducers 0–83 handle indexes from 0–83 respectively, but Reducers 84–89 get no records.

You might want to refer to the discussion on Total Order Sorting earlier in this chapter and relate it to what was just discussed. What is implemented is a simplified variation of the `TotalOrderPartitioner` class in Hadoop. Its simplified nature is useful to explain the key concepts associated with the implementation of Total Order Sorting.

---

■ **Note** In our example there was no need to utilize a Configurable Partitioner. We could have simply calculated the `indexRange` variable in each call to `getPartition()` . This is certainly acceptable. However, it is very common to use MapReduce to handle billions of records. One must avoid creating unnecessary variables. As the JVM's have become efficient, programmers have become used to not worrying about the cost of object creation and garbage collection. However, when applied to batch processing over billions of records, this cost becomes noticeable and can adversely impact t. Pre-calculating `indexRange` in the `setConf()` method is an example of such an optimization.

---

## SortAscMonthDescWeekReducer Class

The Reducer class converts `DelaysWritable` instances received into a comma-separated record by invoking `AirlineDataUtils.parseDelaysWritableToText(...)`. Listing 6-7 shows the `SortAscMonthDescWeekReducer class`.

***Listing 6-7.*** SortAscMonthDescWeekReducer

```
public static class SortAscMonthDescWeekReducer extends Reducer<MonthDoWWritable, DelaysWritable,
NullWritable, Text> {
     public void reduce(MonthDoWWritable key,
                        Iterable<DelaysWritable> values,
                        Context context)
                     throws IOException, InterruptedException {
```

```
        for (DelaysWritable val : values) {
             Text t = AirlineDataUtils.parseDelaysWritableToText(val)
            context.write(NullWritable.get(),t);
        }
    }
}
```

## Run the Sorting Job on the Cluster

Finally, we rebuild using Maven and execute the job in the Hadoop environment as follows:

```
hadoop jar prohadoop-0.0.1-SNAPSHOT.jar \
org.apress.prohadoop.c6.SortAscMonthDescWeekMRJob \
-D mapred.reduce.tasks=21 \
/user/hdfs/sampledata \
/user/hdfs/output/c6/sortbymonthandweek
```

The output directory should have 21 files: `part-r-00000` through `part-r-00020`. The records are guaranteed to be sorted by month (ascending) and day of the week (descending), so every record in the `part-r-{index_i}` file respects the sorting order with respect to the `part-r-{index_j}` file, where `{index_i}<{index_j}`.

## Sorting with Writable-Only Keys

This section examines modifications that have to be made when using only `Writable` instances as keys. Sometimes you are using a third-party library and you have no control over the classes used. Imagine that you are using a library that contains `Writable` classes you want to use as keys. These classes are defined as `final`, which means you cannot extend them to implement `WritableComparable`. This section shows you what to do in this situation.

---

■ **Note** The MapReduce program that demonstrates the use of `Writable` instances as keys is available in the book source code. The class name is `org.apress.prohadoop.c6.SortAscMonthDescWeekWithOnlyWritableKeysMRJob`.

---

Only the key differences are discussed in this section. First, the key class, which is now called `MapDowOnlyWritable`, implements only the `Writable` interface instead of the `WritableComparable` interface. The lines that differ in the `run()` method are these:

```
job.setMapOutputKeyClass(MonthDoWOnlyWritable.class);
job.setSortComparatorClass(ComparatorForMonthDoWOnlyWritable.class);
```

Note that we have now registered a new Sort Comparator class that must implement the `RawComparator` interface. Listing 6-8 shows the source code for this class.

***Listing 6-8.*** Implementation of a RawComparator

```
package org.apress.prohadoop.c6;

import java.io.IOException;
import org.apache.hadoop.io.DataInputBuffer;
import org.apache.hadoop.io.RawComparator;;
```

```
public class ComparatorForMonthDoWOnlyWritable implements RawComparator {
    MonthDoWOnlyWritable first = null;
    MonthDoWOnlyWritable second = null;
    DataInputBuffer buffer = null;
    public ComparatorForMonthDoWOnlyWritable(){
        first = new MonthDoWOnlyWritable();
        second = new MonthDoWOnlyWritable();
        buffer = new DataInputBuffer();
    }

    @Override
    public int compare(Object a, Object b) {
        MonthDoWOnlyWritable first = (MonthDoWOnlyWritable)a;
        MonthDoWOnlyWritable second = (MonthDoWOnlyWritable)b;
        if(first.month.get()!=second.month.get()){
            return first.month.compareTo(second.month);
        }
        if(first.dayOfWeek.get()!=second.dayOfWeek.get()){
            return -1 * first.dayOfWeek.compareTo(second.dayOfWeek);
        }
        return 0;
      }

    @Override
    public int compare(byte[] b1, int s1, int l1, byte[] b2, int s2, int l2) {
        try {
            buffer.reset(b1, s1, l1);
            first.readFields(buffer);

            buffer.reset(b2, s2, l2);
            second.readFields(buffer);

        } catch (IOException e) {
            throw new RuntimeException(e);
        }
        return this.compare(first,second);
    }
}
```

Note the two `compare()` methods:

- `compare(byte[] b1, int s1, int l1, byte[] b2, int s2, int l2)`: Invoked with raw bytes by the framework. This implementation is responsible for calling the `readFields()` on the `Writable` instance created in the constructor to reconstruct `Writable` instances.
- `compare(Object a,Object b)`: Invoked using the reconstructed `Writable` instances in the preceding step. The custom sorting implementation goes into this method.

It is possible to completely bypass the second compare() method. The entire process of reconstructing the Writable instances can be avoided and the comparison can be done on raw bytes. The Hadoop Text.Comparator class does this by using two fields to store information. The first is the length of the string stored in the Text instance, and the second is the field comprising the actual bytes in the string representing the text. The Text.Comparator class uses the following implementation of the first compare() method:

```
@Override
public int compare(byte[] b1, int s1, int l1,
                   byte[] b2, int s2, int l2) {
   int n1 = WritableUtils.decodeVIntSize(b1[s1]);
   int n2 = WritableUtils.decodeVIntSize(b2[s2]);
   return compareBytes(b1, s1+n1, l1-n1, b2, s2+n2, l2-n2);
}
```

Notice that we read the length out of the bytes and completely skip those bytes. The remaining information passed to the compareBytes() method represents *only* the bytes representing the string in the Text instance. The underlying implementation compares the bytes, one at a time. As soon as a byte is encountered that is different between two instances, the decision can be made about the sort order of the two Text instances. Ironically, equal strings require the most work because all bytes need to be compared!

Here are the rules applied by the Hadoop Framework to pick the correct sorting implementation:

- If a comparator class, an implementation of the interface org.apache.hadoop.io.RawComparator, is provided by calling the setSortComparatorClass on the Job instance or by setting the mapred.output.key.comparator (This property is deprecated in Hadoop 2 and replaced with mapreduce.job.output.key.comparator.class. However, as discussed in Chapter 5, the deprecated properties are still valid and widely used).
- If the sort comparator is not set, the default SortingComparator used is the one registered for the key class. Most WritableComparable classes in the Hadoop library such as the IntWritable class register the WritableComparator class (ex. org.apache.hadoop.io.IntWritable.Comparator) during classloading. The following lines of code in Hadoop library IntWritable class performs this function

```
static{
WritableComparator.define(IntWritable.class,
                          new Comparator());
 }
```

The org.apache.hadoop.io.IntWritable.Comparator is an inner class defined in the IntWritable class which extends the WritableComparator class. However if a WritableComparable implementation does not register a comparator class, the default behavior is to use WritableComparator class whose compare method compares two key instances of a WritableComparable class. The WritableComparator.compare method is defined as follows

```
public int compare(WritableComparable a, WritableComparable b) {
   return a.compareTo(b);
}
```

## Recapping the Key Hadoop Features Explored with Sorting

By implementing the Sorting feature, the following new features of Hadoop were discussed:

- Implement a custom key and value as implementations of the `WritableComparable` and `Writable` interfaces, respectively.
- Using a custom `Partitioner` to implement Total Ordering of records across multiple Reducers.
- `TotalOrderPartitioner` is the Hadoop class commonly used to implement Total Ordering across Reducers. It is a more general implementation of the custom Partitioner class implemented in this section.
- Implement sorting when only `Writable` instances are used as the Mapper output key.
- Implement `RawComparator` to provide a very efficient implementation for sorting. We examined how the `Text.Comparator` inner class of the `Text` class from the Hadoop library does this.

# Problem Definition: Analyzing Consecutive Records

This section introduces you to the concept of Secondary Sorting. By default, the Hadoop Framework sorts the Reducer inputs by a key, and sometimes we need sorting by a composite key. In cases such as the one encountered in the previous section, we can simply implement this logic in the `WritableComparator` interface of the key. You have seen how the default `HashPartitioner` decides which Reducer to send the key and value pair, based on the `hashCode()` value of the key class. What if we want to process records together that belong to only part of the key? We will construct a use-case around this concept.

We have created programs that identify arrival and departure delays at an airport. Suppose that along with reporting the arrival delay status for a given flight, we also report the flight details of the previous delayed arriving flight at the airport. It requires us to process all arrival flight records at the airport in a single `reduce` call. We will need to receive the values sorted by arrival time of the flight. This is the first time we are facing the challenge of receiving the values in a sorted order.

But why not simply create a composite key comprising all the attributes (destination airport code + arriving flight date and time) that make up the sort criteria? This question leads to two complexities:

- The Partitioner needs to ensure that the flights referring to the same destination airport code go to the same Reducer. If the hashCode() implementation of our key utilize both attributes of the key (destination airport code and arrival time) and we use the default HashPartitioner, keys having the same destination airport code but different arrival times will go to different reducers.
- Each key is a separate `reduce` call on the Reducer. We have to maintain state variables inside the Reducer class to identify when the destination airport code has changed.

We can implement a more elegant solution. Using the Secondary Sort method, we can not only ensure that all keys belonging to the same arriving airport code go to the same Reducer instance but also that all the values are correctly sorted by arrival date and time, and that these values are processed in the same `reduce` call.

The output is in the following format for a given destination airport port:

```
<ARRIVAL_DELAY_RECORD>|<PREVIOUS_ARRIVAL_DELAY_RECORD>
```

■ **Caution** Why not just use the destination airport code as the key and simply collect all the values for a key in a list by iterating the values `Iterable` instance in the `reduce` call? We can then sort this list in memory by arrival date and time. We can then iterate through the sorted list to produce the required output. Although this process is functionally correct, it will lead to an explosion in the memory usage as there are millions of records per destination airport code. Also we cannot begin sorting until all the records have arrived because the values for a given key (recall, Reducer input is sorted by key. The values for a given key are not sorted). This is the justification for using Secondary Sort; even if it is complicated; it is the only way to achieve the goal.

The `PREVIOUS_ARRIVAL_DELAY_RECORD` is blank for the first arrival delay record for a given destination airport code. The program that implements this functionality is

`org.apress.prohadoop.c6.AnalyzeConsecutiveArrivalDelaysMRJob.`

## Key Components Supporting Secondary Sort

First, we look at the key components needed to implement Secondary Sort; then we examine where in the cluster these components execute (Mapper side or Reducer side).

The four main components required to implement Secondary Sort are shown in Figure 6-3.

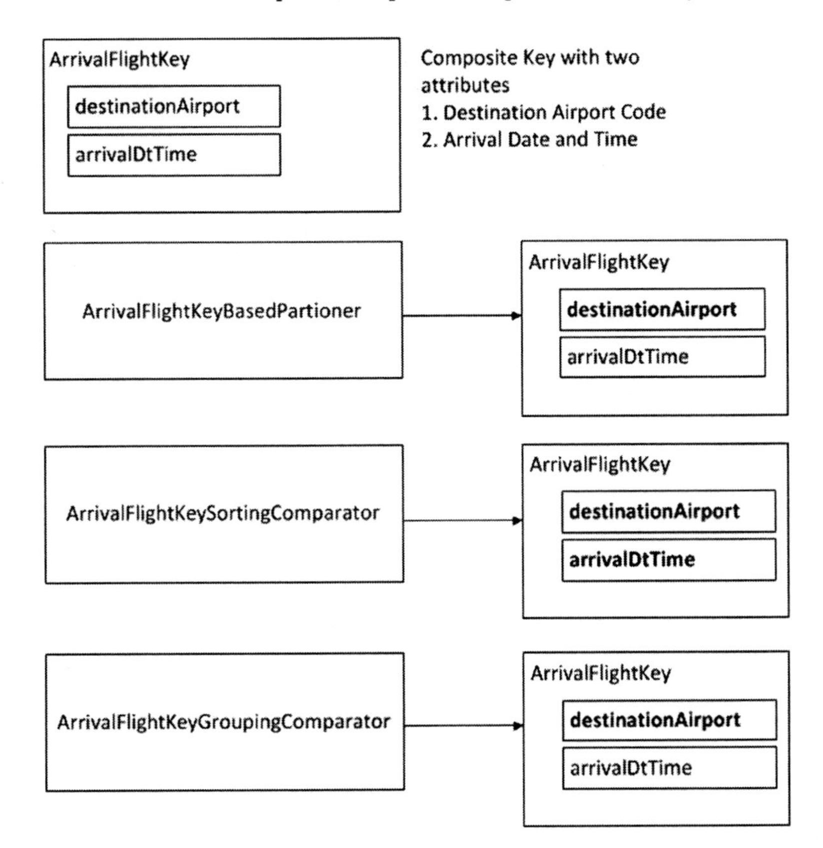


***Figure 6-3.** Components for Secondary Sort and how they use the key attributes*

- Custom `WritableComparable` key class: Used for Mapper output or Reducer input. The way the `compareTo` method is implemented is very subtle (this is discussed soon). The class is `ArrivalFlightKey`.
- Custom `Partitioner` class: Ensures that even if two attributes are used in the custom key class, only the destination airport code is used to choose the Reducer. So even with distinct keys, the keys representing a particular destination airport code go to the same Reducer node. The goal is to have all the values for a given destination code processed by the same `reduce` call. Note the distinction: we have to get the key andvalue pair to the same Reducer node first. Next we process these key and value pairs in the same `reduce` call. We have explicitly defined this component, but it is not needed in the implementation because of the `hashCode` function the custom key uses only the destination airport code. Thus, the default Partitioner (`HashCodePartitioner`) is adequate. However, for illustrative purposes we will define it explicitly. In practice, you might be using a key class over which you have no control. It is important to understand that Partitioner plays a key role in implementing Secondary Sort functionality.
- Sorting Comparator class: Sorts the records. This sorting is complete and uses all fields of the custom `Writable` class `ArrivalFlightKey`. The class name is `ArrivalFlightKeySortingComparator`. We could skip writing this class if we use the `compareTo` operation from `ArrivalFlightKey` to implement sorting based on both the attributes. In the current implementation, we cannot do this because the custom key class sorts only on the destination airport code, and the `SortingComparator` class needs to sort based on both the attributes of the key.
- Grouping Comparator class: Groups the keys on the Reducer side. It is used to determine the Grouping Criteria, which decides which values sent from the Mapper to the Reducer are processed in a single `reduce` call. The Grouping Comparator class we use is `ArrivalFlightKeyGroupingComparator` which groups all the values for a single destination airport code in a single `reduce` call. By default the Grouping Comparator is same as the Sorting Comparator class.
- We can delegate one of the Sorting Comparator or Grouping Comparator functionality to the `compareTo` method of the Reducer input key class. In our example we have chosen to have custom implementations for both the Sorting Comparator class as well as the Grouping Comparator class. However, based on how the `ArrivalFlightKey` class is implemented, we could have delegated the Grouping Comparator function to the `compareTo` method of the ArrvialFlightKey class.

## Custom Key Class: ArrivalFlightKey

The `ArrivalFlightKey` class is shown in Listing 6-9. Only the relevant section of the class is shown for brevity. The `readFields()` and `write()` methods have been omitted because they were discussed in the previous section.

***Listing 6-9.*** ArrivalFlightKey

```
public class ArrivalFlightKey implements WritableComparable<ArrivalFlightKey> {

    public Text destinationAirport = new Text("");
    public Text arrivalDtTime = new Text("");

    public ArrivalFlightKey() {
    }
```

```
    public ArrivalFlightKey(Text destinationAirport,Text arrivalDtTime) {
        this.destinationAirport = destinationAirport;
        this.arrivalDtTime = arrivalDtTime;
    }

    @Override
    public int hashCode() {
        return (this.destinationAirport).hashCode();
    }

    @Override
    public boolean equals(Object o) {
        if (!(o instanceof ArrivalFlightKey))
            return false;
        ArrivalFlightKey other = (ArrivalFlightKey) o;
        return this.destinationAirport.equals(other.destinationAirport) ;
    }

    @Override
    public int compareTo(ArrivalFlightKey second) {
        return this.destinationAirport.compareTo(second.destinationAirport);
    }
...
}
```

The above definition of the compareTo method only uses the destination airport code to implement sorting. As we will soon discover, this implementation is adequate to support the Grouping Comparator function but not the Sorting Comparator function. We could have implemented it differently to support the Sorting Comparator. In this case, neither is necessary because we will implement the Sorting Comparator and Grouping Comparator classes which do not defer their function to the custom key class. However we will demonstrate how the Grouping Comparator can defer its function to the compareTo method of the ArrivalFlightKey class.

## Custom Partitioner: ArrivalFlightKeyBasedPartioner

Listing 6-10 shows the custom Partitioner implementation. The Partitioner partitions on the basis of the destination airport code of the custom key, ignoring the arrival date and time.

***Listing 6-10.*** ArrivalFlightKeyBasedPartioner

```
public static class ArrivalFlightKeyBasedPartioner extends
            Partitioner<ArrivalFlightKey, Text> {
   @Override
    public int getPartition(ArrivalFlightKey key, Text value,
                                              int numPartitions) {
       return Math.abs(key.destinationAirport.hashCode() % numPartitions);
    }
}
```

The hashCode() function of the ArrivalFlightKey class is defined by using only the destination airport code, so we could have skipped defining a custom Partitioner and allowed the Hadoop Framework to use the default HashPartitioner described in the previous chapter. However, the purpose of defining these classes is to enable you to appreciate the components needed to make the Secondary Sorting feature work. Note that despite the two keys being different (the same destination airport code, but a different arrival date and time), they both go to the same Reducer node. Thus the first challenge of Secondary Sort of ensuring distinct keys get routed to the same Partitioner if their destination airport code is the same has been handled. The next challenge is to ensure that these keys and their values are handled by the same reduce invocation on the same Reducer node by using Sorting Comparator and Grouping Comparator (discussed next).

## Sorting Comparator Class: ArrivalFlightKeySortingComparator

The Sorting Comparator class, shown in Listing 6-11, sorts keys by using full information from the keys. It applies the normal sort you saw previously and ensures that the values are correctly sorted by the entire key (ascending with respect to both attributes of the key). The Sorting Comparator class is used in the Sort phase in the Mapper and Reducer. It ensures that Reducer input is sorted in an ascending order by the destination airport code and arrival date time.

***Listing 6-11.*** ArrivalFlightKeySortingComparator

```
public class ArrivalFlightKeySortingComparator extends WritableComparator {
    public ArrivalFlightKeySortingComparator() {
           super(ArrivalFlightKey.class, true);
    }
    /*Default implementation is to invoke a.compareTo(b)*/
    @Override
    public int compare(WritableComparable  a, WritableComparable  b) {
        ArrivalFlightKey first = (ArrivalFlightKey) a;
        ArrivalFlightKey second = (ArrivalFlightKey) b;
        if (first.destinationAirport.equals(second.destinationAirport)) {
            return first.arrivalDtTime.compareTo(second.arrivalDtTime);
        } else {
            return first.destinationAirport
                        .compareTo(second.destinationAirport);
        }
    }
}
```

Note that we need a Sorting Comparator; we cannot use the sorting applied in the `ArrivalFlightKey` class. The `compareTo` method of the `ArrivalFlightKey` class since it only uses the destination code. The constructor in the Sorting Comparator class registers the custom key class. When the second parameter is `true`, the WritableComparator constructor creates two instances of the key class in the constructor using reflection as follows

```
protected WritableComparator(Class<? extends WritableComparable> keyClass,
                                             boolean createInstances) {
   this.keyClass = keyClass;
   if (createInstances) {
      /*key1 and key2 are instance variables of type WritableComparable*/
      /*newKey() is a method which uses reflection to create the
       *custom WritableComparable instance
       */
      key1 = newKey()
      key2 = newKey();
      buffer = new DataInputBuffer();
```

```
    } else {
        key1 = key2 = null;
        buffer = null;
    }
}
```

The Hadoop framework invokes the following method in the WritableComparator class to compare two instances of keys: `compare(byte[] b1, int s1, int l1, byte[] b2, int s2, int l2)`

The two instances created earlier are used to read in the data by using the readFields() method from the `Writable` interface of the key class.

```
public int compare(byte[] b1, int s1, int l1, byte[] b2, int s2, int l2) {
  try {
    buffer.reset(b1, s1, l1); // parse key1
    key1.readFields(buffer);

    buffer.reset(b2, s2, l2); // parse key2
    key2.readFields(buffer);

  } catch (IOException e) {
    throw new RuntimeException(e);
  }
  return compare(key1, key2); // compare them
}
```

The above method finally invokes the `compare(WritableComparable a, WritableComparable b)` using these two instance. We have overridden the `compare(WritableComparable a, WritableComparable b)` method. The default implementation of the `compare` method is as follows:

```
public int compare(WritableComparable a, WritableComparable b) {
  return a.compareTo(b);
}
```

The default implementation makes it clear why we need to override this method. The `ArrivalFlightKey` class compares using on the destination airport code. Contrast the above discussion in the context of the `Text.Comparator` class we examined earlier. That class does not pass true in the constructor as the second parameter. Neither does it invoke the `compare(WritableComparable a, WritableComparable b)` method of the `WritableComparator`. It uses the byte stream to compare bytes directly. Consider this entire discussion from the perspective of our earlier discussion on using only Writable (not WritableComparable) instances as the keys. In such as scenario we cannot invoke the `compare(WritableComparable a, WritableComparable b)` but make the comparison decision in the `compare(byte[] b1, int s1, int l1, byte[] b2, int s2, int l2)` method.

## Grouping ComparatorClass: ArrivalFlightKeyGroupingComparator

The Grouping Comparator class, shown in Listing 6-12, groups key andvalue pairs received by the Reducer only on the basis of the destination airport code. In conjunction with the Sorting Comparator this class allows us to address the second challenge of Seconday Sorting, that of ensuring that flight records for the same destination code are processed in a single reduce call even if their arrival time is different. The Grouping Compartor only cares if its `compare` method returns a zero or non-zero value. Zero implies that two keys are equal from a grouping perspective and non-zero implies that they are not. The sort ordering cannot be changed by the Grouping Comparator.

Recall the discussion on Shuffle and Sort in the last section of Chapter 5. After the key and values are sorted on the Reducer node by the Sorting Comparator, the reduce call is invoked. As each key value pair is encountered a decision needs to be made if it should be processed by the current reduce call or if a new reduce call needs to be invoked. This decision is the responsibility of the Grouping Comparator. Since by default it is the same as the Sorting Comparator, a new reduce invocation is made when the compare operation of the Sorting Comparator returns a non-zero value. However when a Grouping Comparator is configured this behavior can be customized to group values in a single reduce call based on a partial attribute set of the Reducer input key.

The ArrivalFlightKeyGroupingComparator's compare method returns a non-zero value only when the destination code of the ArrivalFlightKey instances are different. Hence a new reduce call is not invoked when new ArrivalFlightKey instance with the same destination code but a different arrival time as compared to the previous instance arrives. Values for keys with the same destination code but different arrival times are processed in the same reduce call. Since the original Mapper output is sorted by destination airport code and arrival time by the Sorting Comparator, the values processed in the reduce call are sorted by destination airport code and arrival time. However, inside the reduce call, the key never changes and remains the first key as sorted by the Sorting Comparator for a given destination airport code. This is also the reason why it is important to include every attribute of the key in the value object; otherwise, those attributes cannot be accessed while iterating the values inside the reduce call.

***Listing 6-12.*** ArrivalFlightKeyGroupingComparator

```
public class ArrivalFlightKeyGroupingComparator extends WritableComparator {
    public ArrivalFlightKeyGroupingComparator() {
        super(ArrivalFlightKey.class, true);
    }
    /* Optional. If not provided the Key class compareTo method is invoked
     * Default implementation is as follows. Hence the custom key could
     * implement the sorting from a Grouping perspective
     *
     * public int compare(WritableComparable a, WritableComparable b) {
     *    return a.compareTo(b);
     * }
     *
     * /
    @Override
    public int compare(WritableComparable a, WritableComparable b) {
        // We could have simply used return a.compareTo(b);  See comment above
        ArrivalFlightKey first = (ArrivalFlightKey) a;
        ArrivalFlightKey second = (ArrivalFlightKey) b;
        return first.destinationAirport.compareTo(second.destinationAirport);
    }
}
```

For illustrative purposes we have overridden the compare method. But we did not need to do so because the ArrivalFlightKey implements its compareTo method to suit the requirements of the Grouping Comparator. The default implementation of the compare method of the WritableComparator class is the following:

```
public int compare(WritableComparable a, WritableComparable b) {
   return  a.compareTo(b)
}
```

The instances of the ArrivalFlightKey would have been passed from the compare(byte[] b1, int s1, int l1, byte[] b2, int s2, int l2) which is responsible for invoking the readFields() method on the instances created in the constructor and invoking the compare(WritableComparable a, WritableComparable b). The only reason we perform explicit type casting in our custom implementation is because we invoke the compareTo on the destinationAirport attribute instead of the ArrivalFlightKey instance.

This is an important point: when implementing Secondary Sort, either the Sorting Comparator or the Grouping Comparator can use the sorting behavior of the custom key class, but not both.

## Mapper Class: AnalyzeConsecutiveDelaysMapper

The Mapper class, which is very straightforward, is shown in Listing 6-13. All it does is determine whether there has been an arrival delay and sends such records to the Reducer after creating the custom key.

***Listing 6-13.*** AnalyzeConsecutiveDelaysMapper

```
public static class AnalyzeConsecutiveDelaysMapper extends
        Mapper<LongWritable, Text, ArrivalFlightKey, Text> {
    public void map(LongWritable key, Text value, Context context)
            throws IOException, InterruptedException {
        if (!AirlineDataUtils.isHeader(value)) {
            String[] contents = value.toString().split(",");
            String arrivingAirport = AirlineDataUtils.getDestination(contents);
            String arrivingDtTime =  AirlineDataUtils
                                        .getArrivalDateTime(contents);
            int arrivalDelay = AirlineDataUtils.parseMinutes(
                    AirlineDataUtils.getArrivalDelay(contents),0);
            if(arrivalDelay>0){
                ArrivalFlightKey afKey =
                     new ArrivalFlightKey(new Text(arrivingAirport),
                                          new Text(arrivingDtTime));
                context.write(afKey, value);
            }
        }
    }
}
```

## Reducer Class: AnalyzeConsecutiveDelaysReducer

Listing 6-14 shows the custom Reducer, which keeps track of the previous arrival delay record for the destination airport inside a *local* variable in the reduce method. This is the key point: when we use Secondary Sort, we can do all our computations using local variables. If we did not use Secondary Sort, we would have to move this variable at the instance level. This is the difference between processing key/value pairs in the same Reducer instance versus the same reduce call.

***Listing 6-14.*** AnalyzeConsecutiveDelaysReducer

```
public static class AnalyzeConsecutiveDelaysReducer extends
        Reducer<ArrivalFlightKey, Text, NullWritable, Text> {
    public void reduce(ArrivalFlightKey key, Iterable<Text> values,
            Context context) throws IOException, InterruptedException {
                Text previousRecord = null;
```

```
        for (Text v : values) {
            StringBuilder out = new StringBuilder("");
            if(previousRecord==null){
                out.append(v.toString()).append("|");
            }
            else{
                out.append(v.toString()).append("|")
                            .append(previousRecord.toString());
            }
            context.write(NullWritable.get(), new Text(out.toString()));
            //Remember to not use references as the same Text instance
            //is utilized across iterations
            previousRecord=new Text(v.toString());
        }
    }
}
```

## run() Method

The key lines in the run() method are shown in Listing 6-15. Note that Sorting Comparator and Grouping Comparator are explicitly defined.

***Listing 6-15.*** AnalyzeConsecutiveArrivalDelaysMRJob.run()

```
job.setMapperClass(AnalyzeConsecutiveDelaysMapper.class);
job.setReducerClass(AnalyzeConsecutiveDelaysReducer.class);
job.setPartitionerClass(ArrivalFlightKeyBasedPartioner.class);
job.setSortComparatorClass(ArrivalFlightKeySortingComparator.class);
job.setGroupingComparatorClass(ArrivalFlightKeyGroupingComparator.class);
```

## Implementing Secondary Sort without a Grouping Comparator

The source code accompanying this book contains the listing for implementing Secondary Sort without the Grouping Comparator. The only difference is that we do not specify the Grouping Comparator in the `run()` method. By default it is the same as the Sorting Comparator. Thus each distinct `ArrivalFlightKey` instance (where one of its two attributes is different) will result in a new reduce invocation. We still use our custom Partitioner which ensures that all the keys for the same destination airport code are processed by the same Reducer instance. The custom Reducer is now shown in Listing 6-16. Notice that the `previousRecord` variable has now moved up from the reduce method to the instance level. The difference is between processing in the same Reducer instance versus processing in the same reduce call.

***Listing 6-16.*** AnalyzeConsecutiveDelaysReducer no Secondary Sort

```
public static class AnalyzeConsecutiveDelaysReducer extends
        Reducer<ArrivalFlightKey, Text, NullWritable, Text> {
    Text previousRecord = null;
  public void reduce(ArrivalFlightKey key, Iterable<Text> values,
            Context context) throws IOException, InterruptedException {
      for (Text v : values) {
         StringBuilder out = new StringBuilder("");
```

```
            if(previousRecord==null){
               out.append(v.toString()).append("|");
            }
            else{
               out.append(v.toString()).
                   append("|").
                   append(previousRecord.toString());
            }
            context.write(NullWritable.get(), new Text(out.toString()));
            //Remember to not use references as the same Text instance
            //is utilized across iterations
            previousRecord=new Text(v.toString());
        }
    }
}
```

---

■ **Note** Based on the previous examples, it appears that Secondary Sort is not so important, and we can easily achieve our goals using a normal sort. This is not the case, however: there are complex use-cases for which you must absolutely use Secondary Sort. Managing state at the instance level (Reducer) is more complicated and has the potential to cause side effects as compared with managing state in a local method: `reduce`. Hadoop implementations are complicated. It pays to be simple and maintainable, especially in the case of Hadoop code, in which data is large, and bugs manifest themselves subtly. Making the code easy to read and review will save you considerable heartburn later.

---

## Running the Secondary Sort Job on the Cluster

Finally we execute a maven build. The result is a JAR file called `prohadoop-0.0.1-SNAPSHOT.jar`.

Execute the job in your Hadoop environment as follows:

```
hadoop jar prohadoop-0.0.1-SNAPSHOT.jar \
org.apress.prohadoop.c6.AnalyzeConsecutiveArrivalDelaysMRJob \
/user/hdfs/sampledata \
/user/hdfs/output/c6/secondarysort
```

## Recapping the Key Hadoop Features Explored with Secondary Sort

As the Secondary Sort was implemented, the following key features of Hadoop were explored:

- How Sorting Comparator and Grouping Comparator work with `WritableComparable` classes.
- How to control which keys are sent to which Reducer using a Partitioner using only a part of the custom key.
- The role a Sorting Comparator plays in ensuring an efficient ordering of records prior to invoking the first reduce call on the Reducer.
- The role a Grouping Comparator plays in grouping multiple values whose keys are only partially equal, in a single reduce call.

# Problem Definition: Join Using MapReduce

Joining two datasets is the most common function invoked with SQL. There are two ways Join can be implemented in the Hadoop Framework:

- **Map-side join**: Used when one of the two datasets is very small. For example, the amount of master data used in this chapter is very small. If we expand the carrier code with the carrier description, we can keep the carrier file in memory in each Mapper. As each flight record flows into the map method, an in memory lookup is performed for a carrier description by carrier code. The record is expanded to include the carrier description and then written out; no Reducer is needed.
- **Reduce-side join**: Used when both datasets are large and one of them cannot be entirely cached in memory, but the join criteria is such that the number of results is very small for any given join instance. In such as scenario we can utilize the Reduce-Side join. Two datasets are read, and the join key is used as the Mapper output key. The Mapper output value also contains a marker attribute that indicates which dataset the record (value part) came from. On the Reducer side, the two records are received together, and the join is performed.For this method to work, the total number of records received for a given join key from both datasets is small enough to be held in memory for the duration of the reduce call. Secondary Sort is often used here if one side of the join is inherently small and the other side is very large. Secondary Sort ensures that the records are ordered by the smaller dataset coming first. The smaller dataset is is retained in memory in the reduce call. As the records of the larger dataset flow in, they are joined with the cached records from the smaller dataset.

This section discusses how to perform a join in MapReduce using Secondary Sort. The implementation in this section can help you appreciate the many uses of Secondary Sort. The characteristics of the join output are as follows:

- We have to produce the same output as the one produced in the previous section on sorting.
- In this case, we have to expand the carrier code into its full description. The final output is similar to the one in the section on Sorting, but it has one extra field at the end: the carrier description.

We will use MapReduce program using Secondary Sort to illustrate this concept, although we could have simply employed Map-side join by caching the entire Carrier Master file in the memory of each Mapper instnace. In fact this method of join is explored when we discuss Map-side join using Distributed Cache in a subsequent section of this chapter.

## Handling Multiple Inputs: MultipleInputs Class

By using MultipleInputs, we can specify multiple input folders along with corresponding Mappers to use for each folder. The only constraint is that their Mapper output definitions should be identical. Their input formats and input definitions can vary. The following lines in the Job definition will configure the input and output paths:

```
String[] args = new GenericOptionsParser(getConf(), allArgs)
                .getRemainingArgs();
Job job = Job.getInstance(getConf());
job.setJarByClass(JoinMRJob.class);
job.setOutputFormatClass(TextOutputFormat.class);
MultipleInputs.addInputPath(job, new Path(args[0]),
                TextInputFormat.class, FlightDataMapper.class);
MultipleInputs.addInputPath(job, new Path(args[1]),
                TextInputFormat.class, CarrierMasterMapper.class);
FileOutputFormat.setOutputPath(job, new Path(args[2]));
```

In this case, we use the same input format for both our inputs. However, if our input files had different formats, MultipleInputs class can accommodate this requirement.

## Mapper Classes for Multiple Inputs

Listing 6-17 shows the definition of CarrierMasterMapper, and Listing 6-18 shows the listing for FlightDataMapper.

***Listing 6-17.*** CarrierMasterMapper

```
public static class CarrierMasterMapper extends
            Mapper<LongWritable, Text, CarrierKey, Text> {

    public void map(LongWritable key, Text value, Context context)
               throws IOException, InterruptedException {

      if (!AirlineDataUtils.isCarrierFileHeader(value.toString()))
      {
         String[] carrierDetails = AirlineDataUtils
                        .parseCarrierLine(value.toString());
         Text carrierCode = new Text(carrierDetails[0]
                                     .toLowerCase().trim());
         Text desc = new Text(carrierDetails[1]
                              .toUpperCase().trim());
          CarrierKey ck = new CarrierKey(CarrierKey.TYPE_CARRIER,
                                        carrierCode,desc);
          context.write(ck,new Text());
       }

    }
 }
```

***Listing 6-18.*** FlightDataMapper

```
public static class FlightDataMapper extends
            Mapper<LongWritable, Text, CarrierKey, Text> {
   public void map(LongWritable key, Text value, Context context)
             throws IOException, InterruptedException {
      if (!AirlineDataUtils.isHeader(value)) {
         String[] contents = value.toString().split(",");
         String carrierCode = AirlineDataUtils
                                  .getUniqueCarrier(contents);

         Text code = new Text(carrierCode.toLowerCase().trim());
         CarrierKey ck = new CarrierKey(CarrierKey.TYPE_DATA,
                                       code);
         DelaysWritable dw = AirlineDataUtils
                      .parseDelaysWritable(value.toString());
         Text dwTxt=AirlineDataUtils.parseDelaysWritableToText(dw);
             context.write(ck, dwTxt);
      }
   }
}
```

The Mapper's output key class is the CarrierKey class, which is shown in Listing 6-19. Note how each of them indicate in the constructor invocation of each CarrierKey, which dataset they respectively process. Also note, the different constructors they utilize. Only the CarrierMasterMapper invokes the constructor which uses the Carrier Key description. Only the relevant section of the CarrierKey class is shown in Listing 6-19 for brevity (the readFields() and write() methods are omitted because they were discussed previously).

***Listing 6-19.*** CarrierKey

```
public class CarrierKey implements WritableComparable<CarrierKey> {
    public static final IntWritable TYPE_CARRIER = new IntWritable(0);
    public static final IntWritable TYPE_DATA = new IntWritable(1);
    public IntWritable type = new IntWritable(3);
    public Text code = new Text("");
    public Text desc = new Text("");

    public CarrierKey() {
    }

    public CarrierKey(IntWritable type, Text code,Text desc) {
        this.type = type;
        this.code = code;
        this.desc = desc;
    }

    public CarrierKey(IntWritable type, Text code) {
        this.type = type;
        this.code = code;
    }

    @Override
    public int hashCode() {
        return (this.code.toString()+
                Integer.toString(this.type.get())).hashCode();
    }
    @Override
    public int compareTo(CarrierKey second) {
        CarrierKey first = this;
        if (first.code.equals(second.code)) {
            return first.type.compareTo(second.type);
        } else {
            return first.code.compareTo(second.code);
        }

    }
...
}
```

We want the records belonging to the same carrier key code (regardless of the Mapper class they were sent from) to be processed by the same reduce call. Recall from Chapter 5 that the default Partitioner in MapReduce is the HashPartitioner which partitions on the basis of the value of the hashCode()invocation on the CarrierKey.

Now look at the hashCode() implementation that uses both the code and type attributes to compute hashcode. Thus, a CarrierKey instance has a different hashcode, depending on which file it came from. Consequently, the master record could go to one Reducer and the flight detail record could go to another.

This problem can be easily fixed. Let us define a custom Partitioner that partitions only on the basis of carrier key code.

## Custom Partitioner: CarrierCodeBasedPartioner

Listing 6-20 shows the implementation of a custom Partitioner: CarrierCodeBasedPartitioner.

***Listing 6-20.*** CarrierCodeBasedPartitioner

```
public static class CarrierCodeBasedPartioner extends
                                 Partitioner<CarrierKey, Text> {
   @Override
   public int getPartition(CarrierKey key, Text value,
                           int numPartitions) {
      return Math.abs(key.code.hashCode() % numPartitions);
   }
 }
```

Add the following line to the run() method:

```
job.setPartitionerClass(CarrierCodeBasedPartitioner.class);
```

Now the records indeed go to the same Reducer instanceIn this instance, we deliberately did not implement the hashCode() method of the custom key to use only the carrier code portion, to illustrate why we cannot use the default HashPartitioner. We are forced to create our own custom Partitioner.

## Implementing the Join in the Reducer

For the join to work we still need all the values for CarrierKey instances having the same carrier code to be processed in the same reduce call regardless of which file they came from (the type attribute of the CarrierKey class). We also want the values to be sorted such that the values corresponding to the master file are processed before the values processed from the flight details file. This would allow the instance of the CarrierKey passed to the reduce call be the one which contains the description of the carrier. Recall from the previous section how the key in the reduce call is the first key encountered by the Grouping Comparator. This would allow the description to be cached in the very first iteration of the values Iterable instance and in the subsequent iterations we can attach the carrier description to the line representing the delayed flight record. To achieve this function we will need to implement a Sorting and a Grouping comparator.

### SortingComparator Class: CarrierSortComparator

Listing 6-21 shows the CarrierSortComparator, which is responsible for ensuring that the records processed by the CarrierMasterMapper are sorted lower than the records processed by the FlightDataMapper. The former assigns a type value of 0 and the latter assigns a type value of 1. See the CarrierType listing for the definitions.

***Listing 6-21.*** CarrierSortComparator

```
public class CarrierSortComparator extends WritableComparator {
    public CarrierSortComparator() {

        super(CarrierKey.class, true);

    }
}
```

We did not override the compare method of the Sorting Comparator class; the custom key is implemented to perform the Sorting Comparator duties. This is different from the previous example, in which the custom key was implemented to perform the Grouping Comparator duties. All we do is register the class in the constructor. The default compare implementation of the WritableComparator invokes the compareTo function of the custom key.

The run() method needs to be updated as follows:

```
job.setSortComparatorClass(CarrierSortComparator.class);
```

## Grouping Comparator Class: CarrierGroupComparator

From the point of view of the Reducer, we want the records for the CarrierKey instance with the same code to be processed together, regardless of whether they came from the carrier master file or the flight details file. In other words, we want records for both of the keys to be processed by the same reduce call in the Reducer. As discussed in the section on Secondary Sort, this goal is met by using GroupingComparator. The implementation of the Sorting Comparator ensures that the master data record comes prior to the flight data record.

Listing 6-22 shows the CarrierGroupComparator. It ensures that the records for the CarrierKey instance with the same carrier code are processed together in a single reduce call, regardless of whether they came from the carrier master file or the flight details file. Note that it uses only the code attribute for making decisions about which records belong together for a CarrierKey instance.

***Listing 6-22.*** CarrierGroupComparator

```
public class CarrierGroupComparator extends WritableComparator {
    public CarrierGroupComparator() {
        super(CarrierKey.class, true);
    }
    @Override
    public int compare(WritableComparable a, WritableComparable b) {
        CarrierKey first = (CarrierKey) a;
        CarrierKey second = (CarrierKey) b;
        return first.code.compareTo(second.code);
    }
}
```

The run() method needs to be updated as follows:

```
job.setGroupingComparatorClass(CarrierGroupComparator.class);
```

This completes the full implementation and we are finally ready to see the Reducer implementation.

## Reducer Class: JoinReducer

Listing 6-23 shows the `JoinReducer class` performing the final join. Note that the Reducer depends on the carrier key records from `CarrierMasterMapper` arriving before the records from `FlightDataMapper`. If the master data does not contain the code, the flight details records use `UNKNOWN` as the carrier description.

***Listing 6-23.*** JoinReducer

```
public static class JoinReducer extends
            Reducer<CarrierKey, Text, NullWritable, Text> {
   public void reduce(CarrierKey key, Iterable<Text> values,
                      Context context)
                      throws IOException, InterruptedException {
        String carrierDesc = "UNKNOWN";
        for (Text v : values) {
           if (key.type.equals(CarrierKey.TYPE_CARRIER)) {
              carrierDesc = key.desc.toString();
              continue;// Coming from the Master Data Mapper
           }
           else{
              //The sorting comparator ensures carrierDesc was already set
              //by the time this section is reached.
              Text out = new Text(v.toString()+","+carrierDesc);
              context.write(NullWritable.get(), out);
           }

        }
     }
}
```

The following modification to the `run()` method configures the Reducer:

```
job.setReducerClass(JoinReducer.class);
```

## Running the MapReduce Join Job on the Cluster

Finally we rebuild using Maven and execute the job in the Hadoop environment as follows:

```
hadoop jar prohadoop-0.0.1-SNAPSHOT.jar \
org.apress.prohadoop.c6.JoinMRJob \
-D mapred.reduce.tasks=12 \
/user/hdfs/sampledata \
/user/hdfs/masterdata/carriers.csv \
/user/hdfs/output/c6/mrjoin
```

The output directory should have 12 files: `part-r-00000` through `part-r-00011`. The lines will look as follows:

```
1,5,1988,29,32,32,19,DFW,CVG,AA,AMERICAN AIRLINES INC.
1,1,1988,18,6,6,0,SNA,SFO,AA,AMERICAN AIRLINES INC.
1,5,1988,1,3,3,-1,HRL,DFW,AA,AMERICAN AIRLINES INC.
```

The last two attributes are the carrier code and carrier description.

## Key Hadoop Features Explored with MapReduce

As we implemented the Join feature using MapReduce, we explored the following new features of Hadoop:

- The role Secondary Sort plays in a reduce-side join.
- The role a custom Partitioner plays in the reduce-side join when using Secondary Sort.
- The role Sorting Comparator plays in ensuring an efficient ordering of records arriving to the Reducer instance.
- The role Grouping Comparator plays in ensuring that even though keys may be physically different, the records for the keys can be processed together in the same `reduce` call, based on logical grouping of value where this grouping is user defined.

Even two large datasets can be joined using this technique. The records of the smaller dataset should be processed first and saved to a local disk. When the iterator eventually arrives at the records from the larger dataset, they can be joined with the records saved to disk for the smaller dataset. Although this process is slower due to the additional disk I/O needed, we will not run out of memory. This method works because records for each side of the join are processed in order and are not mixed with each other on the Reducer side. The Reducer knows exactly when to close the file containing the records for the smaller dataset and start processing the records from the larger dataset.

# Problem Definition: Join Using Map-Only jobs

In the real world, sometimes the amount of master data is very small. For example, suppose that we want to enhance the output of the previous program:

- For each airport, origin, and destination airport code, we want to add details about the airport. For this example, we enhance only the records with the airport name, but we can easily add more attributes. These details come from another file called `airport.csv`. The attributes contained in this file are these:
    - Airport code
    - Airport name
    - Airport city, state, and country
    - Latitude and longitude coordinates of airport location
- For the carrier code, we augment it with the carrier name, similar to the process discussed in the previous section. We do the same here using a map-only job. This method of performing join will be referred to as the map-side join to differentiate it from the method in the previous section known as the reduce-side join.

The advantage of this method is obvious: we are joining based on three separate keys in a single job. With Reducers, this is impossible to do in a single job because the Reducer can have only one join key. Not only are, map-only jobs faster due to the time saved in Sort and Shuffle phases but we can also perform three joins in a single job, saving two passes through a very large number of flight detail records.

## DistributedCache-Based Solution

This is a capability provided by the MapReduce program to cache files (text, archives) needed by applications. Applications can specify the files using the following methods:

- Jobs using the `GenericOptions` parser and implementing the `Tool` interface can be specified on the command line with the `-files <comma separated URI's>` parameter. The files can be in the local filesystem, in HDFS, or even accessible via a HTTP URL. The scheme supplied at the front of the URI is used to determine how to fetch the file. If no scheme is supplied, the files are assumed to be local. Another parameter that can be used is `-archives <comma separated URI's>`. Archives such as ZIP files, TAR files, and JAR files can be passed in this manner. The difference between `-files` and `-archives` is that in the latter case, the files are unarchived on the node where the container runs the Mapper or Reducer tasks.
- Alternatively jobs can specify the files by using `job.addCacheFile(...)` or `job.addCacheArchive(...)` calls.

To ensure performance, the files are copied once at the beginning of the job to the nodes running the tasks of the MapReduce jobs. As each begins, its `setup()` method retrieves the cache files. The class that represents the map-side join using `DistributedCache` is `org.apress.prohadoop.c6.MapSideJoinMRJob`.

## run() method

Listing 6-24 shows the `run()` method of `MapSideJoinMRJob`.

***Listing 6-24.*** MapSideJoinMRJob.run() Implementation

```
public int run(String[] allArgs) throws Exception {
   Job job = Job.getInstance(getConf());
   job.setJarByClass(MapSideJoinMRJob.class);
   job.setInputFormatClass(TextInputFormat.class);
   job.setOutputFormatClass(TextOutputFormat.class);
   job.setOutputKeyClass(NullWritable.class);
   job.setOutputValueClass(Text.class);

   job.setMapperClass(MapSideJoinMapper.class);
   String[] args = new GenericOptionsParser(getConf(), allArgs)
                                      .getRemainingArgs();
   FileInputFormat.setInputPaths(job, new Path(args[0]));
   FileOutputFormat.setOutputPath(job, new Path(args[1]));

       job.addCacheFile((new File(args[2])).toURI());
       job.addCacheFile((new File(args[3])).toURI());

   job.waitForCompletion(true);

   return 0;
}
```

The lines that provide the DistributedCache functionality were highlighted previously. We pass the airlines.csv and carriers.csv files in this manner. The input parameters passed to this program are the following:

- <input path of the file in hdfs>
- <output path in hdfs>
- <local path for airports.csv>
- <local path for carriers.csv>

Next, we examine the Mapper class.

## MapSideJoinMapper Class

The Mapper class is shown in Listing 6-25. The key components of this class are as follows:

- The private variables airports and carriers are used to cache the master data associated with airports and carriers. The lookup key is the airport code and the carrier code.
- The setup() method invokes the context.getCacheFiles() method to get the URIs for the cache files. By the time this call is made, the cache files are already available on the local node via a symbolic link.
- The readAirports and readCarriers methods are responsible for populating the cache variables.
- Next each flight record is processed in a seperate invocation of the map method. The cached variables airports and carriers are available and used to get details for each airport code and carrier code.
- The overridden parseDelaysWritableToText is used to enhance the DelaysWritable instance created earlier. The returned Text instance has the delays line enhanced with the name of the origin and destination airport descriptions as well as the carrier description.

***Listing 6-25.*** MapSideJoinMapper.run() Implementation

```
public static class MapSideJoinMapper extends
            Mapper<LongWritable, Text, NullWritable, Text> {
    //Cache variables for airports and carrier master
    private Map<String, String[]> airports =
                        new HashMap<String, String[]>();
    private Map<String, String[]> carriers =
                        new HashMap<String, String[]>();

    private void readAirports(URI uri) throws Exception {
       //Read airports.csv
       List<String> lines = FileUtils.readLines(new File(uri));
       //populate the airports atttibute. Key is the airport code
       ...
  }

     private void readCarriers(URI uri) throws Exception {
       //Read carriers.csv
       List<String> lines = FileUtils.readLines(new File(uri));
       //populate the carriers atttibute. Key is the carrier code
       ...
  }
```

```
public void setup(Context context) {
  try {
        URI[] uris = context.getCacheFiles();
    for (URI uri : uris) {
       if (uri.toString().endsWith("airports.csv")) {
          this.readAirports(uri);
       }
       if (uri.toString().endsWith("carriers.csv")) {
          this.readCarriers(uri);
       }
    }
  } catch (Exception ex) {
    //IOExceptions while reading master files
    throw new RuntimeException(ex);
  }
}
public void map(LongWritable key, Text value, Context context)
              throws IOException, InterruptedException {
   if (!AirlineDataUtils.isHeader(value)) {
            DelaysWritable dw = AirlineDataUtils
                                          .parseDelaysWritable(value.toString());
      String orginAirportCd = dw.originAirportCode.toString();
      String destAirportCd = dw.destAirportCode.toString();
      String carrierCd = dw.carrierCode.toString();
            String[] originAirport = this.airports.get(orginAirportCd);
            String[] destAirport = this.airports.get(destAirportCd);
            String[] carrier = this.carriers.get(carrierCd);

      String originAirportDesc = "";
      if (originAirport != null)
         originAirportDesc = originAirport[1]
                                .replaceAll(",", "");

      String destAirportDesc = "";
      if (destAirport != null)
         destAirportDesc = destAirport[1]
                                .replaceAll(",", "");

      String carrierDesc = "";
      if (carrier != null)
         carrierDesc = carrier[1].replaceAll(",", "");
            Text outLine = AirlineDataUtils.
                         parseDelaysWritableToText(dw,
                                                   originAirportDesc,
                                                   originAirportDesc,
                                                   destAirportDesc,
                                                   carrierDesc);
       context.write(NullWritable.get(), outLine);
     }
  }
}
```

## Running the Map-Only Join Job on the Cluster

We rebuild using Maven and execute the job in the Hadoop environment as follows:

```
hadoop jar prohadoop-0.0.1-SNAPSHOT.jar \
org.apress.prohadoop.c6.MapSideJoinMRJob \
/user/hdfs/sampledata \
/user/hdfs/output/c6/mapsidejoin \
/user/local/input/master/airports.csv \
/user/local/input/master/carriers.csv
```

The output directory should have several files with names that look like *part-m-nnnnn,* where *n* is a number between 0-9. A sample output record looks like the one that follows (the codes are bold to improve readability). The first code is the origin airport code, the second is the destination airport code, and the last is the carrier code. Each code is followed by a description attribute for the code.

```
1,1,1988,18,89,89,89,STL,Lambert-St Louis International,MSP,Minneapolis-St Paul Intl,NW,Northwest
Airlines Inc.
```

---

■ **Note** The MapSideJoinMRJob may not work as expected in certain versions of Hadoop 2. The method job.getCacheFiles( ) may not return URI instances from the local file system as expected. This bug as well as the fix is described in the following JIRA ticket `https://issues.apache.org/jira/browse/MAPREDUCE-5385`. If you are using a version of Hadoop which does not have the patch described in the above JIRA ticket, utilize the methods adopted in the MapSideJoinMRJob2 or MapSideJoinMRJob3 classes from the book source code. If you use these classes, the only difference from a client perspective is, the master data file paths are HDFS paths: /user/hdfs/masterdata/airports.csv and /user/hdfs/masterdata/carriers.csv. The MapSideJoinMRJob2 uses a deprecated function call job.getLocalCacheFiles( ). This call returns paths from the local file system of the task node on which the Mapper executes. The framework copies these files to the local file system of the Mapper task node before the job initiates. The MapSideJoinMRJob2 class will work in the pseudo distributed or the fully distributed cluster. But unlike the MapSideJoinMRJob, it may not work in your local development environment. Alternatively, utilize the MapSideJoinMRJob3 class. This class is identical to the MapSideJoinMRJob class except for two key differences. Similar to the MapSideJoinMRJob2 class, the master data file paths are HDFS paths and the job.getCacheFiles( ) assumes that the URI instances returned represent the same HDFS paths as passed by the client. In contrast, the MapSideJoinMRJob assumes the URI instances returned by the job.getCacheFiles( ) to map to the files in its local file system (JIRA fix described in the link above). This is the reason why the methods used in the MapSideJoinMRJob and MapSideJoinMRJob3 classes are incompatible. The MapSideJoinMapper inner class from MapSideJoinMRJob3 provides a method readLinesFromJobFS to read the URI instances returned by the job.getCacheFiles( ) method. To recap, there are two key elements to be aware of- the file system (local or HDFS) of master data files from the client perspective and the file system (local or HDFS) of the URI's returned by the job.getCacheFiles( ) method. Based on the Hadoop documentation, the latter is expected to be local. If they are not local, the method described in the MapSideJoinMRJob class will not work. If you want to use cache files from the local file system of the client, use the method described in the MapSideJoinMRJob2 class. If you do not want to use deprecated methods (the job.getLocalCacheFiles( ) method is the Job class is deprecated), use the method described in the MapSideJoinMRJob3 class.

---

## Recapping the Key Hadoop Features Explored with Map-Only Join

As we implemented the Join feature using the distributed cache, we explored the following new features of Hadoop:

- Methods for passing files and archives to a job using the command line arguments as well as programmatically
- How to utilize the `DistributedCache` to perform a Map-Only join on multiple variables in a single `map()` call

Although this section might have given the impression that this method is suitable only if the number of distributed files is small, it is not always true (for example, for a large master data file that is sorted by the Join criteria). Assume we receive a large data file that is also sorted by the Join criteria. Also assume that the records in the master file are sorted by the join criteria. Further assume that the master data file is stored in the HDFS.In the Mapper, we can then cache the first few lines of the master data file before we process lines from the data file. We keep performing the join in the Mapper as the records flow into the map method by doing a lookup on the cached master records. As we continue to process the data file at some point we willsoon as we encounter a value for the join criterion that is not present in the cached master records. At this point we start reading the master data until we either encounter the value in the master data file for the join criterion value equal to the current data record being processed by the Mapper or we reach the end of the master data file. Since both the datasets are sorted by the join criterion, reaching the end of the master data file implies that no more joins (inner) are feasible even if there are data records to process. Depending on the type of join being performed (inner or outer) we handle the remaining data records appropriately. This technique, is called Merge Join, is used extensively in several Hadoop libraries such as Hive and Pig.

# Writing to Multiple Output Files in a Single MR Job

Until now, we discussed jobs that produce only a single set of outputs. The output files have names in the *part-m-nnnnn* format if the files are produced by a Map-only job or *part-r-nnnnn* if the files are produced by a Reducer task. This section examines jobs that produce multiple sets of files.

Let us define the problem first. Suppose that we want to get delay information for the flight records, but we want to produce four separate output files based on the following four criteria:

- Flights that are on time for both arrivals and departures
- Flights that are delayed on arrival but on time for departure
- Flights that are on time for arrival but delayed on departure
- Flights that are delayed on arrival and departure

We will write a Map-only job to achieve these goals. We also suppress the *part-m-nnnnn* files because we do not want any output to those files. The source code for the program is available at `org.apress.prohadoop.c6.MultiOutputMRJob`.

The lines we need to add to the `run()` method to support the previously mentioned features are as follows:

```
job.setOutputFormatClass(NullOutputFormat.class);
MultipleOutputs.addNamedOutput(job,"OnTimeDepOnTimeArr",
                               TextOutputFormat.class,
                               NullWritable.class, Text.class);
MultipleOutputs.addNamedOutput(job, "DelayedDepOnTimeArr",
                               TextOutputFormat.class,
                               NullWritable.class, Text.class);
MultipleOutputs.addNamedOutput(job, "OnTimeDepDelayedArr",
                               TextOutputFormat.class,
                               NullWritable.class, Text.class);
```

```
MultipleOutputs.addNamedOutput(job, "DelayedDepDelayedArr",
                               TextOutputFormat.class,
                               NullWritable.class, Text.class);
```

The first line ensures that there will be no *part-m-nnnnn* files. `NullOutputFormat` is used when you do not want the job to produce any output through the `context.write` invocations. However it does not restrict the ability to create output by invoking the `write` method on the `MultipleOutputs` instance. It is usually used when you want to produce multiple outputs using the `MultipleOutputs` class. The next four lines configure the multiple outputs. The first parameter is the `Job` instance; the second parameter indicates the file name prefix for the named output. The next three parameters define the `OutputFormat` to use as well as the output key and value types to use for the named output. We have not explored formats other than `TextOutputFormat` yet. The value of this feature will become apparent when we discuss output formats such as `SequenceFileOutputFormat` in Chapter 7.

The naming convention for the files is *<name>-m-nnnnn* if the output file is written in the Mapper and *<name>-r-nnnnn* if the output file is written in the Reducer. Listing 6-26 shows the `MultiOutputMapper`. The `MultipleOutputs` instance configured in the `run()` method is initialized as an instance variable in the `setup()` method of the `Mapper`.

The `map` method implementation below demonstrates how various named outputs are written to. Note that the same name is used to recover the handle to the stream, as the one used to configure the named output in the `run()` method.

***Listing 6-26.*** MultiOutputMapper

```
public static class MultiOutputMapper extends Mapper<LongWritable, Text, NullWritable, Text> {

    private MultipleOutputs mos;

    @Override
    public void setup(Context context){
        this.mos = new MultipleOutputs(context);
    }

  public void map(LongWritable key, Text line, Context context)
                      throws IOException, InterruptedException {
     if(!AirlineDataUtils.isHeader(line)){
        DelaysWritable dw =  AirlineDataUtils
                         .parseDelaysWritable(line.toString());
        int arrivalDelay = dw.arrDelay.get();
        int departureDelay =  dw.depDelay.get();
        Text value = AirlineDataUtils
                         .parseDelaysWritableToText(dw);
        if(arrivalDelay<=0 && departureDelay<=0){
             this.mos.write("OnTimeDepOnTimeArr",NullWritable.get(), value);
        }
          if(arrivalDelay<=0 && departureDelay>0){
              this.mos.write("DelayedDepOnTimeArr",NullWritable.get(), value);
          }
          if(arrivalDelay>0 && departureDelay<=0){
              this.mos.write("OnTimeDepDelayedArr",NullWritable.get(), value);
          }
          if(arrivalDelay>0 && departureDelay>0){
              this.mos.write("DelayedDepDelayedArr",NullWritable.get(), value);
          }
         }
  }
}
```

## Collecting Statistics Using Counters

Counters are the mechanism through which various statistics can be collected for a job. The MapReduce job output shows the various counters managed by MapReduce. Examples of such counters include these:

- Map input records
- Map output records
- Reduce output records
- Launched Map tasks
- Failed Map tasks
- Launched Reduce tasks
- Failed Reduce tasks

And there are many more counters that are used by the MapReduce framework to collect statistics on the job.

You can also create custom counters to collect application-specific statistics. There are two ways to create application-specific counters:

- Define a Java enum, which represents a group of counters. The name of the group is the fully qualified name of the enum type.
- Create dynamic counters by dynamically providing a group name for the counter as well as the name for the counter.

Counters are global in nature and can be incremented (or decremented if we pass a negative increment value) from both Mappers and Reducers. The counters are managed at the Application Master level. Information about each increment for the counter is passed to the Application Master through the heartbeat messages between containers running the Mapper and Reducer tasks, and the Application Master.

Let us extend the program developed in the last section on `MultipleOutputs` to collect the following statistics:

- `OnTimeStatistics`: Enum-based statistic. The name of the group is the fully qualified path to the enum definition: `org.apress.prohadoop.c6.MultiOutputMRJob$MultiOutputMapper$OnTimeStatistics`
    - `DepOnTimeArrOnTime`
    - `DepOnTimeArrDelayed`
    - `DepDelayedArrOnTime`
    - `DepDelayedArrDelayed`
- `DELAY_BY_PERIOD`: Dynamically generated for each month or day of the week. Helps to determine how many delays occurred for any given month or day of the week.
    - `MONTH_{MONTH_ID}`
    - `DAY_OF_WEEK_{DAY_OF_WEEK_ID}`

Listing 6-27 shows the modified Mapper class.

***Listing 6-27.*** MultiOutputMapper with Counters

```
public static class MultiOutputMapper
              extends Mapper<LongWritable, Text, NullWritable, Text> {
    enum OnTimeStatistics{
        DepOnTimeArrOnTime,
        DepOnTimeArrDelayed,
        DepDelayedArrOnTime,
        DepDelayedArrDelayed
    }

  private MultipleOutputs mos;
  @Override
  public void setup(Context context){
     this.mos = new MultipleOutputs(context);
  }

  public void map(LongWritable key, Text line, Context context)
                        throws IOException, InterruptedException {
     if(!AirlineDataUtils.isHeader(line)){
      DelaysWritable dw =  AirlineDataUtils
                         .parseDelaysWritable(line.toString());
      int arrivalDelay = dw.arrDelay.get();
      int departureDelay =  dw.depDelay.get();
      Text value = AirlineDataUtils
                         .parseDelaysWritableToText(dw);
      if(arrivalDelay<=0 && departureDelay<=0){
         this.mos.write("OnTimeDepOnTimeArr",
                        NullWritable.get(), value);
                 context.getCounter(OnTimeStatistics
                                   .DepOnTimeArrOnTime).increment(1);
      }
      if(arrivalDelay<=0 && departureDelay>0){
         this.mos.write("DelayedDepOnTimeArr",
                            NullWritable.get(), value);
         context.getCounter(OnTimeStatistics
                            .DepDelayedArrOnTime).increment(1);
      }
      if(arrivalDelay>0 && departureDelay<=0){
         this.mos.write("OnTimeDepDelayedArr",
                        NullWritable.get(), value);
         context.getCounter(OnTimeStatistics
                            .DepOnTimeArrDelayed).increment(1);
      }
      if(arrivalDelay>0 && departureDelay>0){
            this.mos.write("DelayedDepDelayedArr",
                                      NullWritable.get(), value);
            context.getCounter(OnTimeStatistics.
                              DepDelayedArrDelayed).increment(1);
            }
```

```
        if(arrivalDelay>0 || departureDelay>0)
          context.getCounter("DELAY_FOR_PERIOD","MONTH_"+
                                                    Integer.toString(dw.month.get())).increment(1);
                   context.getCounter("DELAY_FOR_PERIOD","DAY_OF_WEEK_"+
                                                    Integer.toString(dw.month.get())).increment(1)
            }
      }
   }
}
```

The counters can be accessed in the run() method after the job completes. The following lines of code in the run() method achieve that:

```
this.printCounterGroup(job,"DELAY_FOR_PERIOD");
this.printCounterGroup(job,"org.apress.prohadoop.c6.MultiOutputMRJob$MultiOutputMapper$OnTime
Statistics");
```

Note the long group name for the counter represented by the enum. Listing 6-28 shows the implementation of the printCounterGroup method. We use the Job instance to retrieve the group of counters and then fetch the iterator for the group which we iterate over to access each counter in that group.

***Listing 6-28.*** printCounterGroup

```
private void printCounterGroup(Job job,String groupName)
                                                    throws IOException{
   CounterGroup cntrGrp = job.getCounters().getGroup(groupName);
   Iterator<Counter> cntIter = cntrGrp.iterator();
   System.out.println("\nGroup Name = " + groupName);
   while(cntIter.hasNext()){
      Counter cnt = cntIter.next();
      System.out.println(cnt.getName() + "=" + cnt.getValue());
   }
}
```

The printCounterGroup method invocation for a run in my development environment for a sampling of the data prints the following output to the console:

```
Group Name = DELAY_FOR_PERIOD
DAY_OF_WEEK_1=25
DAY_OF_WEEK_12=21
MONTH_1=25
MONTH_12=21

Group Name = org.apress.prohadoop.c6.MultiOutputMRJob$MultiOutputMapper$OnTimeStatistics
DepDelayedArrDelayed=27
DepDelayedArrOnTime=2
DepOnTimeArrDelayed=17
DepOnTimeArrOnTime=10
```

# Summary

The last two chapters covered a lot of ground. You learned how to emulate almost every important SQL method using MapReduce. This chapter explored multiple dataset sorting and joining using various methods. In the process, you learned how to create custom I/O classes for the MapReduce framework and how to control the order in which the keys are received and grouped for the Reducers. We discussed complex techniques such as Secondary Sort and distributed cache. You also learned how to handle multiple input folders and write to multiple output files. Finally, you learned how to collect statistics in a job.

The next chapter discusses the MapReduce framework in more detail, especially how it works internally. This understanding is necessary to become a strong Hadoop developer.

# CHAPTER 7

# Hadoop Input/Output

Previous chapters outlined MapReduce concepts in detail and we started delving deeper into the way Hadoop is implemented at the end of Chapter 5. This chapter expands on that theme. First, compression schemes are explained, followed by a detailed discussion on Hadoop I/O. We address various types of files, such as Sequence and Avro files. In the process, you develop a deeper understanding of how the MapReduce framework works internally in the Hadoop engine.

## Compression Schemes

So far, you have learned the basic fundamentals of MapReduce. MapReduce is an I/O intensive process. Reducing or optimizing I/O is the key to improving the performance of MapReduce programs. The Hadoop Framework offers several alternatives to reduce I/O. In Chapter 6 we explored the use of the Combiner to reduce I/O between the Mapper and the Reducer. In this chapter we will explore compression schemes which can be utilized to optimize the I/O performance of MapReduce.

First let us quickly look at various MapReduce steps that involve I/O:

1. Input to the Mapper as files are read from the HDFS.
2. Output from the Mapper that is spilled to local disk.
3. Network I/O between the Reducer and Mapper, as the Reducer's retrieve files from the Mapper nodes.
4. Merge to local disk on the Reducer node as the partitions received from the Mapper nodes are fully sorted on the Reducer node.
5. Reading back from the local disk as records are made available to the `reduce` method on the Reducer instance.
6. Output from the Reducer- this is written back to the HDFS.

I/O is the most expensive operation in any MapReduce program, and anything that can reduce the I/O to disk or over the network is rewarded with better overall throughput. The Hadoop framework allows you to compress output from the Mapper as well as the Reducer. However there is a trade-off: compression is CPU-intensive and consumes CPU cycles. Any CPU cycles used for compression are lost for other processes such as the Mapper, Reducer, Partitioner, Combiner, Sort and Shuffle, and so on. Like most decisions in Hadoop, this one involves resource trade-offs that must be taken seriously at design time.

Compression also has an indirect effect on the number of Mappers that can be started to process the input data. An important aspect of HDFS is that a single file can be stored in multiple blocks. Imagine a large text file consumed by using `TextInputFormat`. Each line of the file is considered a record when using the `TextInputFormat`. But lines do not respect block boundaries, and a line can be split across block boundaries. So when `TextInputFormat` is reading the file, it might have to make a remote call to the subsequent block to fetch the rest of the line. Conversely, the Mapper processing the next block has to skip to the end of the line because it knows that another Mapper will process the partial record at the beginning. This promotes high efficiency because a single file can be allocated to multiple Mappers. As long as the Mapper knows the blocks it is consuming, as well as whether the first block is the starting block and the address of the block subsequent to the last block, it is enough to continue processing.

But what happens when compressed files are used? Even if we know which blocks comprise the compressed file, it is not enough to simply allocate blocks to Mappers. It is not possible to start at any arbitrary point in a compressed file.

Some compression schemes are splittable, which means it is possible to read parts of them and uncompress just those parts. For those that are not splittable, there is no choice: one Mapper consumes one file. Most likely, a large number of blocks of this file are not local to the Mapper node, which has two implications. First, if one file is very large, a single Mapper dominates the performance of the map-side phase. This single long-running Mapper will delay the Reduce phase and consequently the whole job. Second, if the input files are large in size and small in number, only a few Mappers are started. So even if the file sizes are approximately the same, the amount of parallelism is limited and the whole job is delayed. In such a situation, it is probably better to use uncompressed files.

Other file input formats, such as the `SequenceFileInputFormat` discussed later in this chapter, are also splittable in uncompressed and compressed form. The selection of file formats and compression scheme is critical to good performance.

## What Can Be Compressed?

Compression can be performed at the following levels:

- **Input file compression:** If input files are compressed, less I/O has to be performed when reading the input blocks into the Mapper. However, CPU cycles get expended when uncompressing the files.
- **Compressing intermediate Mapper output:** In a MapReduce job, the Mapper produces intermediate files on the Mapper node. These files are partitioned by the destination Reducer and sorted by the Reducer key, and they are downloaded by the Reducer over the HTTP protocol. Compressing this output reduces file I/O as files are written from the Mapper to the local disk and network I/O as partitions are transferred from the Mapper node to the Reducer node.
- **Compressing MapReduce job output:** Applicable regardless of whether the job uses a Reduce phase, it applies to the Mapper output for a Map-only job and the Reducer output for a job using the reducer phase.

As a rule of thumb, it is useful to use compression at each phase. It is very unusual for the Mapper or Reducer to be CPU-intensive, and there are enough cycles available for compression (or de-compression when interpreting compressed).

## Compression Schemes

Compression schemes can be evaluated based on three criteria:

- The size of the compressed file; the smaller the better since less I/O is performed in reading/writing from/to the disk and network.
- The time taken to compress the file; the faster the better.
- Whether the compressed format is splittable. A splittable format helps achieve better parallelism by allowing a single file to be processed by multiple Mappers.

There is a trade-off. You can get very good compression, but it takes more CPU cycles where (as with faster compression schemes) you can get poor compression. You should make a pragmatic choice: to select a compression scheme that is fast and produces moderate to high levels of compression. This scheme enables the free CPU cycles to be used without starving the Mapper or Reducer for the CPU.

Table 7-1 compares various popular compression schemes. Note that the size and speed are relative to one another. GZIP is a good compromise between speed and size, but it is nonsplittable. BZIP2 produces smaller and splittable files, but it is slower. LZO and Snappy are both splittable and fast, but do not compress as much as the other formats.

***Table 7-1.*** *Comparison of Popular Compression Schemes*

| Scheme | Compressed Size | Speed | Splittable |
|---|---|---|---|
| GZIP | Moderate | Medium | No |
| BZIP2 | Small | Slow | Yes |
| LZO | Large | Fast | Yes |
| Snappy | Large | Fast | Yes |

To provide a numerical sense of the trade-offs, Snappy compression can be orders of magnitude faster than GZIP methods, yet the output files are 20–100 percent larger. Snappy originated at Google and is widely used in its internal MapReduce and derivative frameworks.

The Hadoop Framework comes equipped with the codecs to perform file compression and decompression. (A codec is an implementation of the CompressionCodec class). The list of codecs supported is provided by the property io.compression.codecs in the core-site.xml. Its value is a comma separated list of fully qualified CompressionCodec implementation classes. For example, the CompressionCodec implementation class to handle the GZIP scheme is org.apache.hadoop.io.compress.GzipCodec.

A similar class for the Snappy scheme is org.apache.hadoop.io.compress.SnappyCodec.

Compression is not enabled by default. It can be enabled by setting the value true for the following properties in the mapred-site.xml:

- mapreduce.output.fileoutputformat.compress for the job output
- mapreduce.map.output.compress for the intermediate mapper output

If compression is enabled, but no codec is specified, the default codec used is org.apache.hadoop.io.compress.DefaultCodec. The DefaultCodec uses the DEFLATE compression format. This is the same as GZIP without the additional headers that GZIP includes. The default codec can be configured by updating the following properties in the mapred-site.xml:

- mapreduce.output.fileoutputformat.compress.codec, configures the default codec for the job output.
- mapreduce.map.output.compress.codec, configures the default codec for the intermediate mapper output.

## Enabling Compression

How do we indicate to the Hadoop Framework that we want to turn on compression or use specialized compression utilities to read compressed files? For the Mapper input, we do not need to do anything; as long as the required CompressionCodec implementation class is configured in the io.compression.codecs property and present in the classpath the framework infers the compression scheme from the file extension and does what is needed. For the rest of the phases, we need to explicitly tell the framework to use compression.

To turn on compression for the job output (not the intermediate Mapper output), we add the following lines when configuring the job:

```
Configuration conf = new Configuration();
Job job=new Job(conf);

FileOutputFormat.setCompressOutput(job, true);
FileOutputFormat.setOutputCompressorClass(job,GzipCodec.class);
```

The job is an instance of the Job class you have seen before. FileOutputFormat is the base class of output formats that write to files in the HDFS, including TextOutputFormat as well as the SequenceFileOutput format that you will encounter later in the chapter. We used the GzipCodec in the previous example.

To use the Snappy codec, we use this line:

```
FileOutputFormat.setOutputCompressorClass(job,SnappyCodec.class);
```

To turn on compression for intermediate output from the Mapper, we use the following lines:

```
Configuration conf = new Configuration();

//Compress the Mapper Intermediate outoput
//Use this only for jobs with Reduce phase

//or mapred.compress.map.output which is deprecated but widely used
conf.setBoolean("mapreduce.map.output.compress ",true);

//or mapred.map.output.compression.codec is deprecated but widely used
conf.setClass("mapreduce.map.output.compress.codec ",
              GzipCodec.class, CompressionCodec.class);

//Compress job output
FileOutputFormat.setCompressOutput(job, true);
FileOutputFormat.setOutputCompressorClass(job,GzipCodec,class);

Job job=new Job(conf);
```

# Inside the Hadoop I/O processes

This section examines the InputFormat and OutputFormat classes. But before we discuss the details, we explain how these classes read and write data. For this tour, we use two classes we have used before:

- TextInputFormat: This class has been used as an input format to read the airline dataset.
- TextOutputFormat: This class has been used to write the output for all the MapReduce programs until now.

# InputFormat

The InputFormat implementations perform the following functions in a MapReduce job:

- Validate the job specification, which ensures that the job is configured consistently from an input perspective.
- The InputFormat implementations work closely with their respective RecordReader implementation to convert the bytes presented by InputSplit to instances of key and value classes that represent the Mapper class input.

## Anatomy of InputSplit

InputFormat splits the input files into logical InputSplit's, one of which is served to each Mapper. We need to make a distinction between InputSplit and a HDFS block here. The latter represents a physical division of the input data on the disk. The former is a logical division of the data as consumed by the Mapper. InputSplit can comprise multiple blocks in a way that is transparent to the client. For example, consider TextInputFormat: there is no guarantee that each line spans only one block; there are lines at the boundary of the blocks that span multiple blocks. The LineRecordReader, which is the RecordReader for the TextInputFormat, manages this for the client. InputSplit might have to fetch a fragment of a record from a block different from the main block. This block can reside on different node and might incur a communication cost, which is usually inconsequential because this cost is incurred rarely. From the perspective of the Mapper, this remote communication is abstracted out. The Mapper only sees whole records.

An InputSplit has the following two methods:

- int getLength(): This method returns the size in bytes of the InputSplit.
- String[] getLocations(): This method provides the list of hostnames where data for InputSplit is located. These locations are used as hints by the MapReduce framework to schedule Mappers for the splits to maximize data locality for the Mapper.

## Anatomy of InputFormat

InputFormat has two method definitions:

- List<InputSplit> getSplits(): This method is called by the client program to get the list of splits and then writes this list to the HDFS. The Application Master then retrieves this file from HDFS to create resource requests based on the location of each InputSplit
- RecordReader<K,V> createRecordReader(): This method is invoked by the Mapper task in the container executing the Mapper. This method returns the RecordReader instance for the InputSplit processed by the Mapper

The FileInputFormat class, which is a subclass of InputFormat, has the isSplittable() method that indicates whether the file is splittable. The default is true, in which case several splits are produced per file. Recall that one instance of InputSplit is handled by one instance of a Mapper. Thus a splittable InputFormat allows multiple Mappers to handle each input file. Splittable input formats are more complex to develop, but are very efficient because they allow higher degree of parallelism in the map phase. If the return value of the isSplittable() method is false, the entire file is served as a single InputSplit, which implies that all blocks comprising the file are processed by the same Mapper. Some of these blocks may be local to the Mapper, but most of the blocks will not be local if the file is large. A non-splittable InputFileFormat is easier to develop, but it is inefficient compared with a splittable InputFileFormat.

Creating a splittable InputFileFormat is not always possible, however. For example, even if the underlying InputFormat is splittable (for example, TextInputFormat), the InputFormat defaults to non-splittable mode if it needs to process compressed files that do not have a splittable compression scheme. Under the splittable condition, an input file is usually divided into splits based on blocks comprising the file. The number of blocks per input split can be configured via the following properties in the mapred-site.xml

- mapreduce.input.fileinputformat.split.minsize - Default value of 1
- mapreduce.input.fileinputformat.split.maxsize - Usually set to Long.MAX_VALUE
- dfs.blocksize - Default is 128 MB

The three properties work in unison to define the number of blocks per input split such that the following constraint is met:

```
mapreduce.input.fileinputformat.split.minsize< dfs.blocksize<
mapreduce.input.fileinputformat.split.maxsize
```

In most cases this leads to one block per input split. Setting the value of the property mapreduce.input.fileinputformat.split.minsize to be greater than the dfs.blocksize property will result in more than one block per input split. This may not be optimal as it will increase the number of non-local blocks in the input split. The mapreduce.input.fileinputformat.split.maxsize will only have an impact when it less than the dfs.blocksize. This will force splits to be smaller than one block. Recall that the minimum and maximum split size can be configured at run time through the Job configuration and while the dfs.blocksize has a default value (128 MB) it can be configured for each file. Thus the number of blocks per split can be configured for each job run.

## TextInputFormat

The TextInputFormat class is an InputFormat implementation which processes plain text files as lines of text. Lines are assumed to be separated by linefeed and carriage return characters. The TextInputFormat extends the abstract class FileInputFormat (which in turn implements InputFormat) which is the base class of all the file based input formats. TextInputFormat has a complex implementation for the isSplittable method. For uncompressed files the TextInputFormat is splittable. For compressed files it is splittable only if the underlying compression scheme is splittable.

For TextInputFormat, the corresponding RecordReader subclass is LineRecordReader. Each line of text in the input file is considered a record. LineRecordReader is responsible for generating a key that is an offset of each line from the beginning of the file (an instance of LongWritable) and a value that is the entire line of text (an instance of Text). As discussed earlier, it is possible that a line might cross the boundary of the block it originated in. The associated LineRecordReader class is robust enough to handle this condition which often involves a network call because the block of data that completes the line might not be on the same node.

# OutputFormat

The roles of the OutputFormat implementation is two fold

- It validates the output specification of the job. It verifies if the output folder of the job does not exist.
- Provide an instance of the RecordWriter implementation which used to write to the output files. The RecordWriter is responsible for converting a java instance of the output key and the value types into raw bytes.

## TextOutputFormat

The TextOutputFormat class declares the output specification for the job as a text file comprising of lines of text.

Just as an InputFormat class is associated with a RecordReader class, the OutputFormat is associated with a RecordWriter. The RecordWriter implementation utilized by the TextOutputFormat is LineRecordWriter, which converts the key-value output of the job to a line of text, where the key and value are first converted to a string and written out to a line with a tab separator between them. If the key is an instance of NullWritable only the value portion is written out.

Subsequent sections discuss how to create custom OutputFormat and then InputFormat classes. We follow this order because we want to write the results of a MapReduce job using a custom output format. Then we create a custom input format to consume this specialized output format from another MapReduce job. This way, you do not have to create input files for consumption by the specialized custom input format.

As usual, we create a sample problem and solve it using Hadoop APIs.

# Custom OutputFormat: Conversion from Text to XML

This section discusses how to convert flight detail text files into a large XML file comprising only delay information. The sample XML file is shown in Listing 7-1.

***Listing 7-1.*** Sample Output XML File

```
<recs>
 <rec>
    <key>0</key>
    <year>1988</year>
    <month>01</month>
    <date>18</date>
    <dayofweek>1</dayofweek>
    <depdelay>89</depdelay>
    <arrdelay>89</arrdelay>
    <origin>STL</origin>
    <destination>MSP</destination>
    <carrier>NW</carrier>
  </rec>
  <rec>
    <key>98</key>
    <year>1988</year>
    <month>01</month>
    <date>18</date>
    <dayofweek>1</dayofweek>
    <depdelay>0</depdelay>
    <arrdelay>6</arrdelay>
    <origin>SNA</origin>
    <destination>SFO</destination>
    <carrier>AA</carrier>
  </rec>
     <!- More rec tags -->
</recs>
```

Note that the value of the <key> tag is the key of the record as returned by TextInputFormat, which happens to be the byte offset of the start of the line in the file. This is the byte offset inside the text file where the record begins.

The custom `OutputFormat` is defined by the `org.apress.prohadoop.c7.XMLOutputFormat` class.

The MapReduce program using this custom `OutputFormat` is defined in the `org.apress.prohadoop.c7.TextToXMLConversionJob` class.

Listing 7-2 shows the `run()` method of this class. Only the `OutputFormat` definition is different.

***Listing 7-2.*** TextToXMLConversionJob.run()

```
public int run(String[] allArgs) throws Exception {
   Job job = Job.getInstance(getConf());
   job.setJarByClass(TextToXMLConversionJob.class);

   job.setInputFormatClass(TextInputFormat.class);
   job.setOutputFormatClass(XMLOutputFormat.class);

   job.setOutputKeyClass(LongWritable.class);
   job.setOutputValueClass(Text.class);

   job.setMapperClass(TextToXMLConversionMapper.class);
   job.setNumReduceTasks(0);

   String[] args = new GenericOptionsParser(getConf(), allArgs)
              .getRemainingArgs();
   FileInputFormat.setInputPaths(job, new Path(args[0]));
   FileOutputFormat.setOutputPath(job, new Path(args[1]));

   job.waitForCompletion(true);

   return 0;
}
```

This is a Map-only job, and the Mapper class is listed in Listing 7-3. Notice that the `Mapper` class is completely oblivious to the `OutputFormat` used. We can reuse `Mapper` and `Reducer` classes regardless of the `OutputFormat` used.

***Listing 7-3.*** TextToXMLConversionMapper

```
public static class TextToXMLConversionMapper extends
            Mapper<LongWritable, Text, LongWritable, Text> {

   public void map(LongWritable key, Text value, Context context)
                        throws IOException, InterruptedException {
      if (!AirlineDataUtils.isHeader(value)) {
         int month = Integer.parseInt(
                        AirlineDataUtils.getMonth(value
                                   .toString().split(",")));
         context.write(key, value);
      }
   }
}
```

Finally, we can implement the custom `OutputFormat` class. The name of this class is `org.apress.prohadoop.c7.XMLOutputFormat`.

The skeleton implementation of this class is shown in Listing 7-4. Notice that it uses a RecordWriter implementation to actually write the output. The XMLRecordWriter class has all the necessary callbacks to write the XML file. The initial <recs> tag is written out in the constructor, and the close() method closes this outermost <recs> tag. The writeTag() method is repeatedly invoked in response to the context.write() method in the Mapper (or Reducer).

***Listing 7-4.*** XMLOutputFormat and XMLRecordWriter

```
package org.apress.prohadoop.c7;

/*Import Statements*/

public class XMLOutputFormat extends
                            FileOutputFormat<LongWritable,Text> {

   protected static class XMLRecordWriter extends
                             RecordWriter<LongWritable, Text> {
      private DataOutputStream out;
     public XMLRecordWriter(DataOutputStream out) throws IOException
     {
        this.out = out;
        out.writeBytes("<recs>\n");
     }

     private void writeTag(String tag,String value) throws IOException{
        out.writeBytes("<"+tag+">"+value+"</"+tag+">");
     }

     public synchronized void write(LongWritable key, Text value) throws IOException
     {
        out.writeBytes("<rec>");
        this.writeTag("key", Long.toString(key.get()));
        String[] contents = value.toString().split(",");

        String year = AirlineDataUtils.getYear(contents);
        this.writeTag("year", year);
        //Remaining this.write for various tags

        out.writeBytes("</rec>\n");
      }

      public synchronized void close(TaskAttemptContext job)
                                     throws IOException
      {
          try {
             out.writeBytes("</recs>\n");
          } finally {
             out.close();
          }
      }
   }
```

```
    public RecordWriter<LongWritable,Text> getRecordWriter(
                                TaskAttemptContext job)
                                      throws IOException {
        String extension = ".xml";
        Path file = getDefaultWorkFile(job, extension);
        FileSystem fs = file.getFileSystem(job.getConfiguration());
        FSDataOutputStream fileOut = fs.create(file, false);
        return new XMLRecordWriter(fileOut);
    }
}
```

The output writing process is as follows:

1. The Hadoop Framework gets a reference to the `XMLRecordWriter` (subclass of the `RecordWriter`) instance by invoking the `getRecordWriter()` method on the `XMLOutputFormat` (an implementation of the `OutputFormat`) class instance.
2. The `RecordWriter` configures the characteristics of the output, including the file name prefix and extension. The extension is provided in the code listing. The default prefix of the file is *part*, which can be overridden by providing a custom value for the property `mapreduce.output.basename`. It can be configured via passing a generic parameter on the command line as follows:

   ```
   -D mapreduce.output.basename=airlinedata
   ```

   Notice how the `getRecordWriter` method configures the file extension. Until now the jobs we have executed produced output files without any extension. In this example we defined a custom extension by invoking the following two lines in the `getRecordWriter` method of the `XMLOutputFormat`:

   ```
   String extension = ".xml";
   Path file = getDefaultWorkFile(job, extension);
   ```

   These settings will result in the output files having the following naming format:

   ```
   airlinedata-m-nnnnn.xml
   ```

   where m indicates that the Mapper produced this file, and n indicates a number between 0-9 If the job had a reduce phase, the output file name format would be the following:

   ```
   airlinedata-r-nnnnn.xml
   ```

   where r indicates that the Reducer produced the file, and n indicates a number between 0-9
3. The actual output is produced by the framework by repeatedly invoking the `write()` method of the `RecordWriter` instance for each call to `context.write()` in the Mapper for a Map-only job and in the Reducer for a job which utilizes a reduce phase.

## Run the Text-to-XML Job on the Cluster

Finally, we rebuild by using Maven and execute the job in the Hadoop environment as follows:

```
hadoop jar prohadoop-0.0.1-SNAPSHOT.jar \
org.apress.prohadoop.c7. TextToXMLConversionMRJob \
-D mapreduce.output.basename=airlinedata \
/user/hdfs/sampledata \
/user/hdfshdfs/output/c7/xml
```

The output directory has files with the custom naming convention we defined. A sample file name is `airlinedata-m-00000.xml`

# Custom InputFormat: Consuming a Custom XML file

This section shows you how to consume an XML file as input. `TextInputFormat` can support only files that contain one record in one line, but XML files cannot depend on one-line, one-record semantics. We need to write a custom `InputFormat` class; the class that defines it is `org.apress.prohadoop.c7.XMLDelaysInputFormat`.

(For convenience, our custom class `XMLDelaysInputFormat` implements the `FileInputFormat`.)

The `InputFormat` classes rely on a `RecordReader` to read the input bytes and deserialize them into instances of the key and value classes. Our custom `RecordReader` class is `org.apress.prohadoop.c7.XMLDelaysRecordReader`.

Listing 7-5 shows the `XMLDelaysInputFormat` class.

***Listing 7-5.*** XMLDelaysInputFormat

```
public class XMLDelaysInputFormat extends
                  FileInputFormat<LongWritable, DelaysWritable> {
   @Override
   protected boolean isSplitable(JobContext context, Path filename)
   {
      return false;
   }

   @Override
   public RecordReader<LongWritable, DelaysWritable>
           createRecordReader(InputSplit split,
                              TaskAttemptContext context)
   {
      return new XMLDelaysRecordReader();
   }
}
```

The preceding implementation means two things

- For illustrative purposes, we define the custom `InputFormat` as nonsplittable, so one Mapper instance processes one file. As discussed in the earlier section, it is not the most efficient way to process files, however. Making `InputFormat` nonsplittable is not necessary. Similar to `TextInputFormat`, which identifies each line by the presence of a newline character, XML files can discover records by the presence of specific start and end tags. A splittable version of `InputFormat` would need to recognize whether a XML document is split across blocks and fetch a possibly remote block (or blocks) to complete the XML document, starting from the main block for the `InputSplit`. The Mapper processing the `InputSplit` has to process the entire XML document because XML parsers operate only on complete XML documents.
- For each file in the HDFS (`InputSplit`), there is one instance of `XMLDelaysRecordReader`.

Listing 7-6 shows the skeleton of the XMLDelaysRecordReader class.

***Listing 7-6.*** XMLDelaysRecordReader

```
/*Package and import declaration*/

class XMLDelaysRecordReader extends
            RecordReader<LongWritable, DelaysWritable> {
   /*Configure the XML Reader with the file level tag
    * and the record level tag as Job configuration properties
    */
   public static final String FILE_START_TAG_KEY =
                                  "xmlfile.start.tag";
   public static final String RECORD_START_TAG_KEY =
                                  "record.start.tag";
   /*XML specific state variables*/
   private byte[] xmlFileTag;
   private byte[] recordTag;
   private byte[] recordEndTag;

   private String recordTagName = "";
   private long start;
   private long end;

   /*Hadoop API specific variables*/
   private FileSplit fileSplit;
   private Configuration conf;
   private FSDataInputStream in = null;

   /*Mapper input key/value instances*/
   private LongWritable key = new LongWritable(-1);
   private DelaysWritable value = new DelaysWritable();

   private DataOutputBuffer buffer = new DataOutputBuffer();

   /*Initialize the process of XML Parsing.
    *Set the file pointer to the start
    *of the first record
    */
   public void initialize(InputSplit inputSplit,
           TaskAttemptContext taskAttemptContext)
           throws IOException,InterruptedException {
       xmlFileTag = taskAttemptContext.getConfiguration()
                       .get(FILE_START_TAG_KEY).getBytes();
       recordTag=("<"
                + taskAttemptContext.getConfiguration()
                  .get(RECORD_START_TAG_KEY) + ">")
                  .getBytes();
       recordTagName = taskAttemptContext.getConfiguration()
                                  .get(RECORD_START_TAG_KEY);
       recordEndTag = ("</"
                + taskAttemptContext.getConfiguration()
                .get(RECORD_START_TAG_KEY) + ">").getBytes();
```

```
    this.fileSplit = (FileSplit) inputSplit;
    this.conf = taskAttemptContext.getConfiguration();
    start = fileSplit.getStart();
    end = start + fileSplit.getLength();
    FileSystem fs = fileSplit.getPath().getFileSystem(conf);
    this.in = fs.open(fileSplit.getPath());
    this.in.seek(start);
    //Read until start of the first record. Skip the <recs> tag
    readUntilMatch(xmlFileTag, false);
}

/*Method is invoked by the Mapper classes run() method
 *to fetch the next key value class
 */
public boolean nextKeyValue() throws IOException {
   if (this.in.getPos() < this.end &&
       readUntilMatch(recordTag, false)) {
       buffer.write(this.recordTag);
       if (readUntilMatch(this.recordEndTag, true)) {
            key.set(key.get() + 1);
            DelaysWritable dw = this.parseDelaysWritable(this.
                          createInputStream(buffer.getData()));
            value.setDelaysWritable(dw);
            this.buffer = new DataOutputBuffer();
            return true;
       }
    }
    return false;
}

/*The Mapper run() method will call getCurrentKey()
 *and getCurrentValue() if nextKeyValue( method returns true.
 *This indicates that there are more records and the
 *record has been deserialized
 */
@Override
public LongWritable getCurrentKey()
        throws IOException, InterruptedException {
   return key;
}

@Override
public DelaysWritable getCurrentValue() throws IOException,
         InterruptedException {
   return value;
}

/*Used to calculate the progress of the Mapper which is indicated
 *in the MR console
 */
```

```
    @Override
    public float getProgress()
                        throws IOException, InterruptedException {
        float f = (float)(this.in.getPos()-this.start)/
                    (float)(this.end-this.start);
        return f;
    }

    /*When the nextKeyValue() returns false the run() method the mapper
     *will close the RecordReader and call the close() method
     *of the Mapper class
     */
    @Override
    public void close() throws IOException {
         this.in.close();
    }
    /*Logic to parse an XML record from the XML file.
     *Read until the closing tag is encountered.
     */
    private boolean readUntilMatch(byte[] match, boolean withinBlock)
            throws IOException {
       //Skipping implementation. Populates the DataOutputBuffer
    }
    /*Convert an XML record <rec>...</rec> into a
     *DelaysWritable instance
     */
    private DelaysWritable parseDelaysWritable(InputStream
                                                      delaysXML) {
       DelaysWritable dw = new DelaysWritable();
       //Parse into DelaysWritable instance
       Return dw;
    }

    /*Create an InputStream from a byte array representing
     *an XML record string
     */
    private InputStream createInputStream(byte[] bytes)
            throws java.io.UnsupportedEncodingException {
       String xml = (new String(bytes)).trim();
       return new ByteArrayInputStream(xml.getBytes());
    }

}
```

Now we are ready to discuss the workings of the `XMLDelaysRecordReader` class. Its characteristics are as follows:

1. The file level tag and record level tags are required to be passed as configuration parameters. In this case, their values are the following:

   ```
   xmlfile.start.tag=recs
   record.start.tag=rec
   ```

2. The initialize() method is invoked on the XMLDelaysRecordReader by the MapReduce framework before the Mapper is initialized. The InputSplit instance designated to be handled by the XMLDelaysRecordReader instance is passed to the initialize() method. Since our custom InputFormat is a subclass of the FileInputFormat, the InputSplit implementation utilized is the FileSplit class. Since our custom InputFormat is non-splittable, the FileSplit instance represents an entire input file. If the custom InputFormat was splittable, the FileSplit instance would have represented only a section of the input file but it would still need to have set of complete XML documents. The initialization process starts consuming the bytes in the InputSplit and brings the file pointer past the file level tag (<recs>) and at the byte representing the first <rec> tag.

3. The Hadoop runtime engine will now create an instance of the org.apache.hadoop.mapreduce.Mapper.Context class and configure it with an instance of XMLDelaysRecordReader initialized in the previous step.

4. The run() method is invoked on the Mapper by the Hadoop framework using the instance of the org.apache.hadoop.mapreduce.Mapper.Contextconfigured in the previous step. The map method on the Mapper instance will be repeatedly invoked on the custom Mapper instance after retrieving records from the XMLDelaysRecordReader instance. The next section will describe the inner workings of this invocation.

## Anatomy of the org.apache.hadoop.mapreduce.Mapper.run() method

The run() method in the org.apache.hadoop.mapreduce.Mapper class has the following implementation. The highlighted methods have user defined implementation via the user defined Mapper class.

```
public void run(Context context) throws IOException,
                    InterruptedException {
   setup(context);
   try {
      while (context.nextKeyValue()) {
         map(context.getCurrentKey(), context.getCurrentValue(), context);
      }
   } finally {
      cleanup(context);
   }
}
```

The run method performs the following steps

- The setup() method is invoked for the Mapper. The implementation of the setup() method is user defined in the custom Mapper class.
- The methods are invoked on the RecordReader instance through the Context instance to retrieve successive key-value inputs to be passed to the map method.
- The nextKeyValue() method of XMLDelaysRecordReader is responsible for parsing the bytes in the file and converting them to instances of LongWritable (key instance) and DelaysWritable (value instance). If the next record is successfully parsed, this method returns true. If there are no more instances of the <rec> tag, it returns false.
- If the nextKeyValue() method returns true, the getCurrentKey() and getCurrentValue() methods are invoked to fetch the key and value instances that are then used to invoke the user defined map() method.

- If the `nextKeyValue()` method returns `false`, the `cleanup()` method is invoked on the Mapper. This is also a user-defined method.
- Finally, the Hadoop Framework invokes the `close()` method on the `RecordReader` that completes the Mapper life cycle.

If the run() method is overridden, there is no need for the map() method. Imagine an implementation of the Aggregation use case (Aggregation was discussed in Chapter 6) in which the input is known to be sorted by the "GroupBy" criterion. The run() method can perform an intermediate aggregation operation by processing multiple input records. The `map` method invocation can be completely bypassed. The run() method knows exactly when there are no more records to process and can write the final aggregation result to output. An example of such an implementation is shown in Listing 7-7, which assumes that the input is an XML file, and the records in the files are sorted by month. We calculate the total number of records per month. Instead of overriding the `map` method we override the run method from the base `org.apache.hadoop.mapreduce.Mapper` class. This is an example of how multiple records can be processed together in the Mapper.

Notice that we have simulated a more efficient implementation of the Combiner using this technique. You learned in the previous chapter that the Combiner is applied on the output of the Partitioner. In this more efficient version, the output of each Mapper is the aggregation operation applied per month for the entries in the file. No `map()` invocation is needed. Note that this could have been achieved with Mapper-specific instance variables through extending the `map()` method, as we have done so far in the book. But in a `map()` method-based implementation, the last `write` has to be done in the `cleanup()` method as there is no way of knowing in the `map()` method when the last record was processed. The implementation is shown as an illustration of how to override the `run()` method. Also note that we will need a Reducer to produce the final count by month. If we use a splittable `InputFormat` or if records belonging to a single month cross file boundaries, the aggregations computed by the Mapper are only intermediate ones. It is the responsibility of the Reducer to ensure these intermediate aggregations are fully aggregated in the program output.

***Listing 7-7.*** AdvancedGroupBySumMapper

```
public class AdvancedGroupBySumMapper extends
    Mapper<LongWritable, DelaysWritable, IntWritable, LongWritable> {

    public void map(LongWritable key,
                    DelaysWritable value, Context context)
            throws IOException, InterruptedException {
        //Do nothing
    }

    @Override
    public void run(Context context)
                throws IOException, InterruptedException {
       setup(context);
       IntWritable month = new IntWritable(0);
       LongWritable total = new LongWritable(0);
       while (context.nextKeyValue()) {
         LongWritable key = context.getCurrentKey();
         DelaysWritable dw = context.getCurrentValue();
          if(month.get()==dw.month.get()){
              total.set(total.get()+1);
          }
```

```
            else{
                if(month.get()>0){//Skip the first iteration
                  context.write(month, total);//Write intermediate total
                }
                month = new IntWritable(dw.month.get());
                //Reset total to one as it is first record
                total = new LongWritable(1);
            }
          }
          //Write the last aggregation for this file if there were any
          //records which is indicated by month.get() being > 0
          if(month.get()>0){
              context.write(month, total);
          }
          cleanup(context);
      }
}
```

## CompositeInputFormat and Large Joins

This section discusses a special type of InputFormat that can be used to join large tables. CompositeInputFormat can join datasets that are sorted by their join key using a map-side join. As noted in Chapter 6, map-side joins are orders of magnitude faster than map-reduce joins due to avoiding the expensive Sort and Shuffle phase that accompanies the Reducer phase. A map-side operation is a scan through the input files that can be parallelized to a large degree to achieve very high performance.

First, we create a problem definition based on the airline dataset. We want to create a list of connecting flights from the origin to the destination for each day. We assume that connecting flights arrive and depart on the same day. We create the following two datasets:

- Sorted by origin airport code and departure date in *YYYY-MM-DD* format; for example: SFO-2014-12-01.
- Sorted by destination airport code and arrival date in *YYYY-MM-DD* format; for example: SFO-2014-12-01.

Each of the sorted file can be read using KeyValueInputFormat. This implies that there is a key that is followed by a separator and then a value. The default separator is a tab (\t), but it can be any character of choice. The Job needs to configure a custom separator by setting the mapreduce.input.keyvaluelinerecordreader.key.value.separator property. In this case, we assume that the key is the join key that is a combination of airport code and departure/arrival date.

Next, we configure the run() method, which is shown in Listing 7-8.

***Listing 7-8.*** LargeMapSideJoin.run()

```
public  int run(String[] allArgs) throws Exception {
   String[] args = new GenericOptionsParser(getConf(),
                        allArgs).getRemainingArgs();

   Job job = Job.getInstance(getConf());
   job.setJarByClass(LargeMapSideJoin.class);
   //Input Paths
   Path flightByOriginAndDt = new Path(args[0]);
   Path flightByDestinationAndDt = new Path(args[1]);
```

```
    job.setInputFormatClass(CompositeInputFormat.class);
    //Configure join
    String strJoinStmt = CompositeInputFormat.compose("inner",
                                KeyValueTextInputFormat.class,
                flightByOriginAndDt, flightByDestinationAndDt);
    job.getConfiguration().set("mapred.join.expr", strJoinStmt);

    job.setOutputKeyClass(Text.class);
    job.setOutputValueClass(Text.class);
    job.setOutputFormatClass(TextOutputFormat.class);

    job.setMapperClass(LargeMapSideJoinMapper.class);
    job.setNumReduceTasks(0);

    FileOutputFormat.setOutputPath(job, new Path(args[2]));
    job.waitForCompletion(true);
    return 0;
}
```

The key features of this method are the following:

1. First, we define input paths to both datasets which need to be joined. We will assume that each path represents a single file produced by the `TextOutputFormat`. Each file has many lines with, each line comprising of a key and value pair separated by tab ("\t") and the lines are sorted by the key. The goal is to perform the join on the two inputs based on this key.
2. Next, we define `CompositeInputFormat` as the input format and configure it by invoking the `compose` method. The configuration parameters are:
   a. **Type of join:** The "`inner`" value configures an inner join. A value of "`outer`" should be used to configure a full outer join. Note the use of the `mapred.join.expr` property to configure the join criteria on the `Job` instance in the next line.
   b. **Delegate `InputFormat` to consume inputs:** The `CompositeInputFormat` delegates reading for each input path to another InputFormat suitable for consuming the input. We utilize the `KeyValueTextInputFormat`. This InputFormat assumes the input to be lines of text. Each line comprises of a key and value pair separated by a character defined by the property `mapreduce.input.keyvaluelinerecordreader.key.value.separator`. The default separator character is tab ("\t"). Since our input files have key and value separated by a tab character we will utilize this default setting. If the separator byte does not exist in a line, the entire line will be returned as the key and the value will be empty. The underlying `RecordReader` implementation used by the `KeyValueTextInputFormat` is `KeyValueLineRecordReader` which returns the key and value of the underlying input as instances of `Text`.
   c. **Input `Path` instances:** The last two parameters to `compose()` are the paths to the datasets which need to be joined.

---

■ **Note** for the delegate InputFormat utilized (KeyValueTextInputFormat in our example). We assumed The CompositeInputFormat forces one set of files (two files, one from each path) to be processed by one mapper by configuring the minimum split size to be equal to Long.MAX_VALUE one file for each dataset being joined. However, if there are multiple files in the paths representing each side of the join, each path must contain files such that the records contained in them are total order sorted by their key. We discussed total order sorting in Chapter 6. It means that if the files are sorted by the file name in lexicographical order and appended into a single file, the records in the single file thus created must still remain sorted by their keys. Yet another constraint which needs to be enforced in order to use the CompositeInputFormat is, the paths representing each side of the join must have an equal number of files and each of the file in one path must contain the same key range as the corresponding file in the other path.

---

The unique feature of using a CompositeInputFormat is the Mapper we configure. The Listing 7-9 shows the LargeMapSideJoinMapper class. The Mapper is configured with key-value classes as follows:

- Text: This is the common join key from both inputs. Each of the underlying KeyValueLineRecordReader instances consuming its respective input path will return a Text instance of the key. Since we use the Inner join in our example, the CompositeInputFormat utilizes the InnerJoinRecordReader class as its RecordReader. InnerJoinRecordReader is the subclass of the JoinRecordReader which implements the join functionality. InnerJoinRecordReader consumes the key instances returned by the two underlying KeyValueLineRecordReader's and if they are equal, serves this instance as the key to the map invocation of the LargeMapSideJoinMapper class.
- TupleWritable: The InnerJoinRecordReader will return instances of this class as the value parameter to the map invocation. It is a list of two Writable instances, one from each input path. The sort order of the Writable instances in the list is identical to the order of input paths in the CompositeInputFormat.compose() invocation in the run() method. Since our underlying input format was the KeyValueTextInputFormat these are instances of the Text class. Each Text instance is the value portion of the record returned by the KeyValueLineRecordReader.

***Listing 7-9.*** LargeMapSideJoinMapper

```
public static class LargeMapSideJoinMapper
                   extends Mapper<Text, TupleWritable, Text, Text> {
    public void map(Text key, TupleWritable value, Context context)
                          throws IOException, InterruptedException {
       Text connectingFlight = new Text("");
       if (value.toString().length() > 0) {
          String originAtts[] = value.get(0).toString().split(",");
          String destAtts[] = value.get(1).toString().split(",");
          String oFN = AirlineDataUtils.getFlightNum(originAtts);
          String oDT = AirlineDataUtils.getDepartureTime(destAtts);
          String dFN = AirlineDataUtils.getFlightNum(destAtts);
          String dAT = AirlineDataUtils.getArrivalTime(destAtts);
```

```
            //Add logic to ensure originDepTime > depArrTime
            connectingFlight.set(key+","+oFN+","+oDT+","+dFN+","
                                  +dAT);
            context.write(key, connectingFlight);
        }
    }
}
```

Note the following calls on the TupleWritable instance inside the Mapper:

- value.get(0): Fetches the flight data line for the origin side flight that will depart from the origin.
- value.get(1): Fetches the flight data line for the arriving side flight that will arrive at the origin (or destination, from the arriving flight's point of view) location.

---

■ **Note** To simplify the code, we skipped the logic to ensure that the departing flight time (at the origin side) must be greater than the arriving flight time (at the departure time).

---

Finally, we merge the lines from the two datasets and send them to the output. Note that the Mapper receives lines from both datasets so that all permutations are accounted for. For example, if we receive three arriving flights and two departing flight for the same airport and date combinations, we have six pairings, and the map() method receives all six pairings through six separate calls to the Mapper.

### How Map-Side Join Works with Sorted Datasets

This section discusses at a high level how datasets sorted by the join key can be joined on the map side.

To perform the inner join, we need to start scanning both datasets from the start. We either encounter the same key on both datasets or we do not. In the former case, we produce a pairing of records with the same key. In the latter case, we either start skipping the records on the dataset whose current key is behind (from a sort order perspective) the key in the other dataset. If the dataset being skipped ends up with a key higher than the key in the other dataset, we start skipping the other dataset. We do this until both keys are the same again and we start producing permutations.

The preceding discussion is for inner joins, but it can also be easily adapted to full outer, left outer, or right outer joins. In those cases, instead of skipping, we produce output records with one side of the join empty, depending on the type of join requested.

## Hadoop Files

This section discusses specialized file formats. We discuss the Sequence files, which is a binary format to store binary key value pairs. Next we briefly discuss the Map files which is a sorted and indexed Sequence file. Finally, we will discuss the Avro format, a compact and fast binary format which relies on schema's to support multiple languages. Thus while both, Sequence and Avro file formats are binary, the former only supports the Java language, while the latter can be used with any language for which Avro bindings exist.

## SequenceFile

We have frequently used the TextInputFormat, which requires each record to be a line of Text. In the last two chapters, we have been working with datasets that work only with delayed flight records. We created a custom Writable class called DelaysWritable, and to work with this class, we had to write it as Text using TextOutputFormat and then read it back using TextInputFormat and reconvert it to a DelaysWritable instance in the Mapper class. If we use a SequenceFile, we do not need to do this. We can simply store a DelaysWritable in the file and read it back as a DelaysWritable instance from the SequenceFile. The InputFormat used to read a SequenceFile is SequenceFileInputFormat, and the corresponding OutputFormat is the SequenceFileOutputFormat. SequenceFileInputFormat extends FileInputFormat and SequenceFileOutputFormat extends FileOutputFormat.

A SequenceFile is a binary format that stores binary key and value pairs. The key and value instances do not necessarily have to be Writable instances, but they do have to be at least Serializable However in practice, key and values to Sequence files are usually Writable instances. Despite being a key-value store, Sequence files do not have map-like semantics. The keys do not need to be unique and can also be an instance of NullWritable if you cannot come up with sensible keys. In this case, it functions like a collection of Writable (or Serializable) instances.

## SequenceFileOutputFormat: Creating a SequenceFile

We illustrate creating a SequenceFile by writing a program that converts the XML file shown in Listing 7-1 to a SequenceFile. The key of the SequenceFile is a LongWritable instance that contains the value of the <key> tag. The value is an instance of DelaysWritable, and the class that performs this conversion is org.apress.prohadoop.c7.XML ToSequenceFileConversionJob.

Listing 7-10 shows the run() method of this class. The key characteristics of this method are the following:

- The use of the XMLDelaysInputFormat class defined earlier.
- The use of the SequenceFileOutputFormat to save the LongWritable instance of the key and DelaysWritable instance of the value. The value portion of the SequenceFile does not have to be a custom Writable instance. In fact, it does not need to be Writable at all; it can also be a java.io.Serializable instance. However, when you encounter a SequenceFile in Hadoop, the value portion is usually a Writable instance.

***Listing 7-10.*** XMLToSequenceFileConversionJob.run()

```
public int run(String[] allArgs) throws Exception {
   String[] args = new GenericOptionsParser(getConf(), allArgs)
                              .getRemainingArgs();
   Job job = Job.getInstance(getConf());
   job.setInputFormatClass(XMLDelaysInputFormat.class);
   job.setOutputFormatClass(SequenceFileOutputFormat.class);

   job.setOutputKeyClass(LongWritable.class);
   job.setOutputValueClass(DelaysWritable.class);

   job.setMapperClass(XMLToSequenceFileConversionMapper.class);
   job.setNumReduceTasks(0);

   FileInputFormat.setInputPaths(job, new Path(args[0]));
   FileOutputFormat.setOutputPath(job, new Path(args[1]));
   //Could also use SequenceFileOutputFormat.setOutputPath
   job.waitForCompletion(true);

   return 0;
}
```

The Mapper is shown in Listing 7-11. Notice how uncluttered it is because we can write out the key and value instances directly without making all the Text conversions.

***Listing 7-11.*** XMLToSequenceFileConversionMapper

```
public static class XMLToSequenceFileConversionMapper extends
            Mapper<LongWritable, DelaysWritable,
                   LongWritable, DelaysWritable> {
    public void map(LongWritable key,
                    DelaysWritable value,
                    Context context)
                    throws IOException, InterruptedException {
       context.write(key, value);

    }
 }
```

You can also create a SequenceFile from a Java program. Listing 7-12 shows how to do it when creating an uncompressed SequenceFile. The method for creating compressed files is similar.

***Listing 7-12.*** SequenceFileWriter.writeFile() Method

```
public static void writeFile(String fName) throws IOException {
   SequenceFile.Writer writer = null;
   try {
      Configuration conf = new Configuration();
      FileSystem fs = FileSystem.getLocal(conf);
      Path seqFilePath = new Path(fName);
      LongWritable key = new LongWritable();
      DelaysWritable val = new DelaysWritable();

      writer = SequenceFile.createWriter(fs,conf,
                                         seqFilePath,
                                         key.getClass(),
                                         val.getClass());

      for (int i = 1; i <= 12; i++) {
        key.set(i);
        //We just set the month for this illustration
        val.month = new IntWritable(i);
        writer.append(key, val);
      }
   }
   finally {
      writer.close();
   }
}
```

■ **Note** Refer to the Apache documentation on SequenceFile for techniques on creating compressed SequenceFile from a stand-alone Java program. The program to create an uncompressed SequenceFile is included in the book's source code (www.apress.com) as SequenceFileWriter.

## SequenceFileInputFormat: Reading from a SequenceFile

This section discusses how to read a SequenceFile using a MapReduce program. We will write a program that converts the SequenceFile we created in the last section using the SequenceFileInputFormat to a text file using TextOutputFormat. The program that does this is org.apress.prohadoop.c7.SequenceToTextFileConversionJob.

Listing 7-13 shows this program. The only unique feature about it is the use of the SequenceInputFormat.

***Listing 7-13.*** SequenceToTextFileConversionJob

```
/*Package and import declarations*/
public class SequenceToTextFileConversionJob
                      extends Configured implements Tool {

   public static class SequenceToTextConversionMapper extends
            Mapper<LongWritable, DelaysWritable, NullWritable, Text> {
      public void map(LongWritable key,
                      DelaysWritable value,
                      Context context)
                   throws IOException, InterruptedException {
         Text line = AirlineDataUtils.parseDelaysWritableToText(value);
         context.write(NullWritable.get(), line);
      }
   }

   public int run(String[] allArgs) throws Exception {
      Job job = Job.getInstance(getConf());
      job.setInputFormatClass(SequenceFileInputFormat.class);
      job.setOutputFormatClass(TextOutputFormat.class);
      job.setOutputKeyClass(NullWritable.class);
      job.setOutputValueClass(Text.class);
      job.setMapperClass(SequenceToTextConversionMapper.class);
      job.setNumReduceTasks(0);
      String[] args = new GenericOptionsParser(getConf(), allArgs)
               .getRemainingArgs();
      FileInputFormat.setInputPaths(job, new Path(args[0]));
      FileOutputFormat.setOutputPath(job, new Path(args[1]));
      job.waitForCompletion(true);
      return 0;
   }

   public static void main(String[] args) throws Exception {
      Configuration conf = new Configuration();
      ToolRunner.run(new SequenceToTextFileConversionJob(), args);
   }
}
```

You can also read a SequenceFile from a Java program. Listing 7-14 shows how to do that. The SequenceFile.Reader works with both compressed and uncompressed SequenceFiles. (The program is included in the book source code as SequenceFileReader.java.) Notice that we created empty key and value instances that are populated in the reader.next() call in each iteration.

***Listing 7-14.*** SequenceFileReader.readFile()

```
public static void readFile(String fName) throws IOException {
   SequenceFile.Reader reader = null;
   try {
      Configuration conf = new Configuration();
      FileSystem fs = FileSystem.getLocal(conf);
      Path seqFilePath = new Path(fName);

      reader = new SequenceFile.Reader(conf,
                          Reader.file(seqFilePath));

      LongWritable key = new LongWritable();
      DelaysWritable val = new DelaysWritable();

      while (reader.next(key, val)) {
         System.out.println(key + "\t"
         + val.year + "/" + val.month + "/" + val.date
         + " from/to==" + val.originAirportCode + " to "
         + val.destAirportCode);
      }

    } finally {
       reader.close();
    }
}
```

## Compression and SequenceFiles

This section discusses the structure of a SequenceFile and the types of compression it supports. We examine the structure from a conceptual point of view. (The details of the format can be found at the Apache web site at http://hadoop.apache.org/docs/current/api/org/apache/hadoop/io/SequenceFile.html.)

### Sequence File Header

Every SequenceFile has a header. The header comprises the following information:

- Version no of the SequenceFile format.
- Key and value class names. This is an example where the output key and value class names configured using the invocations to the setOutputKeyClass and setOutputValueclass methods on the Job instance are utilized.
- A Boolean value indicating whether the key and value instances are compressed.
- A Boolean value indicating whether BLOCK compression is turned on. When it is turned on, multiple records are compressed together.
- Additional metadata and finally a sync marker to denote the end of the header. This is the marker that a SequenceFile reader looks for when it needs to start processing records after skipping the header portion.

## Sync Marker and Splittable Nature of SequenceFiles

A SequenceFile has a sync marker every few records, which enables a SequenceFile to be splittable. The SequenceFileInputFormat can break up a SequenceFile into multiple InputSplits based on this marker. It is guaranteed that only whole records (key-value pairs) are contained between sync records. Similar to text files, records can traverse block boundaries, so the underlying SequenceFileInputFormat and its corresponding reader must be capable of interpreting the bytes across block boundaries when this happens.

## RECORD Compression

A SequenceFile with a RECORD level of compression consists of the following:

- Header
- Groups of records separated by a sync marker with each record structured as follows:
    - Record length
    - Key length
    - Key
    - Compressed value (uncompressed when compression is not turned on)
- Sync marker every few records

Note how the value length does not have been stored in the record as it can be inferred from the record length and the key length. To turn on compression at a RECORD level for a job using SequenceFileOutputFormat, configure the job with the following:

```
SequenceFileOutputFormat.setCompressOutput(job, true);
SequenceFileOutputFormat.setOutputCompressorClass(job,
                                        SnappyCodec.class);
SequenceFileOutputFormat.setOutputCompressionType(job,
                                   CompressionType.RECORD);
```

The value portion of each record is compressed if compression is turned on via the SequenceFileOutputFormat.

## BLOCK Compression

A SequenceFile with a BLOCK level of compression consists of the following:

- Header
- Sync marker-separated blocks, each of which comprises several records structured using the following fields:
    - Number of records
    - Compressed key lengths
    - Compressed keys
    - Compressed value lengths
    - Compressed values
- Sync marker between blocks

To turn on compression at the BLOCK level for a job using SequenceFileOutputFormat, configure the job with the following:

```
SequenceFileOutputFormat.setCompressOutput(job, true);
SequenceFileOutputFormat.setOutputCompressorClass(job,
                                    SnappyCodec.class);
SequenceFileOutputFormat.setOutputCompressionType(job,
                                    CompressionType.BLOCK);
```

## MapFiles

The goal of a MapFile, which is a special form of SequenceFile, is to support random access to data stored in a sorted SequenceFile. The idea is to provide an index for a sampling of keys from a sorted SequenceFile. The index marks the byte offset in the sorted SequenceFile for the keys stored in the index. When a specific key is requested from a MapFile, the nearest key from the index is sought, and the sequential search proceeds from that byte offset until the requested key is either found or a key higher than the requested key is found.

A MapFile is a set of two SequenceFiles stored in a folder which has the name of the map file (denoted by the variable ${map.file.name} below):

- ${map.file.name}/data is a SequenceFile containing the data sorted by key. This condition is checked during creation of the MapFile using the MapFile.Writer.append() call and throws an IOException if the condition is violated.
- ${map.file.name}/index is a smaller SequenceFile that contains a fraction of the keys. For each key in the index file, a byte offset within the data file for that key is stored. When a key lookup is performed on a MapFile, the index file is used to go to the nearest key prior to the key being looked up. The byte offset enables the MapFile reader to go to the start of this key's record. From that point, the keys are sequentially traversed until the requested key is found or a key larger than the looked-up key is encountered. The number of keys stored in the index file can be set by invoking the setIndexInterval() method on the MapFile.Writer instance. This interval defines the number of keys to be skipped before setting a key in the index file. The smaller the interval, the more keys that will be stored in the index file. Consequently, the MapFile becomes more randomly accessible. Remember that the index file is completely loaded in memory, so too small an interval (leading to a large index file) leads to an OutOfMemoryException while reading the MapFile. Thus we need to strike a balance between the size and random accessibility of the MapFile.

Two restrictions of the MapFile are that the key class must be an instance of WritableComparable (to ensure key ordering), and there is no enforcement of the duplicate keys. If you make multiple appends with the same key, they are all accepted. If you iterate the MapFile, you should be able to retrieve all the value instances for the duplicate key. However, if you invoke a get() call on the MapFile.Reader instance, only the first one is retrieved. The MapFile.Reader has a seek(WritableComparable key) method; if you invoke it with a key that has duplicates, it sits at the end of the record with the first instance of the key. So if there is a MapFile in which every key is duplicated twice (three instances of the same key), a seek call puts it at the end of the first instance of the key. Starting to iterate the MapFile.Reader returns two instances of the key (used to invoke seek) and three instances of every subsequent key. The book source code has example code to write and read a MapFile from a Java program. The classes are the following:

```
org.apress.prohadoop.utils.MapFileReader
org.apress.prohadoop.utils.MapFileWriter
```

MapFile can be used as input to MapReduce programs. The InputFormat class to use is the SequenceFileInputFormat, which is smart enough to ignore the index file and consider only the data file when processing the input folders.

Producing a MapFile as the output of the MapReduce program is also feasible. It is usually a natural thing to do in a Reduce phase because input to a Reducer is naturally sorted by the key instance. Each Reducer produces a separate MapFile, so this is applicable even if multiple Reducers are used. The OutputFormat to use is the MapFileOutputFormat. It is also possible to use this OutputFormat for Map-only jobs as long as we know that the output of the Mapper is sorted by the MapFile key. This is usually feasible only when the input to the Mapper is sorted by the attribute that will be used as the key of the MapFile.

## Map Files and Distributed Cache

The last chapter discussed how Distributed Cache (represented by an instance of DistributedCache and managed by the Job instance as demonstrated in Chapter 6)is used to propagate small master data files to each of the task nodes. But a Distributed Cache does not have to be used only with small files; a relatively large MapFile can also be distributed using Distributed Cache.

The individual Mappers and Reducers can then apply lookup on the MapFile to get the records they want. This method is very appropriate when each Mapper and Reducer needs keys that are relatively close to each other.

In fact, we can do a large table join using this method, simulating the CompositeInputFormat functionality. If the input SequenceFile is sorted by the join key (or just a MapFile with sort criteria based on the join key on the dataset), and the other dataset is a MapFile (stored in the Distributed Cache) sorted by a join key, we can have a Map-only job that scans the first SequenceFile (or MapFile). In each Mapper, we collect a bunch of keys from the input SequenceFile (or MapFile) and then perform a seek in the Mapper on those keys on the second MapFile from the Distributed Cache. Due to the sorted nature of the MapFile, these keys can be scanned very quickly after the first record is fetched. This is another method in which a large table join can be performed using a Map-only job.

# Avro Files

Avro is a binary format that allows Hadoop files to be supported by multiple languages. It is similar to SequenceFile, but SequenceFile is limited to Java programming, whereas Avro is supported by Java as well as other programming languages. Using Avro, data can be serialized to a very compact binary format based on the schema specified. Avro is rich in data types. All the major primitive data types are supported. Complex data types such as records, enums, arrays, and maps are also supported. This section is not a detailed exposure to Avro; instead, we demonstrate how the Avro file format can be used to work with MapReduce jobs.

## FlightDelay.avsc: Avro Schema File

Listing 7-15 shows an example of an Avro schema file that mimics the DelaysWritable class. We store this file in the src/main/avro folder of our Maven project.

***Listing 7-15.*** FlightDelay.avsc

```
{"namespace": "org.apress.prohadoop.avro",
 "type": "record",
 "name": "FlightDelay",
 "fields": [
     {"name": "year", "type": "int"},
     {"name": "month",  "type": "int"},
     {"name": "date", "type": "int"},
     {"name": "dayOfWeek", "type": "int"},
```

```
        {"name": "depDelay", "type": ["int", "null"]},
        {"name": "arrDelay", "type": ["int", "null"]},
        {"name": "origin", "type": "string"},
        {"name": "destination", "type": "string"},
        {"name": "carrier", "type": "string"}

 ]
}
```

Although it is not essential to perform any code generation, helper libraries exist to use strongly typed classes with Avro schemas. We want to convert the preceding schema to a Java class via code generation; the POM file updates needed to do so are shown in Listing 7-16.

***Listing 7-16.*** Maven Setup in pom.xml

```
<dependencies>
...
  <dependency>
     <groupId>org.apache.avro</groupId>
      <artifactId>avro</artifactId>
      <version>1.7.6</version>
  </dependency>
  <dependency>
      <groupId>org.apache.avro</groupId>
      <artifactId>avro-mapred</artifactId>
      <version>1.7.6</version>
      <classifier>hadoop2</classifier>
  </dependency>
...
</dependencies>
<build>
 <plugins>
   <plugin>
     <groupId>org.apache.maven.plugins</groupId>
     <artifactId>maven-compiler-plugin</artifactId>
   </plugin>
   <plugin>
      <groupId>org.apache.avro</groupId>
      <artifactId>avro-maven-plugin</artifactId>
      <version>1.7.6</version>
      <executions>
         <execution>
            <phase>generate-sources</phase>
            <goals>
               <goal>schema</goal>
            </goals>
            <configuration>
               <sourceDirectory>
                  ${project.basedir}/src/main/resources/avro/
               </sourceDirectory>
```

```
                <outputDirectory>
                   ${project.basedir}/src/main/java/
                </outputDirectory>
             </configuration>
          </execution>
       </executions>
    </plugin>
  </plugins>
</build>
```

When we do the Maven build, this class is created: `org.apress.prohadoop.avro.FlightDelay`. It can be used as a helper class to perform factory operations on instances of the schema from Listing 7-15.

## Job to Convert from Text Format to Avro Format

We now look at how to use the Avro format in MapReduce programs, specifically to convert flight input files to Avro files. Listing 7-17 shows this program.

***Listing 7-17.*** TextToAvroFileConversionJob

```
/*Package and Import declarations*/

public class TextToAvroFileConversionJob
                           extends Configured implements Tool {

    public static class TextToAvroFileConversionMapper extends
              Mapper<LongWritable, Text,
                    AvroKey<NullWritable>,
                    AvroValue<FlightDelay>> {
      public void map(LongWritable key, Text value, Context context)
               throws IOException, InterruptedException {
         if(!AirlineDataUtils.isHeader(value)){
           String[] contents = value.toString().split(",");
               FlightDelay.Builder fd =  FlightDelay.newBuilder();

           int year = Integer.parseInt
           AirlineDataUtils.getYear(contents));
               fd.setYear(year);

           int month = Integer.parseInt(
            AirlineDataUtils.getMonth(contents));
               fd.setMonth(month);

           int date = Integer.parseInt(
              AirlineDataUtils.getDateOfMonth(contents));
               fd.setDate(date);

           int dow = Integer.parseInt(
              AirlineDataUtils.getDayOfTheWeek(contents));
               fd.setDayOfWeek(dow);
```

```
            int arrDelay = AirlineDataUtils.parseMinutes(
               AirlineDataUtils.getArrivalDelay(contents),0);
                fd.setArrDelay(arrDelay);

            int depDelay = AirlineDataUtils.parseMinutes(
               AirlineDataUtils.getDepartureDelay(contents),0);
                fd.setDepDelay(depDelay);

            String origin = AirlineDataUtils.getOrigin(contents);
            fd.setOrigin(origin);

            String dest = AirlineDataUtils.getDestination(contents);
                fd.setDestination(dest);

            String carrier = AirlineDataUtils
                             .getUniqueCarrier(contents);
                fd.setCarrier(carrier);

                context.write(
                  new  AvroKey<NullWritable>(NullWritable.get()),
                  new AvroValue<FlightDelay>(fd.build()));
            }
        }
    }

    public int run(String[] allArgs) throws Exception {
        Job job = Job.getInstance(getConf());
        job.setJarByClass(TextToAvroFileConversionJob.class);
        job.setInputFormatClass(TextInputFormat.class);
           job.setOutputFormatClass(AvroKeyValueOutputFormat.class);
           AvroJob.setOutputKeySchema(job,
               Schema.create(org.apache.avro.Schema.Type.NULL));
           AvroJob.setOutputValueSchema(job, FlightDelay.SCHEMA$);
        job.setMapperClass(TextToAvroFileConversionMapper.class);
        job.setNumReduceTasks(0);
        String[] args = new GenericOptionsParser(getConf(), allArgs)
                .getRemainingArgs();
        FileInputFormat.setInputPaths(job, new Path(args[0]));
        FileOutputFormat.setOutputPath(job, new Path(args[1]));
        job.waitForCompletion(true);
        return 0;
    }

    public static void main(String[] args) throws Exception {
        ToolRunner.run(new TextToAvroFileConversionJob(), args);
    }
}
```

Following are the key features of Listing 7-17:

- The output key and value classes in the signature of the Mapper are a `NullWritable` class and the custom Avro class `FlightDelay` wrapped in the `AvroKey` and `AvroValue` classes. The underlying `OutputFormat` performs the mapping between the standard Hadoop `Writable` classes and the underlying Avro schema for standard primitive Avro types.
- The `OutputFormat` is an `AvroOutputFormat` class. But in addition to specifying the output format, we also specify the schema. We access the `Schema` instance directly from the `FlightDelay` helper class, which is how `FlightDelay` acts as the helper class. In its absence, we would have to construct a `Schema` instance. Providing the schema is necessary because the output Avro file contains the schema along with the data, enabling its use by any program without any additional information. The data and schema are contained in the file.
- We used the `FlightDelay.Builder` from the generated class to act as the factory method to create instances of the objects based on the Avro schema. We work with them as with any other Java class. Again, the `FlightDelay` class hides all the Avro-specific details. Without it, we would have to work with instances of `org.apache.avro.generic.GenericRecord`. Refer to the Avro documentation for more details.
- The `context.write()` call writes the `AvroKey` and `AvroValue` instances, which are wrappers for the `NullWritable` and `FlightDelay` instances.

## Job to Convert from Avro Format to Text Format

We now examine how to consume Avro files using `AvroInputFormat`. Listing 7-18 shows a program to perform a reverse conversion from Avro to text format.

***Listing 7-18.*** AvroToTextFileConversionJob

```
/*Package and import declarations*/
public class AvroToTextFileConversionJob
                    extends Configured implements Tool {

    public static class AvroToTextFileConversionMapper
            extends Mapper<AvroKey<NullWritable>,
                           AvroValue<FlightDelay>,
                           NullWritable, Text> {

        public void map(AvroKey<NullWritable> key,
                        AvroValue<FlightDelay> value,
                        Context context)
                throws IOException, InterruptedException {
            FlightDelay fd = value.datum();
            StringBuilder output = new StringBuilder("");
            output.append(fd.getYear().toString()).append(",")
                    .append(fd.getMonth().toString()).append(",")
                    .append(fd.getDate().toString()).append(",")
                    .append(fd.getDayOfWeek().toString()).append(",")
                    .append(fd.getArrDelay().toString()).append(",")
                    .append(fd.getDepDelay().toString()).append(",")
```

```
                .append(fd.getOrigin().toString()).append(",")
                .append(fd.getDestination().toString());
            context.write(NullWritable.get(),
                        new Text(output.toString()));
        }
    }

    public int run(String[] allArgs) throws Exception {
        Job job = Job.getInstance(getConf());
        job.setJarByClass(AvroToTextFileConversionJob.class);
        job.setInputFormatClass(AvroKeyValueInputFormat.class);
        AvroJob.setInputKeySchema(job,
                Schema.create(org.apache.avro.Schema.Type.NULL));
        AvroJob.setInputValueSchema(job,FlightDelay.SCHEMA$);
        job.setOutputFormatClass(TextOutputFormat.class);
        job.setMapperClass(AvroToTextFileConversionMapper.class);
        job.setNumReduceTasks(0);
        job.setOutputKeyClass(NullWritable.class);
        job.setOutputValueClass(Text.class);

        String[] args = new GenericOptionsParser(getConf(), allArgs)
                .getRemainingArgs();
        FileInputFormat.setInputPaths(job, new Path(args[0]));
        FileOutputFormat.setOutputPath(job, new Path(args[1]));

        job.waitForCompletion(true);
        return 0;
    }

    public static void main(String[] args) throws Exception {
        Configuration conf = new Configuration();
        ToolRunner.run(new AvroToTextFileConversionJob(), args);
    }
}
```

Following are the key features of Listing 7-18:

- Note how `AvroInputFormat` is configured. It is also required to configure the Avro schema for the key class and the Avro schema for the value class. We used `NullWritable` as our key in the previous example, but now we have to use an instance of `org.apache.avro.Schema.Type.NULL` because `NullWritable` is mapped to this schema when the underlying `RecordWriter` class of `AvroOutputFormat` wrote the keys for the Avro records in the Avro file.
- Note the input key and value classes in the Mapper definition. We again wrap the standard `NullWritable` instance in the `AvroKey` class and the generated `FlightDelay` instance in the `AvroValue` class. The underlying `RecordReader` class of `AvroInputFormat` assumes the responsibility of performing the translations from bytes to the respective class instances.

# Summary

In this chapter, we conclude our journey inside the MapReduce framework. Chapter 5 and Chapter 6 showed how to implement various data-intensive programming patterns in MapReduce. This chapter examined MapReduce in detail from an underlying implementation perspective.

This book's first seven chapters were foundational. Armed with this knowledge, you can now appreciate the higher-level frameworks built on top of MapReduce such as Hive, Pig, HBase, and many more. We start examining those frameworks in Chapter 10.

The next two chapters discuss unit testing and monitoring of MapReduce programs.

# CHAPTER 8

# Testing Hadoop Programs

This chapter explores how Hadoop programs can be unit tested in your local Integrated Development Environment (IDE). Although unit-testing of Hadoop programs has come a long way since the early days of Hadoop, it continues to be challenging because Hadoop components such as Mappers and Reducers run in a distributed environment.

We discuss the MapReduce unit-testing API called MRUnit, which enables us to unit test Mapper and Reducer classes independently. After discussing the limitations of MRUnit, we explore the LocalJobRunner class that handles these limitations. The chapter closes by looking at exploring MiniMRCluster classes, which enable us to run an entire MapReduce framework in memory, making it suitable for integrated testing of MapReduce components in the context of the overall MapReduce framework.

## Revisiting the Word Counter

Listing 8-1 is the Reducer for the word counter that you saw in Chapter 3.

***Listing 8-1.*** Reducer Portion of WordCountNewAPI.java

```
public static class MyReducer extends
            Reducer<Text, IntWritable, Text, IntWritable> {
   public void reduce(Text key, Iterable<IntWritable> values,
                    Context context) throws IOException,
                                          InterruptedException
   {
      int sum = 0;
      for (IntWritable val : values) {
         sum += val.get();
      }
      context.write(key, new IntWritable(sum));
   }
}
```

Although this program serves the very simple function of counting the words in a set of files, it is very hard to unit-test because it needs to be executed in a MapReduce cluster due to its dependencies on various Hadoop classes such as the Context class. A typical design pattern used by Hadoop programmers is to move all the important logic into services as Plain Old Java Objects (POJOs) because these POJO classes can be easily unit-tested. The Mappers and Reducers act as an orchestration mechanism between various services and are responsible for the I/O from/to the HDFS.

The complete code of UnitTestableReducer, shown in Listing 8-2, demonstrates this concept. There is a Reducer that is appropriately called UnitTestableReducer. Inside the reduce invocation we instantiate an instance of WordCountingService. Note that each invocation of the reduce method creates its own instance of WordCountingService, which is stateful and maintains the state of the count for the given word. This program works even if a Combiner is used because the incrementCount method of the service takes an int parameter. Without a Combiner, val.get() returns 1; with a Combiner, val.get() can return a value greater than 1.

***Listing 8-2.*** UnitTestableReducer.java

```
public static class UnitTestableReducer extends
            Reducer<Text, IntWritable, Text, IntWritable> {

  public void reduce(Text key, Iterable<IntWritable> values,
                    Context context)
                    throws IOException,
                           InterruptedException {
      WordCountingService service = new WordCountingService(key);
      for (IntWritable val : values) {
          service.incrementCount(val.get());
      }
      IntWritable wrdCnt = IntWritable(service.getCount());
      context.write(service.getWord(),wrdCnt);
  }
}
```

Listing 8-3 shows the entire WordCountingService class.

***Listing 8-3.*** WordCountingService.java

```
/*Package and Import declarations*/
public class WordCountingService {
    private Text word=null;
    private int count=0;

    public WordCountingService(Text word){
        this.word = word;
    }
    public void incrementCount(int incr){
        this.count=this.count+incr;
    }
    public Text getWord() {
        return word;
    }
    public int getCount() {
        return this.count;
    }
}
```

# Introducing MRUnit

Although the development pattern discussed in the previous section should enable you to test many MapReduce jobs, sometimes your input will be subjected to multiple state changes in your Mapper or Reducer. There might also be dependency relationships between values received by the Reducer for a given key. One example of such a dependency is a Reducer performing some calculations as the values are evaluated for a given key in the Reducer. It then performs additional calculations when all the values are received for a key. Such scenarios also require certain values to be stored until all the values are received. In such complex scenarios, using the development pattern in the previous example can prove tedious and might even lead to duplication of code between the Reducer and the testing unit.

MRUnit is a library that enables us to test Mappers, Reducers, and MapReduce programs as a whole. MRUnit, which was developed and open-sourced by Cloudera, enables a convenient integration between MapReduce and standard testing libraries such as JUnit and Mockito. The next few sections demonstrate how MRUnit enables unit-testing of MapReduce jobs.

## Installing MRUnit

Because we use Maven to build our code, we also use it to install MRUnit. To install MRUnit add the following lines to your pom.xml file.

```
<!--  Start MR Unit -->
<dependency>
    <groupId>junit</groupId>
    <artifactId>junit</artifactId>
    <version>4.10</version>
</dependency>

<dependency>
    <groupId>org.apache.mrunit</groupId>
    <artifactId>mrunit</artifactId>
    <version>1.0.0</version>
    <classifier>hadoop2</classifier>
</dependency>
<!--  End MR Unit -->
```

The `<classifier>` tag indicates that the MRUnit JAR file being downloaded is for Hadoop 2. The resulting JAR file has the name `mrunit-1.0.0-hadoop2.jar`, and the naming format for the file is `<artifactId>-<version>-<classifier>.jar`.

## MRUnit Core Classes

The core classes of MRUnit are as follows:

- MapDriver: The driver class responsible for calling the Mapper's map() method.
- ReducerDriver: The driver class responsible for calling the Reducer's reduce() method.
- MapReduceDriver: The combined MapReduce driver responsible for calling the Mapper's map() method, followed by an in-memory Shuffle phase. The in-memory Shuffle phase is then followed by invoking the Reducer's reduce() method.

Each of the drivers has methods that allow it to provide inputs and expected outputs for those inputs around the run() method. The JUnit API's setup() method is responsible for creating new instances of the Mapper, Reducer, and appropriate MRUnit drivers mentioned previously.

As you learned in Chapter 3, there are two versions of MapReduce (old API and new API), and MRUnit has drivers that work with each one. The drivers that correspond to the MapReducerV1 API (old API) can be found in the org.apache.hadoop.mrunit package; the drivers that correspond to the MapReduceV2 API (new API) are in the org.apache.hadoop.mrunit.mapreduce package.

---

■ **Note** This chapter uses only the new API (MapReduceV2), so we use the latter package. If you are using one of the older APIs that are still very prevalent, use drivers from the former package.

---

## Writing an MRUnit Test Case

It is time to write an MRUnit test case. Because the MapReduce version of HelloWorld is WordCount, you will write a unit-test for the WordCountNewAPI program from Chapter 3.

In the source code (which can be downloaded from the book's web site), you find the following listings in the org.apress.prohadoop.c8 package, for which we write MRUnit test cases:

- WordCountMapper.java
- WordCountReducer.java

The file that contains the MRUnit test case is WordCountMRUnitTest, which can also be found in the org.apress.prohadoop.c8 package.

The skeleton of this class is shown in Listing 8-4.

***Listing 8-4.*** WordCountMRUnitTest.java

```
/*Package and Import Declarations*/
public class WordCountMRUnitTest extends TestCase {
    private Mapper mapper;
    private Reducer reducer;
    private MapDriver mapDriver;
    private ReduceDriver reduceDriver;
    private MapReduceDriver mapReduceDriver;

   @Before
    public void setUp() {
      /*Initialize Mappers, Reducers and MRUnit Drivers*/
    }
    @Test
    public void testWordCountMapper() {
    {
        /*Test only the Mapper*/
    }
    @Test
    public void testWordCountReducer() {
    {
        /*Test only the Reducer*/
    }
```

```
    @Test
    public void testWordCountMapReducer() {
    {
        /*
         Test both Mapper and Reducer together where Mapper Inputs are
         passed to the Reducer after an in-memory Shuffle
       */
    }
```

A typical MRUnit program follows the skeleton pattern shown in Listing 8-4. (We assume that you are familiar with JUnit4 for the rest of this section.) Methods tagged with @Before are invoked before every method tagged @Test is invoked. The setUp() method ensures that new instances of Mappers, Reducers, and Drivers are available for each test. Listing 8-5 shows the setUp() method.

***Listing 8-5.*** setUp() Method

```
@Before
public void setUp() {
   mapper = new WordCountMapper();
   mapDriver = new MapDriver(mapper);

   reducer = new WordCountReducer();
   reduceDriver = new ReduceDriver(reducer);

   mapReduceDriver = new MapReduceDriver(mapper,reducer);
}
```

Next, we demonstrate how easily the Mapper can be tested. Earlier versions of MRUnit allowed only a single input to the Drivers, but version 1.0.0 removed this limitation, so multiple inputs can now be passed to the Drivers (which expands the type of tests that can be run using MRUnit). Multiple inputs for a given key are often needed to formulate comprehensive tests. For example, a graph problem in which the key is the common vertex and the values are edges that share the common vertex cannot be tested comprehensively if only a single input is allowed for the Mappers.

Listing 8-6 shows how the MapDriver instance can be used to test only the Mapper.

***Listing 8-6.*** testWordCountMapper() Method

```
@Test
public void testWordCountMapper() {
{
   mapDriver.withInput(new LongWritable(1), new Text("map"))
            .withInput(new LongWritable(2), new Text("reduce"))
            .withOutput(new Text("map"), new IntWritable(1))
            .withOutput(new Text("reduce"), new IntWritable(1))
            .runTest();
}
```

This code takes two inputs through the withInput() method invocation on the MapDriver instance. Note that we physically hard-coded the typical inputs to this program that come from TextInputFormat (see Listing 8-1). The first parameter (LongWritable instance) is the byte offset in the file for the start of each line, and the second parameter is the actual line. Because we ignore the first parameter, we added a value of the appropriate type.

The output pair is a Text instance of the word passed as input followed by an instance of IntWritable with a value of 1. The word order in the output pair must match the ordering of the words in the input, or else the unit test fails.

Next we test only the Reducer with the relevant `testWordCountReducer()` method, as shown in Listing 8-7.

***Listing 8-7.*** testWordCountReducer()

```
@Test
public void testWordCountReducer() throws Exception {
   Text key1 = new Text("map");
   List<IntWritable> values1 = new ArrayList<IntWritable>();
   values1.add(new IntWritable(1));
   values1.add(new IntWritable(1));

   Text key2 = new Text("reducer");
   List<IntWritable> values2 = new ArrayList<IntWritable>();
   values2.add(new IntWritable(1));
   values2.add(new IntWritable(1));
   values2.add(new IntWritable(1));
   reduceDriver.withInput(key1, values1)
               .withInput(key2, values2)
               .withOutput(key1, new IntWritable(2))
               .withOutput(key2, new IntWritable(3))
               .runTest();
}
```

We are simulating that the Reducer receives two inputs for the word `map` and three for the word `reducer`. The expected output in this case are the `<map, 2>` and `<reduce, 3>` pairs, in that order. The output types for key and value instances for the Reducer output are `Text` and `IntWritable`, respectively.

Finally, the Mapper and Reducer are tested together, which is a true test of the complete MapReduce program. The output of the Mapper is shuffled by the in-memory Shuffler that comes with the `MRUnit` program and finally consumed by the Reducer. The expected output provided to this program (via the `withOutput()` method) is the expected output of the MapReduce program as a whole for the Mapper inputs provided through the `withInput()` method.

Listing 8-8 shows how the MapReduce program can be tested as a whole.

***Listing 8-8.*** testWordCountMapReducer() Method

```
@Test
public void testWordCountMapReducer() throws Exception {
  mapReduceDriver.withInput(new LongWritable(1), new Text("map"))
                 .withInput(new LongWritable(2), new Text("map"))
                 .withInput(new LongWritable(3), new Text("reduce"))
                 .withOutput(new Text("map"), new IntWritable(2))
                 .withOutput(new Text("reduce"), new IntWritable(1))
                 .runTest();
}
```

# Testing Counters

MapReduce counters (introduced in Chapter 6) are used extensively in Map Reduce programs. In certain cases, counters are used as a testing artifact. For example, a complex graph algorithm used in a social networking or fraud detection application might have extremely complex logic for interconnecting events. Counters are often used in such scenarios to ensure that the actual count of certain named event-interactions matches the expected count. This serves as a high-level check to ensure that all the expected event-interactions were consumed as expected.

The way MRUnit can be used to test counters can be demonstrated with a very simple example. We enhance the WordCountReducer to count the number of words starting with a certain letter, which gives potentially 26 counts when the first letter is case-insensitive.

Listing 8-9 shows the new WordCountWithCounterReducer that manages this counter.

***Listing 8-9.*** WordCountWithCounterReducer

```
public class WordCountWithCounterReducer
             extends Reducer<Text, IntWritable, Text, IntWritable> {
    public void reduce(Text key, Iterable<IntWritable> values,
            Context context) throws IOException,
                           InterruptedException
  {
     int sum = 0;
     for (IntWritable val : values) {
        sum += val.get();
     }
     String firstLt = key.toString().substring(0,1).toLowerCase();
     context.getCounter("FIRST_LETTER", firstLt).increment(sum);
     context.write(key, new IntWritable(sum));
  }
}
```

The source code that tests the MapReduce program with the previous Reducer can be found on the book's web site in the WordCountMRUnitCountersTest.java source file in the following package: org.apress.prohadoop.c8. The only components that vary are the setUp() method shown in Listing 8-10. The new Reducer is used in place of the older one, which does not manage the counters.

***Listing 8-10.*** setUp() Method for WordCountWithCounterReducer

```
/*Other Import Statements*/
/*JUnit specific import statements start here*/
import org.junit.Before;
import org.junit.Test;

import junit.framework.Assert;
import junit.framework.TestCase;
@Before
public void setUp() {
  mapper = new WordCountMapper();
  mapDriver = new MapDriver(mapper);

  reducer = new WordCountWithCounterReducer();
  reduceDriver = new ReduceDriver(reducer);

  mapReduceDriver = new MapReduceDriver(mapper,reducer);
}
```

The two methods that test the counter are testWordCountReducer(), shown in Listing 8-11, and testWordCountMapReducer(), shown in Listing 8-12. Note the lines in bold for the way counters are retrieved and tested.

***Listing 8-11.*** testWordCountReducer() Method

```
@Test
public void testWordCountReducer() throws Exception {
   Text key = new Text("map");
   List<IntWritable> values = new ArrayList<IntWritable>();
   values.add(new IntWritable(1));
   values.add(new IntWritable(1));
   reduceDriver.withInput(key, values)
               .withOutput(key, new IntWritable(2))
               .runTest();
   Assert.assertEquals(2,
               reduceDriver.getCounters()
                          .findCounter("FIRST_LETTER", "m")
                          .getValue());
   Assert.assertEquals(0,
               reduceDriver.getCounters()
                          .findCounter("FIRST_LETTER", "n")
                          .getValue());
}
```

***Listing 8-12.*** testWordCountMapReducer() Method

```
@Test
public void testWordCountMapReducer() throws Exception {
  this.mapReduceDriver
              .withInput(new LongWritable(1), new Text("map"))
              .withInput(new LongWritable(2), new Text("map"))
              .withInput(new LongWritable(3), new Text("reduce"))
              .withOutput(new Text("map"), new IntWritable(2))
              .withOutput(new Text("reduce"), new IntWritable(1))
              .runTest();

  Assert.assertEquals(2,
             mapReduceDriver.getCounters()
                         .findCounter("FIRST_LETTER", "m")
                         .getValue());
  Assert.assertEquals(0,
             mapReduceDriver.getCounters()
                         .findCounter("FIRST_LETTER", "n")
                         .getValue());

}
```

The preceding programs tested only the counters from ReducerDriver, but the testing process of counters managed by the Mappers from the MapDriver is similar. The method call on the MapDriver (assuming that mapDriver is an instance of MapDriver) is the following:

```
mapDriver.getCounter().findCounter(...)
```

The MapReduceDriver can also verify counters managed by the Mapper and the Reducer.

---

■ **Note** Assert is a JUnit class that has a series of overloaded assertEquals() methods that compare the expected value (first parameter) with the actual value (second parameter). This example uses the assertEquals(long expected,long actual) method to compare with long values.

---

## Features of MRUnit

We have shown you some key features of MRUnit and tested Mappers, Reducers, MapReduce jobs, and MapReduce Counters. There is more to MRUnit, however. For example, if a program uses distributed cache (refer to Chapter 6), MRUnit can accommodate this requirement.

Another very useful type of scenario supported by MRUnit is testing of chained MapReduce jobs. In this scenario jobs are chained one after the other where output of one job is the input to the next one in the chain. The PipeLineMapReduceDriver class is the MRUnit driver that allows a chain of such jobs to be tested. The Hadoop library comes equipped with two classes, IdentityMapper and IdentityReducer.When used in the context of a single MapReduce program the output is identical to the input. Following is an example of how PipeLineMapReduceDriver can be used with a series of three MapReduce jobs using the IdentityMapper and IdentityReducer:

```
PipelineMapReduceDriver<Text, Text, Text, Text> driver =
PipelineMapReduceDriver.newPipelineMapReduceDriver();
driver.withMapReduce(new IdentityMapper<Text, Text>(),
                     new IdentityReducer<Text, Text>())
      .withMapReduce(new IdentityMapper<Text, Text>(),
                     new IdentityReducer<Text, Text>())
      .withMapReduce(new IdentityMapper<Text, Text>(),
                     new IdentityReducer<Text, Text>())
      .withInput(new Text("one"), new Text("two"))
      .withOutput(new Text("one"), new Text("two")).runTest();
```

The real value of PipeLineMapReducerDriver is that it allows developers to chain MapReduce jobs in memory very quickly. Contrast that with running a MapReduce job that writes to disk, followed by another job that reads this output from disk and produces another output on disk. The latter approach is very slow due to the increased I/O, even for a small dataset that is amenable for unit-testing. The former approach allows a series of MapReduce jobs in a chain to be unit-tested without writing a single record to disk, making it very suitable for the develop, unit-test, fix, develop cycle.

The MRUnit source code that can be downloaded from the MRUnit web site has a comprehensive test suite with excellent test cases. You should download the source code and go through the test cases in the following packages (depending on whether MR v1 or MR v2 is used):

- MapReducev1: org.apache.hadoop.mrunit package
- MapReducev2: org.apache.hadoop.mrunit.mapreduce package

The distributed cache-based test case is a class named TestDistributedCache. The pipeline-based test cases are in the TestPipelineMapReduceDriver class. The code snippet shown previously for PipelineMapReduceDriver has been borrowed from TestPipelineMapReduceDriver.

## Limitations of MRUnit

Traditionally, MapReduce programs have been very hard to test. MRUnit is a fantastic library that enables MapReduce developers to easily test MapReduce jobs. It has a very intuitive interface and smooth learning curve. The library is nonintrusive and does not require developers to modify Mapper or Reducer code in any library-specific manner. MRUnit operates completely in memory, which enables tests to execute very quickly.

MRUnit is not without limitations, however. For example, it can test only certain Hadoop artifacts such as Mappers, Reducers, MapReduce Counters, and Distributed Cache. Note that Combiner testing is identical to Reducer testing because they share the same interface.

As discussed in Chapter 7, sometimes we have to write a custom InputFormat or custom OutputFormat, and neither can be tested using MRUnit. Also, MRUnit operates under the assumption of a single Mapper and Reducer. Any aspect of the program that depends on the multithreaded or multiprocess nature of MapReduce, such as Partitioner, cannot be tested using MRUnit.

All these features also need to be tested. Hadoop comes equipped with the LocalJobRunner, which was implicitly used in Chapter 3 when we ran our programs in our local IDE environment. LocalJobRunner is discussed in the next section.

# Testing with LocalJobRunner

MapReduce jobs can be tested end to end by using LocalJobRunner. Let us test our WordCount program with this method.

Listing 8-13 shows the skeleton of the test case program using LocalJobRunner with only the setUp() and testXXX() methods populated.

***Listing 8-13.*** WordCountLocalRunnerUnitTest Class

```
/*Package and Import Declarations*/

public class WordCountLocalRunnerUnitTest extends TestCase {
   private Job job = null;
   private Path inputPath  = null;
   private Path outputPath = null;

   @Before
   public void setUp() throws Exception{
      this.inputPath =
              new Path("src/main/resources/input/wordcount/");
      this.outputPath =
              new Path("src/main/resources/output/wordcount/");
      Configuration conf = new Configuration();
      conf.set("mapred.job.tracker", "local");
      conf.set("fs.default.name", "file:////");

      FileSystem fs = FileSystem.getLocal(conf);
      if (fs.exists(outputPath)) {
         fs.delete(outputPath, true);
      }
      this.job = this.configureJob(conf, inputPath, outputPath);
   }
```

```
    private Job configureJob(Configuration conf,
                             Path inputPath,
                             Path outputPath)
            throws Exception {
        /*
          Configure the Job instance with input/output paths and
          input/output formats
        */
        this.job = Job.getInstance(conf);
        job.setJarByClass(WordCountLocalRunnerUnitTest.class);
        job.setOutputKeyClass(Text.class);
        job.setOutputValueClass(IntWritable.class);
        job.setMapperClass(WordCountMapper.class);
        job.setReducerClass(WordCountReducer.class);
        job.setInputFormatClass(TextInputFormat.class);
        job.setOutputFormatClass(TextOutputFormat.class);
        FileInputFormat.setInputPaths(job, inputPath);
        FileOutputFormat.setOutputPath(job, outputPath);
    }

    private Map<String, Integer> getCountsByWord(File outDir)
                                 throws Exception {
           /*
             Return a Map instance of counts by word for the job output
          */
    }

    @Test
    public void testWordCount() throws Exception {
       boolean status = job.waitForCompletion(true);
         assertTrue(job.isSuccessful());
       File outFile = new File(outputPath.toUri() + "/");
       Map<String, Integer> countsByWord = getCountsByWord(outFile);
         assertEquals(new Integer(1),countsByWord.get("java"));
         assertEquals(new Integer(1),countsByWord.get("job"));
         assertEquals(new Integer(3),countsByWord.get("test"));
         assertEquals(new Integer(1),countsByWord.get("word"));
    }
}
```

The following sections explain the key components of Listing 8-13.

## setUp() method

First, we configure the job and define the input and output paths. Because this test uses `LocalJobRunner`, we do not want to use the HDFS for this test; we want to use the local filesystem instead. The `Configuration` instance with the following characteristics is defined as follows:

```
Configuration conf = new Configuration();
conf.set("mapred.job.tracker", "local");
conf.set("fs.default.name", "file:////");
```

The first line instantiates the Configuration instance, the second line configures the JobRunner as local, and the last line configures FileSystem as the local filesystem.

The next three lines delete the output path if it already exists (a MapReduce job does not run if it already exists).

```
FileSystem fs = FileSystem.getLocal(conf);
if (fs.exists(outputPath)) {
   fs.delete(outputPath, true);
}
```

The key line to notice is FileSystem.**getLocal(conf)**, which retrieves the local filesystem.

Finally, we configure the Job instance with call configureJob(). By now, you are familiar with the method listing, so we do not describe it in detail here.

We run the job and then verify whether the job ran successfully. We invoke getCountsByWord(), which reads the output files from the output directory and constructs a map of word count by word, and verify the word count.

Listing 8-14 shows the source code for getCountsByWord(), and Listing 8-15 shows the sample file used to calculate word counts.

***Listing 8-14.*** getCountsByWord

```
private Map<String, Integer> getCountsByWord(File outDir)
                              throws Exception {
   Map<String, Integer> countsByWord =
                         new HashMap<String, Integer>();
   Collection<File> files = FileUtils.listFiles(outDir,
                              TrueFileFilter.INSTANCE,
                              TrueFileFilter.INSTANCE);

   for (File f : files) {
      if (f.getName().startsWith("part")
                && !f.getAbsolutePath().endsWith("crc")) {
         List<String> lines = FileUtils.readLines(f);
         System.out.println(lines.size());
         for (String l : lines) {
            String[] counts = l.split("\\t");
            countsByWord.put(counts[0],
                              Integer.parseInt(counts[1]));
         }
         break;
      }
   }
   return countsByWord;
}
```

***Listing 8-15.*** Test input file

```
test
test
word
test
java
job
```

## Limitations of LocalJobRunner

One of the limitations of using `LocalJobRunner` is that it is single-threaded, and only one Reducer can be started. This is true even as `setNumReducers` is invoked with a value higher than `1`. Map-only jobs with no Reducers are supported as well.

Single threading is the limitation it shares with `MRUnit`: any code that depends on a multithreaded or multiprocess nature cannot be tested. One example of such code is a custom Partitioner class. We developed several custom Partitioners in Chapter 6.

However, because `LocalJobRunner` runs in a proper Hadoop environment, we can test all the other features of MapReduce, such as custom `InputFormat` and `OutputFormat`. We can now test our MapReduce programs end to end.

The other main limitation compared to `MRUnit` should be emphasized. With `LocalJobRunner`, there is a real file I/O, even if the filesystem used is local, so unit tests slow down. There is no overhead with `MRUnit`; all testing occurs in memory. Consequently, unit tests based on `MRUnit` run considerably faster. The test suite should prefer `MRUnit` test cases over `LocalJobRunner` cases and use the latter only when necessary.

# Testing with MiniMRCluster

So far, we have unit-tested the job using `MRUnit`. We used `LocalJobRunner` to test every aspect of MapReduce, but it has limitations, and we want a testing method that completely tests MapReduce jobs.

A feature of the Hadoop API known as Mini Hadoop clusters can be used to test the test suite in an environment that is a true simulation of a fully distributed Hadoop environment. Hadoop provides classes such as `MiniMRCluster`, `MiniYarnCluster`, and `MiniDFSCluster` to simulate a true HDFS and MapReduce environment in memory. Separate tasks are started in separate JVMs, so we finally have an environment that is a true replica of the actual runtime Hadoop environment that enables us to test features that depend on multiple Reducers or Partitioners. This feature comes at a price, of course: because multiple JVMs are started to run the job tasks, executing tests is slower and debugging is more complex.

Mini-Clusters are used extensively in the Hadoop test suite. As developers, we should avail ourselves of it, too. Hadoop provides an abstract base class to abstract away the details of starting/stopping up the HDFS and the MapReduce framework in the `setup()` and `teardown()` methods. The class is `ClusterMapReduceTestCase`, but the documentation for this class is very limited. This section guides you in setting up the development environment to enable this feature for Hadoop 2.x.

## Setting up the Development Environment

First we need to setup our development environment to use the `MiniMRCluster`. Follow these steps to enable the `MiniMRCluster` feature:

1. The following JAR files are used by the `MiniMRCluster`:
   a. `hadoop-mapreduce-client-jobclient-2.2.0-tests.jar`
   b. `hadoop-common-2.2.0-tests.jar`
   c. `hadoop-hdfs-2.2.0-tests.jar`
   d. `hadoop-yarn-server-tests-2.2.0-tests.jar`
   e. `hadoop-yarn-server-nodemanager-2.2.0.jar`
   f. `hadoop-mapreduce-client-hs-2.2.0.jar`

2. Enable these JAR files in your POM by adding dependencies for each of the previous JAR files. (Note the extensive use of the <classifier> tag.)

```
<dependency>
   <groupId>org.apache.hadoop</groupId>
   <artifactId>
      hadoop-mapreduce-client-jobclient
   </artifactId>
   <version>2.2.0</version>
   <classifier>tests</classifier>
</dependency>
<dependency>
   <groupId>org.apache.hadoop</groupId>
   <artifactId>hadoop-hdfs</artifactId>
   <version>2.2.0</version>
   <classifier>tests</classifier>
</dependency>
<dependency>
   <groupId>org.apache.hadoop</groupId>
   <artifactId>hadoop-common</artifactId>
   <version>2.2.0</version>
   <classifier>tests</classifier>
</dependency>
<dependency>
   <groupId>org.apache.hadoop</groupId>
   <artifactId>hadoop-yarn-server-tests</artifactId>
   <version>2.2.0</version>
   <classifier>tests</classifier>
</dependency>
<dependency>
   <groupId>org.apache.hadoop</groupId>
   <artifactId>hadoop-yarn-server-nodemanager</artifactId>
   <version>2.2.0</version>
</dependency>
   <dependency>
   <groupId>org.apache.hadoop</groupId>
   <artifactId>hadoop-mapreduce-client-hs</artifactId>
   <version>2.2.0</version>
</dependency>
```

3. Ensure that the following dependency is also configured:

```
<dependency>
   <groupId>org.apache.hadoop</groupId>
   <artifactId>hadoop-client</artifactId>
   <version>2.2.0</version>
</dependency>
```

This step completes the setup for the MiniMRCluster in the development environment.

## Example for MiniMRCluster

The source file that contains an example of a `TestCase` using `MiniMRCluster` is `WordCountTestWithMiniCluster`. Listing 8-16 shows the key sections of this `TestCase`.

***Listing 8-16.*** WordCountTestWithMiniCluster

```
/*Package and import declarations*/

public class WordCountTestWithMiniCluster
                         extends ClusterMapReduceTestCase {
    public static final String INPUT_STRING =
            "test\ntest\nword\ntest\njava\njob\n";
    Path inputPath = null;
    Path outputPath = null;
    Configuration conf = null;
    /*
     *Read output file and construct a Map of
     *word count by word
     */
    protected Map<String, Integer> getCountsByWord()
            throws Exception {
       Map<String, Integer> results = new HashMap<String, Integer>();
        FileSystem fs = FileSystem.get(conf);
        FileStatus[] fileStatus = fs.listStatus(outputPath);
        for (FileStatus file : fileStatus) {
            String name = file.getPath().getName();
            if (name.contains("part")) {
                Path outFile = new Path(outputPath, name);
                BufferedReader reader = new BufferedReader(
                        new InputStreamReader(fs.open(outFile)));
                String line;
                line = reader.readLine();
                while (line != null) {
                    String[] vals = line.split("\\t");
                    results.put(vals[0],
                               Integer.parseInt(vals[1]));
                    line = reader.readLine();
                }
                reader.close();
            }
        }
        return results;
    }

    private void createFile(FileSystem fs, Path filePath)
                                     throws IOException {
        FSDataOutputStream out = fs.create(filePath);
        out.write(INPUT_STRING.getBytes());
        out.close();
    }
```

```
    private void prepareEnvironment() throws Exception {
        this.conf = this.createJobConf();
        this.inputPath = new Path("input/wordcount/");
        this.outputPath = new Path("output/wordcount/");
        FileSystem fs = FileSystem.get(conf);
        fs.delete(inputPath, true);
        fs.delete(outputPath, true);
        fs.mkdirs(inputPath);
        this.createFile(fs, new Path(inputPath, "test.txt"));
    }

    public Job configureJob() throws Exception {
        Job job = Job.getInstance(conf);
        job.setJarByClass(WordCountTestWithMiniCluster.class);
        job.setOutputKeyClass(Text.class);
        job.setOutputValueClass(IntWritable.class);
        job.setMapperClass(WordCountMapper.class);
        job.setReducerClass(WordCountReducer.class);
        job.setInputFormatClass(TextInputFormat.class);
        job.setOutputFormatClass(TextOutputFormat.class);
        FileInputFormat.setInputPaths(job, inputPath);
        FileOutputFormat.setOutputPath(job, outputPath);
        return job;
    }

    @Test
    public void testWordCount() throws Exception {
        this.prepareEnvironment();
        Job job = this.configureJob();
        boolean status = job.waitForCompletion(true);
        assertTrue(job.isSuccessful());
        Map<String, Integer> countsByWord = getCountsByWord();
        assertEquals(new Integer(1), countsByWord.get("java"));
        assertEquals(new Integer(1), countsByWord.get("job"));
        assertEquals(new Integer(3), countsByWord.get("test"));
        assertEquals(new Integer(1), countsByWord.get("word"));
    }
}
```

The following key features of this program require further explanation:

- This TestCase class extends ClusterMapReduceTestCase, which is the class that comes with the Hadoop distribution and has the implementation to start up and shut down the Hadoop cluster around each test case.
- We create HDFS directories and files as needed for our TestCase. The prepareEnvironment() method implements this functionality.

- Do not create a `setUp()` method as is typical in unit tests. The `setUp()` method is provided by the `ClusterMapReduceTestCase` class that the custom unit test extends. The super class `setUp()` method is invoked before each test executes. This method starts the YARN cluster in memory for the tests to execute. The custom `prepareEnvironment()` method has to be called at the start of the test.
- Because each test starts the mini-YARN cluster via the `setUp()` method of `ClusterMapReduceTestCase` and stops the minicluster with the `tearDown()` method, it is not a good idea to have too many test methods. When using `MiniMRCluster`, put all your tests into one consolidated test method to optimize test performance.

## Limitations of MiniMRCluster

This discussion of `MiniMRCluster` finishes up the topic of testing toolkits for Hadoop. Although `MiniMRCluster` `enables us` to test every single aspect of MapReduce jobs, there are a few limitations. Bear the following in mind:

- `MiniMRCluster` starts up the Hadoop cluster in memory and is a true multithreaded simulation of Map Reduce framework. Be aware that running out of memory is a real risk in this approach. Make sure that you keep the size of your test data small.
- The `ClusterMapReduceTestCase` base class has the implementation to start up and shut down `MRCluster` for every test case, but it can be a slow process. It is a good idea to bundle test cases together in a single method to ensure that the test cases run quickly. Traditional unit-testing methods require that each test case is isolated in its own test function. However, tests using `MiniMRCluster` are not unit tests; they are more like integration tests. They take longer than a typical unit test to execute due to the overhead associated with starting and stopping the Hadoop cluster in memory. Bundling them is a good idea.

# Testing MR Jobs with Access Network Resources

Eventually, you will write MapReduce jobs that access network resources. I have written MapReduce jobs that write to Apache Solr, the open-source faceted search engine. I have also written jobs that write to HBase, the Hadoop NoSQL database that you will encounter in Chapter 14. Similarly, I had to write to Cassandra, which is another distributed NoSQL database. Using any of the methods mentioned in this chapter will be inadequate for such tests unless you are willing to write to these network resources from your MapReduce test suites or your external services provide a class similar to `MiniMRCluster`. It can also considerably increase the time it takes to run your test cases.

I have adopted the following process to test jobs that depend on such network resources:

1. Use a separate service class to communicate with network resources. This service class can be separately tested outside of the Hadoop MapReduce environment.
2. Initialize this service class in the `configure()` method of your Mapper or Reducer, and shut down the network resources in the `close()` methods.
3. Provide a configuration parameter (using the `-D <name> <value>`) method to your MapReduce job. This parameter is a Boolean value that indicates whether the external service should be used. For test suites, set this parameter to `false`.
4. Read this parameter in your Mapper and Reducer instance and determine whether to initialize the service in the `configure()` method or shut down the service in the `close()` method.

5. Provide an alternative code path that writes the data meant for the network resources to the HDFS when unit testing.

6. Read the HDFS output to test the data meant for the network resource.

# Summary

This chapter covered all aspects of testing Hadoop jobs. First and foremost, you learned that it is a good practice to use some form of service-oriented style of programming in which services are invoked from the Mapper or Reducer. Services are far easier to test than Mappers and Reducers are.

We discussed `MRUnit` in considerable detail. `MRUnit` is a fantastic open-source library that enables Mapper and Reducer class testing. However, it allows only a limited number of Hadoop classes to be tested.

Testing of custom `InputFormat`, `OutputFormat`, and `RecordReader` classes need more than `MRUnit`. `LocalJobRunner` enables us to test MapReduce jobs end to end and in memory in a local development environment. Yet `LocalJobRunner` is a single-threaded process; aspects of MapReduce that depend on multiple threads, such as instances of a Reducer or Partitioner, cannot be tested using `LocalJobRunner`.

Finally, we examined `MiniMRCluster`, which is a true `MRCluster` started in memory in the local development environment. Multiple tasks are created in separate threads, which makes it a true simulation of the real-world distributed Hadoop cluster. This method finally allows us to test every aspect of MapReduce jobs.

CHAPTER 9

# Monitoring Hadoop

This chapter discusses how to monitor the Hadoop cluster. Monitoring covers the whole gamut of operations: interpreting log files, provisioning the Hadoop cluster components, and monitoring services running on the Hadoop cluster. This chapter covers various aspects of monitoring the Hadoop cluster including the following:

- Writing and interpreting log messages using web-based user interface (UI) management tools that come packaged with the Apache Hadoop Framework.
- Improvements in Hadoop log management in YARN.
- How to monitor the performance of the Hadoop cluster.
- Vendor-based tools for provisioning, monitoring, and managing Hadoop clusters.

## Writing Log Messages in Hadoop MapReduce Jobs

First, we need to be able to log messages in our MapReduce jobs, including using standard libraries such as `log4j` and writing to standard output streams with `System.out.println()` or `System.err.println()`.The main challenge is that processes such as Mappers and Reducers run on various machines in the cluster. Hadoop comes equipped with a web interface to view the logs.

There are three types of logs in Hadoop for every Mapper and Reducer:

- `stdout - System.out.println()` is directed to this file.
- `stderr - System.err.println()` is directed to this file.
- `syslog - log4j` logs are directed to this file. The stack traces of all exceptions that occur during job execution and not handled in the job also show up in the `syslog`.

Listing 9-1 shows how to write these logs using the familiar WordCount program. Although WordCount is a simple Hadoop program, it illustrates how to log messages without getting distracted by what the program is trying to do.

***Listing 9-1.*** WordCountWithLogging.java

```
/*Package and Import Declarations*/

public class WordCountWithLogging {
    public static class MyMapper extends
            Mapper<LongWritable, Text, Text, IntWritable> {
        public static Logger logger =
                     Logger.getLogger(MyMapper.class);
```

```
        public void map(LongWritable key,
                        Text value,
                        Context context) throws IOException,
                                                 InterruptedException {
            String w = value.toString();
            logger.info("Mapper Key =" + key);
            if(logger.isDebugEnabled()){
            logger.info("Mapper value =" + value);
            }
            context.write(new Text(w), new IntWritable(1));
        }
    }

    public static class MyReducer extends
            Reducer<Text, IntWritable, Text, IntWritable> {
        public static Logger logger =
                          Logger.getLogger(MyReducer.class);
        public void reduce(Text key,
                           Iterable<IntWritable> values,
                           Context context) throws IOException,
                                                   InterruptedException {

            int sum = 0;
            for (IntWritable val : values) {
                sum += val.get();
            }
            logger.info("Reducer Key =" + key);
            if(logger.isDebugEnabled()){
            logger.info("Reducer value =" + sum);
            }
            System.out.println("Reducer system.out >>> " + key);
            System.err.println("Reducer system.err >>> " + key);
            context.write(key, new IntWritable(sum));
        }
    }

    public static void main(String[] args) throws Exception {
        Job job = Job.getInstance(new Configuration());
        job.setJarByClass(WordCountWithLogging.class);
        job.setOutputKeyClass(Text.class);
        job.setOutputValueClass(IntWritable.class);
        job.setMapperClass(MyMapper.class);
        job.setReducerClass(MyReducer.class);
        job.setInputFormatClass(TextInputFormat.class);
        job.setOutputFormatClass(TextOutputFormat.class);
        FileInputFormat.setInputPaths(job, new Path(args[0]));
        FileOutputFormat.setOutputPath(job, new Path(args[1]));
        boolean status = job.waitForCompletion(true);
```

```
        if (status) {
            System.exit(0);
        } else {
            System.exit(1);
        }
    }
}
```

Note the bold lines in the listing. To view the logs, we have to access the JobHistory service. In the browser address bar, enter the following URL (assuming that you are running this in a pseudo-distributed environment):

```
http://localhost:19888/jobhistory
```

You see a screen like the one shown in Figure 9-1, which provides summary information about the job.

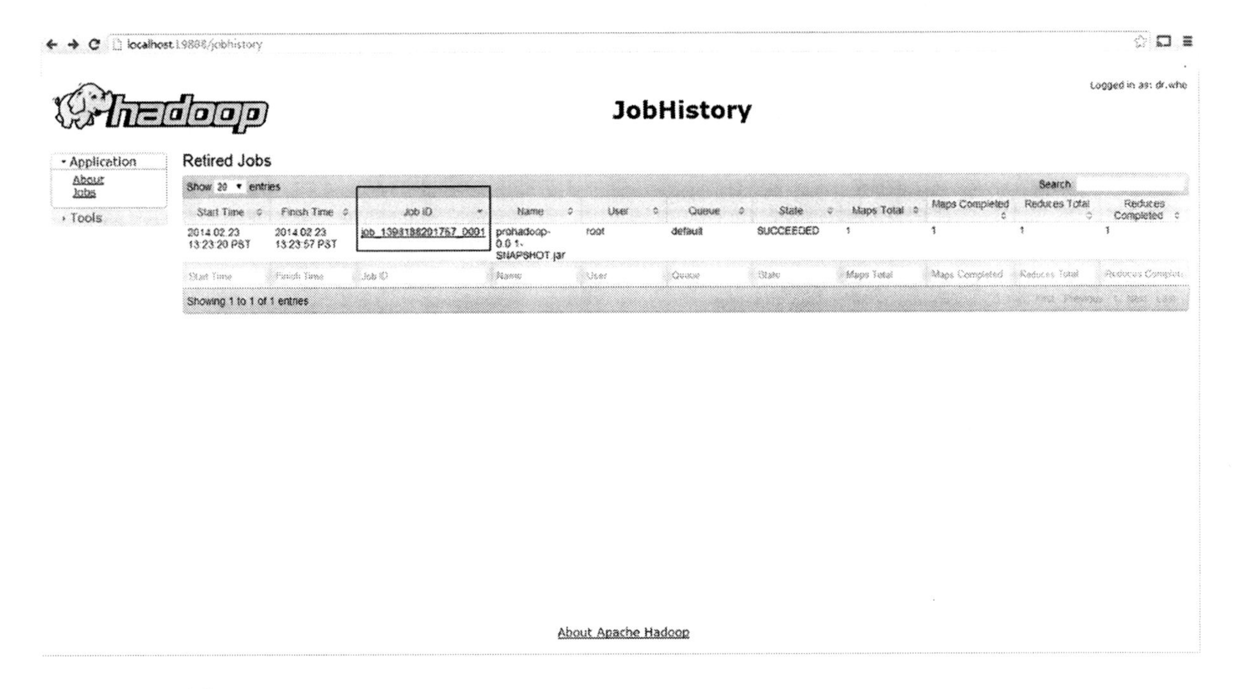


*Figure 9-1. Job history viewer*

Click the Job ID link (boxed for emphasis in Figure 9-1), which displays the screen shown in Figure 9-2. This screen displays additional information about the job, including basic information about its execution status, its start and stop times, and the queue in which the job ran (refer to the section "Scheduler" in Chapter 4). We can identify how many Mappers and Reducers were used in the job execution. Note the average execution time of the Map, Shuffle, Sort, and Reduce phases. We can now see links to drill into details for the Mappers and Reducers.

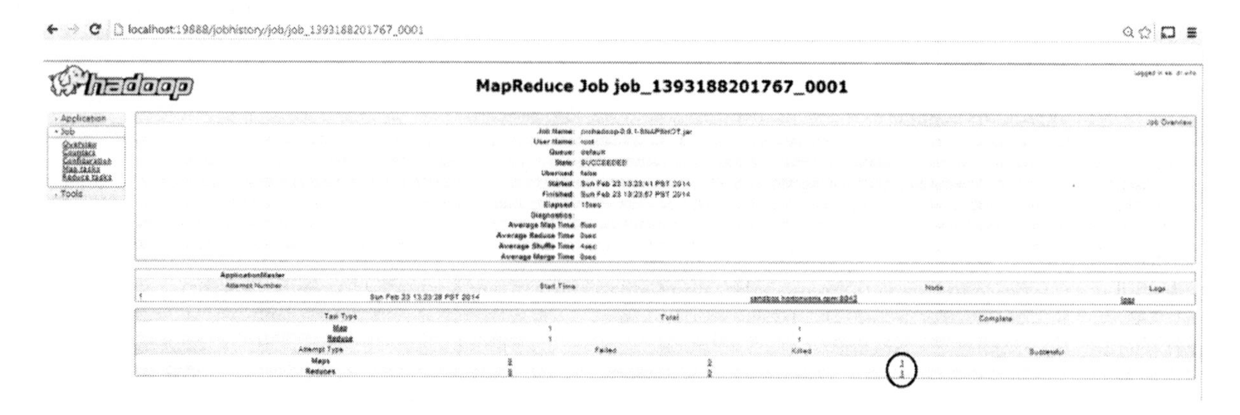


*Figure 9-2.* Job details viewer

# Viewing Log Messages in Hadoop MapReduce Jobs

Let us drill down into the Mapper logs by clicking the `1` link for the `Maps`. The view shown in Figure 9-3 is displayed.

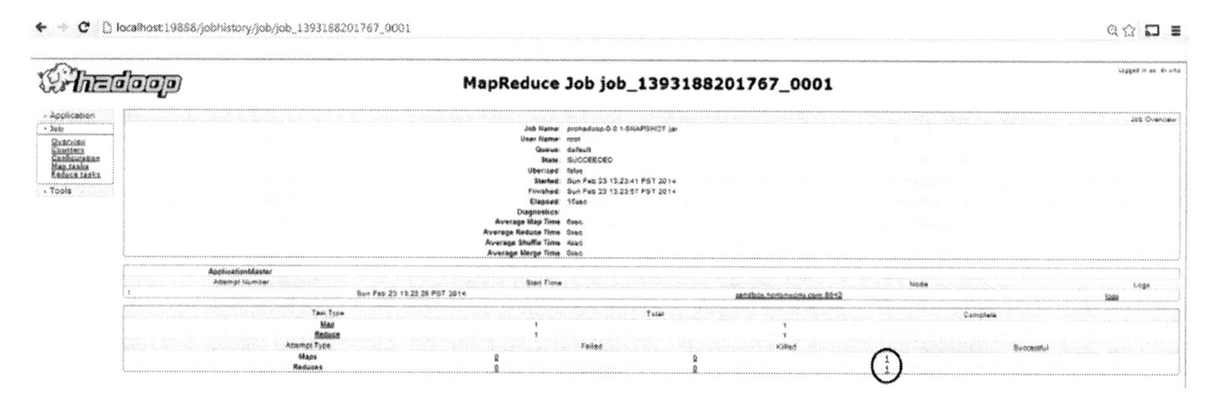


*Figure 9-3.* Map details viewer

Now we can view the logs for this specific instance of the Mapper. We can click the logs that take us to the view shown in Figure 9-4. Notice that you can now view all three logs:

- `stdout`
- `stderr`
- `syslog`

*Figure 9-4.* *Mapper view logs*

The Mapper includes only `log4j` logging, so `stdout` and `stderr` are empty. There are additional custom messages in the `syslog` that are boxed for emphasis in Figure 9-5. This is the result of the following line in the Mapper code:

```
logger.info("Mapper Key =" + key);
```

```
2014-02-23 13:23:50,175 INFO [main] org.apache.hadoop.mapred.MapTask: bufstart = 0; bufvoid = 209715200
2014-02-23 13:23:50,225 INFO [main] org.apress.prohadoop.c8.WordCountWithLogging$MyMapper: Mapper Key -0
2014-02-23 13:23:50,226 INFO [main] org.apress.prohadoop.c8.WordCountWithLogging$MyMapper: Mapper Key -6
2014-02-23 13:23:50,226 INFO [main] org.apress.prohadoop.c8.WordCountWithLogging$MyMapper: Mapper Key -12
2014-02-23 13:23:50,226 INFO [main] org.apress.prohadoop.c8.WordCountWithLogging$MyMapper: Mapper Key -18
2014-02-23 13:23:50,226 INFO [main] org.apress.prohadoop.c8.WordCountWithLogging$MyMapper: Mapper Key -24
2014-02-23 13:23:50,226 INFO [main] org.apress.prohadoop.c8.WordCountWithLogging$MyMapper: Mapper Key -30
2014-02-23 13:23:50,242 INFO [main] org.apache.hadoop.mapred.MapTask: Spilling map output
```

*Figure 9-5.* *Mapper syslog*

Because DEBUG mode is not enabled by default, the following line never gets executed:

```
logger.info("Mapper value =" + value);
```

We follow the same set of steps to get to the Reducer logs, and the full scope of the logging is shown in Figure 9-6. There is a section for `stdout`, `stderr`, and `syslog`.

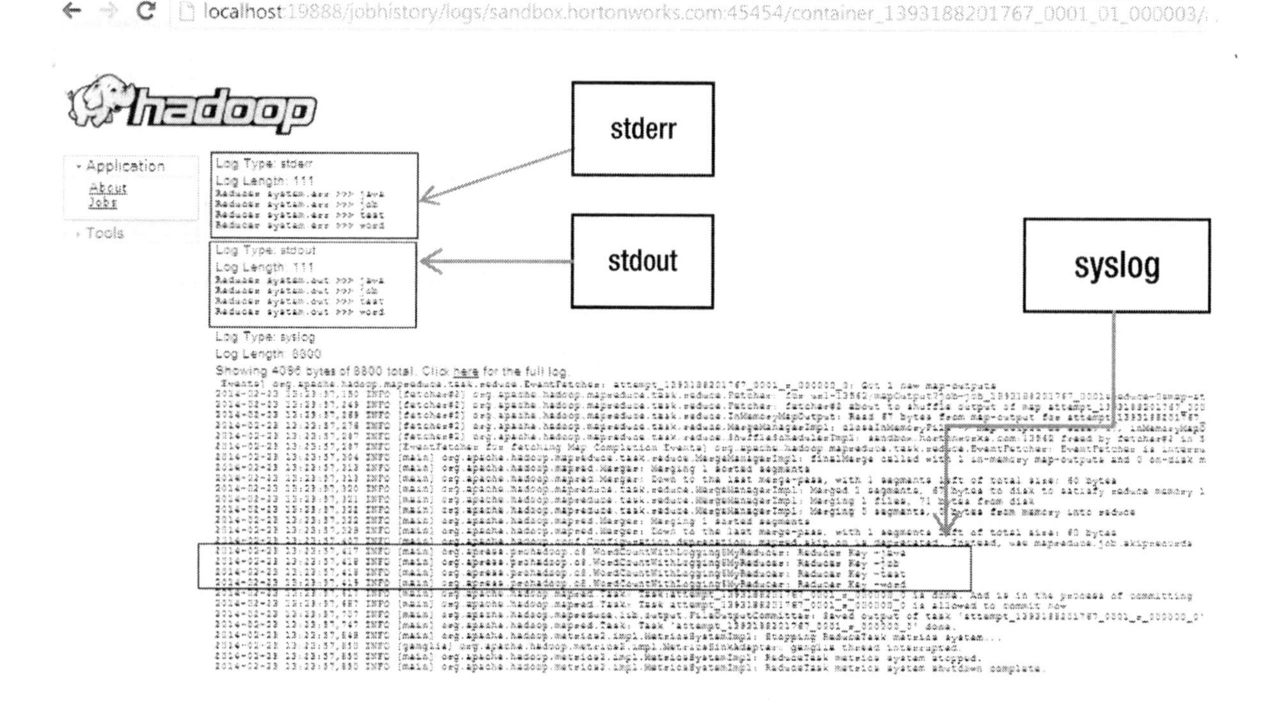


*Figure 9-6.* *View logs for the Reducer*

The `stdout` file output is the result of the following line in the Reducer:

```
System.out.println("Reducer system.out >>> " + key);
```

The `stderr` output is the result of the following line in the Reducer:

```
System.err.println("Reducer system.err >>> " + key);
```

The `syslog` output is the result of the following line in the Reducer:

```
logger.info("Reducer Key =" + key);
```

Similar to the Mapper, the DEBUG mode is disabled for the `log4j` logger, so the following line does not get executed:

```
logger.info("Reducer value =" + sum);
```

The default location of the `log4j.properties` file that can be used to configure custom logging characteristics is `$HADOOP_CONF_DIR/log4j.properties`. No restart of the Hadoop cluster is needed; each new job picks `log4j.properties`.

# User Log Management in Hadoop 2.x

We now explore the internals of logging in Hadoop. Logging is one area that has been substantially improved in Hadoop 2.x, but you need to understand a bit about logging in Hadoop 1.x and its limitations in practice to understand the improvements and what motivated them.

In Hadoop 1.x, logs were stored on the individual task nodes that ran the Mapper and Reducer. Hadoop 2.x also supports this behavior; in fact, it is the default behavior in the Apache distribution. However, there is now an option to aggregate these logs after the job completes and move them to the HDFS, which alleviates a lot of the log-management problems found in Hadoop 1.x.

In Hadoop 1.x, logs were truncated after they reached a predetermined size to save disk space on task nodes. This size was driven by the following set of parameters:

- Logs were cleaned after a predetermined time interval, which was decided by the value specified by the `mapred.userlog.retain.hours` parameter (pre-Hadoop 0.21) or the `mapreduce.job.userlog.retain.hours` parameter (from Hadoop 0.21 and before Hadoop 0.23). The default value of this parameter was 24 (hours), which implied that logs were deleted after this threshold age was reached.
- The other main problem was that every log file got truncated every time it reached a certain length, which was driven by the value of the following properties:
  - `mapred.userlog.limit.kb`: In this approach (prior to Hadoop 0.21), the `stdout` and `stderr` streams are piped to the UNIX tail program for each task, and only the specified size is retained at the end of each job. The main challenge with this approach is that it increases memory usage for a job with lots of logging.
  - `mapreduce.cluster.map.userlog.retain-size`: With this option (from Hadoop 0.21 and prior to Hadoop 0.23), truncation occurred after the job was complete. This method did not have the same memory problem as the prior method, but a runaway job could easily fill up the disk on the data node.

The configuration of the log-retention parameters is global to the cluster. Hadoop clusters are rarely used by a single group with homogenous needs; they are used by various heterogeneous groups in the Enterprise. Although different groups have different purposes for log files, a global configuration needs to meet the requirements of the group that needs the log files for the longest duration. It leads to increased operations cost for managing disks on the data nodes.

## Log Storage in Hadoop 2.x

The log management-related properties belong to the `yarn-site.xml` file. The two main properties are the following:

- `yarn.log-aggregation-enable`: It is `false` by default in the Apache Hadoop distribution. If `false`, the logs are kept on the nodes running the Node Manager. This behavior is similar to Hadoop 1.x, but when aggregation is enabled, all the logs are aggregated and moved to the HDFS after the job completes. Check the default property of the Hadoop distribution you are using; some distributions use a value of `true` by default.
- `yarn.nodemanager.log-dirs`: The local directory on the node running the Node Manager. The container executing within the Node Manager stores the logs in this directory. The default value is `${yarn.log.dir}/userlogs`. Note that `yarn.log.dir` is not an environment variable of the operating system; it is a Java system property configured through the `yarn-env.sh` (or `yarn-env.cmd` in Windows) file. By default, it is set to the value of the operating system environment variable `YARN_LOG_DIR`.

The application-level log directory is maintained as follows (${appid} represents the application ID of the job):

```
${ yarn.nodemanager.log.dirs}/application_${appid}/
```

All the logs for this application instance on the given node are written to subdirectories of the previous directory. Those subdirectories have the following naming convention:

```
container_${contid}
```

${contid} represents the container ID of the container running the task. Each of the container directories contains three files: stdout, stderr, and syslog, produced by the container. A sample hierarchy is shown in Figure 9-7.

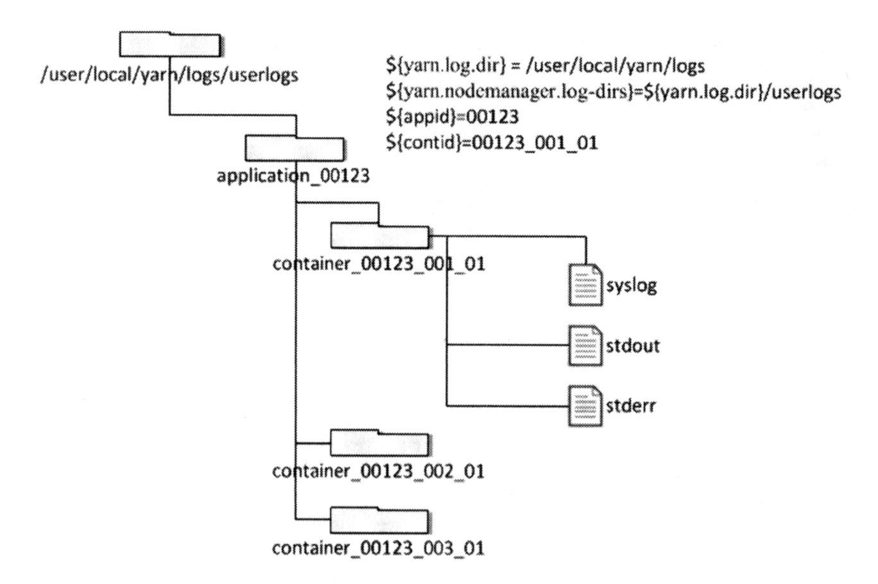


***Figure 9-7.*** *Node-level log hierarchy*

${contid} is determined by the specific implementation of the org.apache.hadoop.yarn.api.records.ContainerId class. Figure 9-7 assumes that it is a concatenation of application ID, container ID, and node ID.

When log aggregation is enabled, all the logs of the application are concatenated into a single file and moved to a single directory based on the following properties (the local logs on the nodes are deleted after the aggregated logs are moved to the HDFS):

- yarn.nodemanager.remote-app-log-dir: The root HDFS directory in which the aggregated log files are exported when log aggregation is enabled. The logs are maintained in a user specific sub-folder. Each user will find the logs for the jobs they executed in the following folder ${ yarn.nodemanager.remote-app-log-dir}/${user.name}/
- yarn.nodemanager.remote-app-log-dir-suffix: The subdirectory that will be created in the folder ${ yarn.nodemanager.remote-app-log-dir}/${user.name}/. This property allows the user to group jobs based on user defined criteria. For example, if a user is running jobs for two separate departments, the user can provide a value for this property as the identifier of the department on whose behalf the job is being executed. This would allow the user to manage the job logs in a department specific sub-folder. ${yarn.nodemanager.remote-app-log-dir-suffix}/${user.name}/ ${ yarn.nodemanager.remote-app-log-dir-suffix}/${appid}/, where ${user.name} is the user running the job and ${appid} is the ID of the application.

This aggregated log file is not a text file; it is a binary file. It can be viewed only through the Job History Manager, which abstracts away the aggregated nature of the logs or the command-line tools (discussed later).

## Log Management Improvements

Hadoop 2.x introduced log-management improvements, including the following:

- Log aggregation is performed in the HDFS. Logs no longer have to be deleted or truncated because they now reside in the distributed file system (DFS), in which it is more tolerable to keep them around for longer durations.
- The logs for all the containers that executed on a given node manager are written to a single log file at a configured location. This log file can be possibly compressed.
- The aggregated logs are maintained at multiple levels. The log directory hierarchy is as follows:
    - Application-level log directory.
    - Per-node log file that contains the logs for all the containers of the application that ran on a given Node.
- Support for command-line tools with fine-grained configurations to access logs in a fine-grained manner (by application ID and container ID). It enables usage of standard tools such as `grep` to search log files from the command line without having to move them to the local disk.
- Web-based UI that is agnostic to log aggregation. Users can view and download logs, whether they are aggregated or not.

## Viewing Logs Using Web-Based UI

The section "Viewing Log Messages in Hadoop MapReduce Jobs" showed that for completed jobs, the logs can be viewed using the `JobHistoryManager` web interface. By default, it can be accessed through the following URL. When we are running in a pseudo-distributed mode, the `job_history_manager` is `localhost`:

```
http://job_history_manager_host:19888
```

As mentioned earlier, whether logs are aggregated or not is transparent to the user. If logs aggregation is enabled, the Job History Manager fetches them from the HDFS. If logs aggregation is disabled, the logs are fetched from the individual nodes by delegating the requests to the Node Managers.

While the job is running, the logs can be viewed through the Application Master web interface that in turn defers it to the Node Manager web interface. The Application Master web interface in turn can be accessed through the "RUNNING" jobs link on the left bar of the Resource Manager web interface. By default, the Resource Manager web interface can be accessed through the following URL.

```
http://resource_manager_host:8088
```

When we are running in a pseudo-distributed mode, the `resource_manager_host` is `localhost`.

## Command-Line Interface

The command-line utility to interact with logs is `$HADOOP_YARN_HOME/bin/yarn logs`.

The usage to obtain the logs for an entire job is as follows:

```
$HADOOP_YARN_HOME/bin/yarn logs -applicationId <application i>
```

The preceding command retrieves logs for the entire application. To retrieve logs only for a container, the usage is the following:

```
$HADOOP_YARN_HOME/bin/yarn logs -applicationId <application id> -containerId <container id>
-nodeAddress <nodemanager ip_add:port>
```

The default user is assumed to be the current user. However, to retrieve logs for another user, the command line is this:

```
$HADOOP_YARN_HOME/bin/yarn logs -applicationId <application id> -appOwner <user id> -containerId
<container id> -nodeAddress <nodemanager ip_add:port>
```

The command-line interface enables regular scripting utilities such as `grep` to be used to search the log files.

## Log Retention

When log aggregation is enabled, the following properties determine how long the logs are retained in the HDFS:

- `yarn.log-aggregation.retain-seconds`: Number of seconds after which the aggregated logs are deleted. A negative number disables this feature, and the logs are maintained forever. The default value is `-1`.
- `yarn.log-aggregation.retain-check-interval-seconds`: Determines the frequency in seconds when the aggregated log-retention checks are applied. If the value is `0` or negative, the value defaults to one-tenth of the value for `yarn.log-aggregation.retain-seconds`. Default value for this property is `-1`. Do not make this value too small or you will spam the name node.
- `yarn.nodemanager.log.retain-seconds`: When log aggregation is disabled, this property determines the duration in seconds for which the logs are retained in the local directories of the node. The default value is 10800, which is 3 hours.

One of the main pain points of Hadoop 1.x is that log retention was a global setting. Unfortunately, it is still a global setting in Hadoop 2.x. The Hadoop cluster is rarely used by one group in the organization; there are often several sponsors, and they have various service-level agreement requirements. It is conceivable for one group to want longer retention on their logs as compared with another group using Hadoop. Hadoop 2.x alleviates the problem because longer retention is more tolerable when the logs are maintained in the HDFS when compared to being maintained on local task nodes, as it used to be in Hadoop 1.x. In my opinion, a number of Hadoop design decisions do not take into account the fact that a Hadoop cluster will have multiple sponsors with different requirements. Hadoop is still in the early stages of being adopted on a mass scale in diverse production environments. As more feedback is gained from such environments, the software will evolve in the latter versions to address the feedback gained.

# Hadoop Cluster Performance Monitoring

Managing the Hadoop cluster usually boils down to managing resources on the cluster nodes. Two common tools used for monitoring large clusters of machines are:

- **Ganglia:** A performance-monitoring tool for distributed systems. Ganglia agents (called Ganglia monitors) collect metrics on individual nodes of a cluster and report back to a Ganglia collector that aggregates these metrics for reporting. Ganglia is designed from the ground up to be integrated into applications running in a cluster to support collection of vital metrics. Hadoop can be integrated with Ganglia, which enables Ganglia to be used as a performance-monitoring tool for Hadoop. The setup instructions are provided at `http://wiki.apache.org/hadoop/GangliaMetrics`.

- **Nagios:** A scheduling and notification engine for monitoring large clusters. While Ganglia will monitor Hadoop specific metrics the health of the cluster is monitored using Nagios. Nagios will monitor and collect metrics from network, disks and CPU's and sends notifications to the system administrators regarding important events.

Today, there is no need to install and configure monitoring tools from scratch, however. Most Hadoop vendors provide excellent web-based monitoring tools for the Hadoop cluster that use these products under the covers while providing a very intuitive web interface to the user. Hadoop 2.x also provides REST-based APIs to develop monitoring utilities. We will discuss both options in the next few sections.

# Using YARN REST APIs

The Hadoop YARN service provides REST APIs that enable the user to access the cluster, nodes, and historical job information conveniently through a familiar set of tools. The syntax of these URIs follows this pattern:

- `http://${service_host}:${service_port}/ws/${version}/${resource_path}`, where `service_host` is the DNS host name or service host IP address.
- `service_port` is the port number in which the service is running.
- `version` is the version of the API. For Hadoop 2.2.0. it is `v1`.
- `resource_path` defines the resource requested. Depending on the resource, it can be either a collection or a singleton.

Currently, only `GET` operations are supported by the API. The web REST APIs respect the same security supported by the web UI. If security is enabled, the user has to authenticate via the mechanism specified.

Two headers are supported by the REST API:

- `Accept`: Accepts values of XML or JSON and determines the response format.
- `Accept-Encoding`: If the value of Gzip is provided, the response is returned compressed in Gzip format.

The response error code should be checked to determine whether the response was successful (an HTTP code of 200 indicates success). In case of an error, the details are contained in the response body. The details include the exception type, the Java class name of the exception, and a detailed error message.

Specific examples of how to use the REST API can be obtained from `http://hadoop.apache.org/docs/r2.2.0/hadoop-yarn/hadoop-yarn-site/WebServicesIntro.html`.

# Managing the Hadoop Cluster Using Vendor Tools

Earlier sections showed the low-level APIs and tools available to monitor Hadoop. Ganglia enables metrics collection, Nagios provides notification capability, and REST APIs allow monitoring jobs through simple web-based APIs.

Major vendors have combined these three capabilities and now provide user-friendly tools to manage Hadoop clusters. Intuitive web interfaces now complement APIs to enable administrators to provide the entire range of activities, from provisioning/extending clusters to managing clusters. Developers can now use these tools to monitor their jobs.

Examples of such web-based monitoring tools include the following:

- **Cloudera Manager:** A Hadoop management application from Cloudera. It comes in two flavors: Cloudera Express, which is free; and Cloudera Enterprise, which is a paid version. It supports remote Hadoop provisioning, deployment, configuration, management, monitoring, and has diagnostic tools. The Enterprise edition comes with advanced management features such as Simple Network Management Protocol (SNMP) support, automated disaster recovery, and support for the Cloudera distribution.
- **Ambari:** Hortonworks has developed the open-source Apache Ambari product, which is software that enables Hadoop management by supporting activities such as provisioning, managing, and monitoring Hadoop clusters. Ambari uses Ganglia, Nagios, and Hadoop REST-based APIs and provides a very intuitive web-based user interface that allows administrators to manage Hadoop clusters.

The following section discusses the Ambari architecture, which is an example of how such tools work. Because Apache Ambari is an open-source product, it is discussed in the appendix of this book as an illustration. You should use the appendix to become familiar with the nature of these third-party monitoring tools. These tools are often maintained by system administrators. This book is written primarily from the perspective of a Hadoop developer, and it is important to be familiar with tools used to monitor Hadoop jobs in an organization.

## Ambari Architecture

The high-level architecture of Ambari is shown in Figure 9-8.

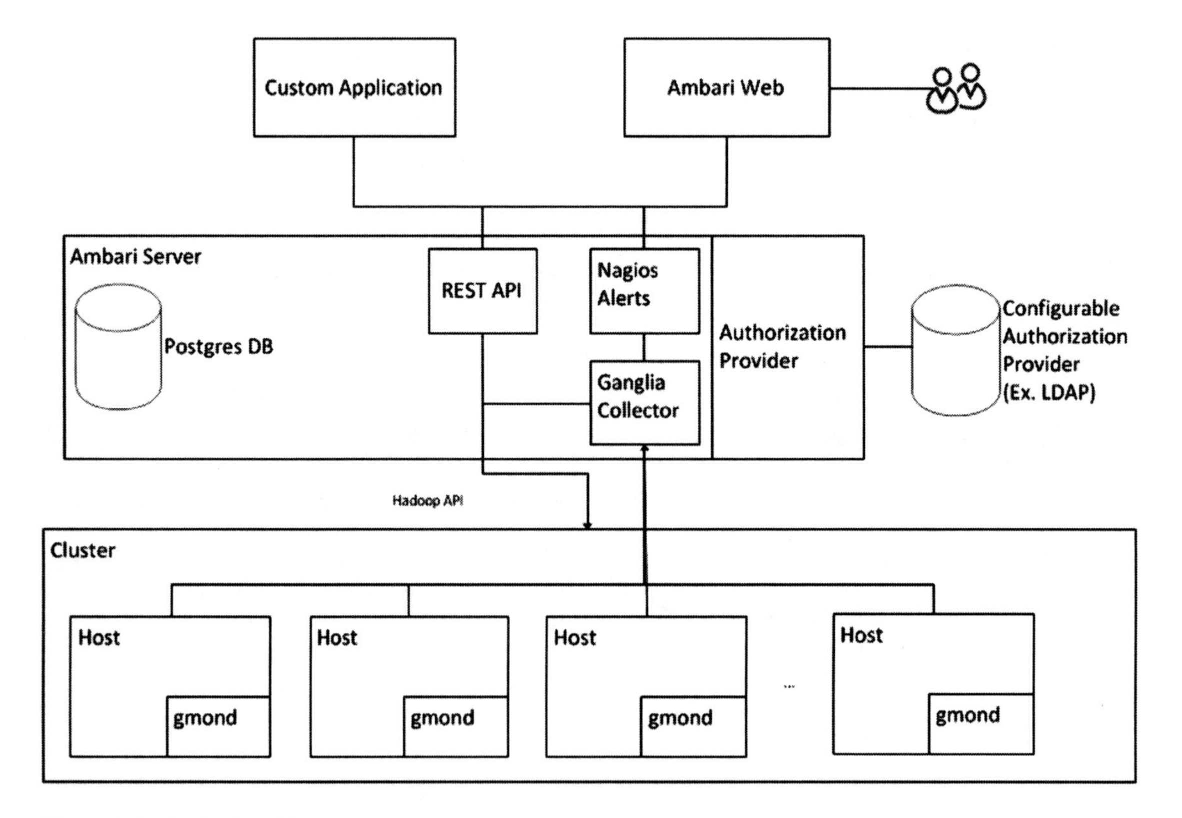


***Figure 9-8.*** *Ambari architecture*

The following are the three main components of Ambari:

- **Ambari agents** are the Ganglia monitors (`gmond`) running on individual hosts that collect metrics.
- **Ambari server** is responsible for collecting information received from each host through the Ganglia monitors running on them. The Ambari server has the following components:
  - **Postgres RDBMS** stores the cluster configurations.
  - **Authorization provider** can integrate with an organizational authentication/authorization provider such as the LDAP service.
  - **Optional Nagios service** supports alerts and notifications.
  - **REST API** integrates with the web front-end Ambari Web. The REST API can also be used by custom applications.
- **Ambari Web** is the web-based interface for the users.

Tools such as Ambari and Cloudera Manager are now widely used to manage Hadoop clusters. They improve your productivity as a developer because you can more quickly track down the source of any problems.

# Summary

Monitoring long and complex Hadoop jobs can be challenging, and this chapter introduced various techniques used to monitor them. From a developer's point of view, the most basic aspect of monitoring a Hadoop job is the ability to write custom logs and read the `syslog`. The first section of this chapter introduced you to the methods used to do this.

Hadoop YARN comes with standard web-based UIs that can be used to monitor the cluster and the jobs running on the cluster. The Resource Managers, Node Managers, Containers, and tasks running in the containers can all be monitored through the web front-ends that are packaged with the Hadoop distribution. You were introduced to these interfaces.

Log management has been significantly improved in Hadoop 2.x. These improvements were based on lessons learned in production from Hadoop 1.x implementations. Features such as log aggregation allow Hadoop users to retain logs as long as they need to. They also provide convenient multifaceted interfaces. Regular users can use web-based UIs to read and download the logs. Power users can utilize command-line tools to analyze the logs using tools such as `grep`, `sort`, and so on.

Hadoop provides basic web-based interfaces to manage and monitor the cluster. Vendors have used these basic interfaces and integrated them with standard cluster management tools such as Ganglia and Nagios to provide complex yet intuitive Hadoop cluster-management tools. Cloudera provides Cloudera Manager; and Hortonworks has developed an open-source, cluster-management tool called Apache Ambari. We discussed Ambari architecture as an illustration of how cluster management is performed in typical production environments. You should be familiar with tools that are in use in your organization. Cluster Management tools, like Cloudera Manager and Apache Ambari, are a powerful yet intuitive way to monitor Hadoop jobs.

Monitoring Hadoop jobs is a large part of what a Hadoop developer does, even in the development phase. It is a tedious and painstaking process to debug a complex job, and tools such as Cloudera Manager and Apache Ambari make the task easier.

When a system is in production, these tools enable troubleshooting. The distributed nature of Hadoop ensures that there will be many challenges in production. Jobs fail sometimes, and tools such as Cloudera Manager and Apache Ambari can help you perform root cause analysis of such events with ease.

CHAPTER 10

# Data Warehousing Using Hadoop

The Hadoop platform supports several data warehousing solutions, including Apache Hive, Impala, and Shark. These solutions are conceptually similar to relational databases at much larger scale but differ in their implementation and usage model. Relational databases are often used in transactional systems in which single row inserts, updates, and deletes must be executed atomically. Efficient indexing and referential integrity with primary/foreign keys allow modern relational databases to find records quickly and guarantee that all data satisfies a strict schema. Relational databases try to avoid full table scans whenever possible because I/O bandwidth is limited and tends to be the bottleneck in these systems.

Data warehousing solutions are designed to solve various problems. Hadoop ecosystem solutions achieve their performance by using a cluster of many networked nodes in which I/O bandwidth is considerably higher. The primary advantage of clusters is the capability to scale to very large datasets where conventional relational databases fail to perform. High-speed atomic operations are difficult to implement in a distributed system, so cluster-based solutions do not support transactions or even single row inserts/updates/deletes. Data warehousing solutions are preferred for analytics over very large datasets. In these usage models, full table scans are common. Indexing, transactions, and related features are not needed in these cases.

Hive, Impala, and Shark share many common features; in fact, they are designed to interoperate and access the same data in the HDFS. From a user's perspective, they all work the same way. The primary difference is in their implementation and the types of workloads they support, so most of this chapter will focus on Hive. Implementation differences between Hive, Impala, and Shark are described later in the chapter.

## Apache Hive

Hive is an open-source warehousing solution that runs on top of the Hadoop platform. Like relational data stores, Hive supports familiar concepts such as databases, tables, and views. It can be accessed interactively through a command-line interface or programmatically through Java using Java Database Connectivity (JDBC).

Hive executes SQL-like queries known as HiveQL, which can perform selects, inserts, joins, subqueries, and even windowing and analytic functions. Hive queries are parsed and compiled into MapReduce jobs that execute just like Java MapReduce jobs. The Hive compiler converts simple selects into Map-only jobs. Complex queries that include multiple table joins and subqueries are converted into multistage MapReduce jobs.

HiveQL can be extended through UDFs and user-defined aggregate functions (UDAFs). In fact, many Hive functions are implemented as UDFs or UDAFs. Source code for these functions is available and not difficult to follow and build on.

Although HiveQL looks like SQL, it is not the same. SQL is a mature language with a well-defined standard that is widely implemented. The nature of MapReduce and commodity clusters constrains what HiveQL can do well. Complex multitable joins can require many MapReduce stages whose startup time and intermediate data writes to disk make Hive appear sluggish for these queries.

Hive is not a transactional data store; it does not support updates, deletes, or single row inserts. Non-equal joins are not supported because they are difficult to implement with MapReduce. Hive is not always the most efficient solution. It has minimal knowledge of the data, so it can't exploit optimizations that an experienced MapReduce developer might implement. Even with these limitations, Hive solves a significant class of problems using familiar query syntax. Hive is best suited for long-running batch queries for which reliability is critical and startup time is not important. If a problem can be expressed efficiently in Hive, it is often preferred over hand-coded MapReduce jobs.

---

■ **Note** The Hive architecture, capabilities, and usage models are described in this chapter, but specific HiveQL syntax and other details are not discussed. That information is available online in the Hive language manual, which is the best and most up-to-date reference for language specifics. The Hive manual can be found here: `http://hive.apache.org/`.

---

## Installing Hive

Hive can be used in conjunction with a fully distributed Hadoop cluster or in local mode. In local mode, data is stored in the local filesystem, and MapReduce jobs are executed locally. Local mode, which can be useful for querying small datasets, runs on a single node that does not scale to large datasets.

Hive runs on Linux and Mac OS X, and requires Java 1.6. Basic installation is straightforward; the steps are as follows:

1. Download a release from `hive.apache.org`.
2. Unpack the release into a suitable directory.
3. Set the `HIVE_HOME` environment variable to the installation directory.
4. Add `$HIVE_HOME/bin` to the `PATH` environment variable.

At this point, Hive operates in local mode, and the built-in Derby database is used as the Hive metastore. The Hive command-line interpreter (CLI) can be started by using the Hive command from a UNIX shell prompt, which creates a `metastore_db` directory in the current directory for the metastore.

To run Hive queries on a Hadoop cluster, the following additional configuration is necessary:

1. Hadoop must be in the `PATH` or `HADOOP_HOME` environment variable set to the Hadoop installation directory.
2. Create `/tmp` and `/user/hive/warehouse` directories in the HDFS with g+w mode (use Hadoop `fs -chmod g+w`).
3. Configure the metastore for multiple users.

Detailed installation details are available on the Apache Hive home page.

## Hive Architecture

The Hive architecture is shown in Figure 10-1.

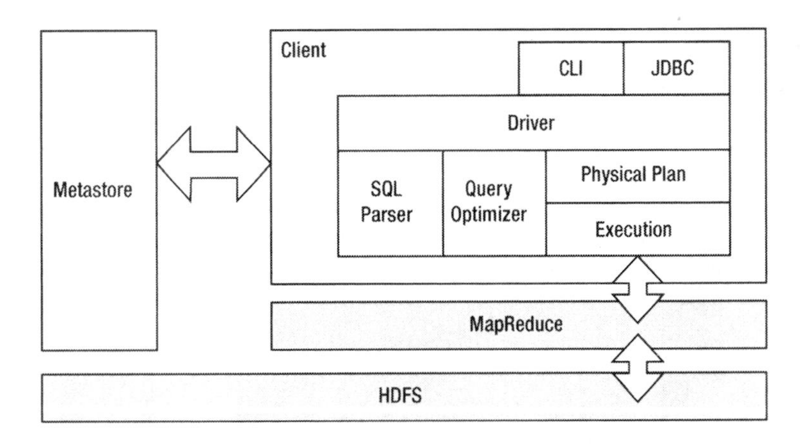


***Figure 10-1.*** *HIVE architecture*

## Metastore

Hive requires a metastore that contains describing objects such as databases, tables, columns, views, and so on. The metastore is a central repository of Hive metadata. HiveQL Data Definition Language (DDL) statements such as `CREATE TABLE` update the metastore behind the scenes.

By default, Hive uses the built-in Derby database for the metastore. It is adequate for local testing and development, but MySQL is often used for a production cluster with multiple users.

## Compiler Basics

HiveQL statements issued from a client are ultimately turned into one or more MapReduce jobs that are executed within Hadoop. Simple `SELECT` statements without joins or aggregations can be implemented as a single stage Map-only job. Aggregation queries using a `group by` clause can be implemented as a single stage MapReduce job. Complex queries with large multitable joins require multiple MapReduce jobs executed in sequence.

The HiveQL compiler operates in multiple phases. In the first phase, a query is parsed and converted into an abstract syntax tree. The syntax tree is passed to the semantic analyzer, in which column names and types are verified using the information from the metastore. The semantic analyzer generates an internal representation that is sent to the logical plan generator, in which a tree of logical operations is created. The logical plan is optimized over several passes and sent to the physical plan generator. The physical plan generator creates a DAG that is used to generate MapReduce jobs.

Achieving high performance requires some knowledge of how Hive translates HiveQL into MapReduce jobs. Improperly defined tables or queries can result in unacceptably poor performance. More details on the compiler and how HiveQL statements are converted into MapReduce jobs are covered later in this chapter.

## Hive Concepts

Hive supports concepts similar to conventional databases, such as databases, tables, and views. Hive also adds features such as partitions and buckets to manage large-scale data. There is also some support for indexes, although not as sophisticated as with conventional databases.

## Databases

A database in Hive is the top-level container for tables and views, and it can be created using the CREATE DATABASE statement. Once created, a database can be selected for use with the use command. A table in a database can be referenced using the database name as a prefix, very much like SQL. If no database is used, the default database, default, is used.

The following examples illustrate creating, using, and dropping a database. Like SQL, HiveQL is not case-sensitive, and lines prefixed with hyphens (--) are comments. The trailing semicolon is required by the Hive command-line interface, but not allowed when executed through JDBC.

```
create database mydb;

-- if not exists clause is optional
create database if not exists mydb;

-- use the database that was just created
use mydb;

-- if exists is optional
drop database if exists mydb;
```

## Tables

Hive supports two types of tables: managed and external. Managed tables are stored in the Hive warehouse, which is an HDFS directory under /user/hive/warehouse. Managed tables can be populated using the LOAD or INSERT INTO AS SELECT statements. Dropping a managed table deletes the table structure as well as any data in the table. A managed table is created when the EXTERNAL keyword is not used. Data storage details such as column delimiters and file formats can be specified if needed. By default, the \u0001 (Control+A) character is used as the column delimiter.

Here is an example that creates a managed table and loads it with data from a directory in the HDFS:

```
create table videos (producer string, title string, category string, year int) row format
delimited fields terminated by ",";
LOAD DATA LOCAL INPATH 'data/videos.txt' OVERWRITE INTO TABLE videos;
```

External tables function just like managed tables, but the location of the underlying data is specified at creation time. The location can be any HDFS file or directory that is accessible by the client. The external table location must be an absolute HDFS path when used with a Hadoop cluster.

Here is an example of an external table:

```
CREATE EXTERNAL TABLE videos_ex (producer string, title string, category string, year int) ROW
FORMAT DELIMITED FIELDS TERMINATED BY "," LOCATION '/user/hadoop/external/videos_ex';
```

External tables are a convenient way of accessing or modifying structured data that already exists in the HDFS. Selecting from an external table returns data in the underlying HDFS location. Any changes in the underlying data from another process, a MapReduce job for example, immediately appear in the external table. Similarly, any data that is inserted into an external table appears in the corresponding HDFS location.

Dropping an external table is different from a managed table. When an external table is dropped, information about the table is deleted from the metastore, but the underlying data in the HDFS remains. To modify the schema for an external table, it can be safely dropped and re-created as needed. If the data types for a column do not match the underlying data in an HDFS file, Hive returns a NULL for that column.

Both managed and external tables can be created as a SELECT statement. In this case, the table is immediately populated with data from the results of a select query.

Hive supports several file formats, including text files, sequence files, and Gzip files. Note that text and sequence files can span multiple HDFS blocks, whereas Gzip files cannot. If text file storage is used, each line in a file represents a row. If sequence files are used, each value represents a row, and the key is ignored. Columns are delimited by the Control+A character (\u0001) by default. Alternate delimiters can be specified when a table is created.

The online Hive language manual describes the create table syntax, options, and properties in great detail.

## Views

Hive views are logical objects with no underlying storage. The data for a view is the result of a select query. The select query is executed only when the view is selected. Note that views are read-only; it is not possible to load views or insert into them.

Here is a simple example of creating and using a view:

```
create view group_by_year_vw
as
select year, count(*) as video_ct from videos group by year;

select * from group_by_year_vw;
```

Views are commonly used as an abstraction layer to publish data to external clients. A view can be created with stable public column names and data types. The view allows private tables to change structure without affecting downstream consumers.

## Partitions

Very large datasets can take a considerable amount of time to process. In many cases, data can be divided or partitioned to reduce the amount of data that must be scanned, which improves performance dramatically. Data is partitioned based on one or more columns. Data is usually partitioned over time, geographic region, or some business dimension. The video table, for example, could be partitioned by movie category or rating, although production year is likely to yield a more uniform distribution. To achieve consistent performance, partition columns are chosen so data is more or less evenly distributed. It is best to choose a partition that is usually part of a where clause.

Hive partitions are implemented as subdirectories in the HDFS. Each subdirectory name contains the name of the partitioned columns and the value for each column. This does impose a limitation because the HDFS is not designed to support a very large number of subdirectories. It is important to estimate the number of expected partitions in a table to avoid serious problems as more partitions are added.

Consider the video table partitioned by production year. The DDL for this table looks like this:

```
CREATE EXTERNAL TABLE videos_part (producer string, title string, category
string)
PARTITIONED BY (year int)
ROW FORMAT DELIMITED
FIELDS TERMINATED BY "," LOCATION
'/user/hadoop/external/hive/partitioned';
```

Partition columns are logically part of the table, but cannot be included in the standard column list. Partition columns appear at the end of the standard column list:

```
hive> describe videos_part;
OK
producer                string                  None
title                   string                  None
category                string                  None
year                    int                     None
```

Files in the HDFS representing the partitioned data might look like these:

```
/user/hadoop/external/hive/partitioned/year=1991/part-r-00001
/user/hadoop/external/hive/partitioned/year=1991/part-r-00002
...
/user/hadoop/external/hive/partitioned/year=1992/part-r-00001
...
/user/hadoop/external/hive/partitioned/year=1993/part-r-00001
...
```

Notice that the partitions are subdirectories under the directory specified by the external table location. Each partition is of the form *column=value.*

A Hive query often contains the partitioned column as a condition. In that case, the underlying MapReduce job can restrict the scan to specific subdirectories in the HDFS. The HDFS directory structure is effectively used as an index.

## Buckets

Data in partitions can be further divided into buckets, which are used to sample data based on a hash of a column other than partition columns. Buckets are preferred when the number of partitions might be large enough to overwhelm the filesystem. The number of buckets is fixed. Hive distributes rows across buckets by invoking a hash function on the column used for bucketing as follows:

```
hash_function(bucket_column)%num_of_buckets
```

The choice of hash function depends on the data type of the column used for bucketing. Apart from sampling, bucketing can also be used to implement very efficient map-side joins. We discussed map-side joins in Chapter 6 and Chapter 7.

Consider the following example, in which a partitioned table is subdivided into 100 buckets:

```
CREATE EXTERNAL TABLE videos_b (producer string, title string, category
string)
PARTITIONED BY (year int)
CLUSTERED BY (title) INTO 100 BUCKETS;
```

Next we populate the table as follows,

```
set hive.enforce.bucketing = true;
FROM videos
INSERT OVERWRITE TABLE videos_b
PARTITION (year=1999)
SELECT producer, title, string WHERE year=2009;
```

If we do not invoke the `set hive.enforce.bucketing = true` property, we will need to explicitly set the number of reducers via the call `set mapred.reduce.tasks=100`. In addition to that call, the `INSERT` query would need a `CLUSTER BY` clause after the `SELECT` statement as follows:

```
set mapred.reduce.tasks = 100;
FROM videos
INSERT OVERWRITE TABLE videos_b
PARTITION (year=1999)
SELECT producer, title, string WHERE year=2009 CLUSTER BY title;
```

Refer to the Hive documentation for more details on bucketing.

## Indexes

Hive supports indexes, but not in the same way that relational databases do. Hive does not support primary or foreign keys, for example. Indexes can be created on table columns to improve some operations, such as reducing the number of blocks read in underlying MapReduce jobs. Indexes are sometimes preferred over partitions when a large number of partition values are expected. Refer to the HIVE documentation for more details on Indexing of HIVE tables.Keep in mind that Hive is not a transactional database in which individual rows are selected, updated, or deleted. Those types of operations require efficient indexing to achieve high performance. Hive is a batch-processing tool designed to scan large volumes of data at high speed using many processing nodes. Relational databases are typically designed to run on single machines that are I/O-bound. Hive achieves performance through parallelism, so full table scans are expected rather than discouraged as with relational databases.

## Serializer/Deserializer interface

In this section we will briefly review how data stored in HDFS is managed by HIVE using the table abstraction. SerDe is short for Serializer/Deserializer. SerDe allows even a highly customized data format to be supported via a familiar Database Table abstraction. A SerDe is an implementation of the interface `org.apache.hadoop.hive.serde2.SerDe`.

A good description of how SerDe's work is provided at the following link:

```
https://cwiki.apache.org/confluence/display/Hive/SerDe
```

A SerDe implementation class defined for the table (in the `CREATE TABLE` call), tells HIVE how raw data record should be processed. The Deserializer portion of the SerDe takes the raw data and translates it into a Java instance that HIVE can work with. Likewise, the Serializer portion of the SerDe converts an object that HIVE is using into raw bytes. Serializers are used to execute the `INSERT` statement. Deserializers are used to execute the `SELECT` statement. If your data is in a custom format and you want to use HIVE to process it, you can implement your own SerDe. An excellent introduction to developing a custom `SerDe` implementation can be found at the following link:

```
http://blog.cloudera.com/blog/2012/12/how-to-use-a-serde-in-apache-hive/
```

# HiveQL Compiler Details

Compiler operation can be observed using the EXPLAIN statement, which accepts a query and produces the following output:

- A syntax tree for the query
- Dependencies between stages of the plan
- A description of each of the stages

Here are a few examples of the EXPLAIN statement. First is a DDL statement that creates a managed table:

```
hive> explain create table videos (producer string, title string, category string, year int) row
format delimited fields terminated by ",";
OK
ABSTRACT SYNTAX TREE:
  (TOK_CREATETABLE (TOK_TABNAME videos) TOK_LIKETABLE (TOK_TABCOLLIST (TOK_TABCOL producer TOK_
STRING) (TOK_TABCOL title TOK_STRING) (TOK_TABCOL category TOK_STRING) (TOK_TABCOL year TOK_INT))
(TOK_TABLEROWFORMAT (TOK_SERDEPROPS (TOK_TABLEROWFORMATFIELD ","))))

STAGE DEPENDENCIES:
  Stage-0 is a root stage

STAGE PLANS:
  Stage: Stage-0
      Create Table Operator:
        Create Table
          columns: producer string, title string, category string, year int
          field delimiter: ,
          if not exists: false
          input format: org.apache.hadoop.mapred.TextInputFormat
          # buckets: -1
          output format: org.apache.hadoop.hive.ql.io.IgnoreKeyTextOutputFormat
          name: videos
          isExternal: false
```

The syntax tree is displayed in a format resembling LISP S-expressions. The nodes of the tree contain a value and a list of children. The values include tokens representing keywords, such as TOK_CREATETABLE; and literals, such as the table name (videos) or columns (producer, title, and so on).

This statement executes in a single root stage named Stage-0. Because there is only one stage, there are no dependencies.

The stage description is straightforward: create a table given a set of columns with data types. Notice that a number of properties, such as I/O formats, are displayed. No specific properties were supplied in this statement, so default values were selected by the compiler.

Next is a basic SELECT statement:

```
hive> explain select * from videos;
OK
ABSTRACT SYNTAX TREE:
  (TOK_QUERY (TOK_FROM (TOK_TABREF (TOK_TABNAME videos))) (TOK_INSERT (TOK_DESTINATION (TOK_DIR
TOK_TMP_FILE)) (TOK_SELECT (TOK_SELEXPR TOK_ALLCOLREF))))

STAGE DEPENDENCIES:
  Stage-0 is a root stage

STAGE PLANS:
  Stage: Stage-0
    Fetch Operator
      limit: -1
      Processor Tree:
        TableScan
          alias: videos
```

```
        Select Operator
          expressions:
                expr: producer
                type: string
                expr: title
                type: string
                expr: category
                type: string
                expr: year
                type: int
          outputColumnNames: _col0, _col1, _col2, _col3
          ListSink
```

This example also contains a single root stage. The stage plan contains a table scan containing a selection of all columns in the table. Each of the columns and their types has been identified from the metastore. Notice the alias and output column names.

Now for a slightly more complex select query:

```
hive> explain select year, count(*) from videos where category = 'Comedy' group by year limit 10;
OK
ABSTRACT SYNTAX TREE:
  (TOK_QUERY (TOK_FROM (TOK_TABREF (TOK_TABNAME videos))) (TOK_INSERT (TOK_DESTINATION (TOK_DIR
TOK_TMP_FILE)) (TOK_SELECT (TOK_SELEXPR (TOK_TABLE_OR_COL year)) (TOK_SELEXPR (TOK_FUNCTIONSTAR
count))) (TOK_WHERE (= (TOK_TABLE_OR_COL category) 'Comedy')) (TOK_GROUPBY (TOK_TABLE_OR_COL year))
(TOK_LIMIT 10)))

STAGE DEPENDENCIES:
  Stage-1 is a root stage
  Stage-0 is a root stage

STAGE PLANS:
  Stage: Stage-1
    Map Reduce
      Alias -> Map Operator Tree:
        videos
          TableScan
            alias: videos
            Filter Operator
              predicate:
                  expr: (category = 'Comedy')
                  type: boolean
              Select Operator
                expressions:
                      expr: year
                      type: int
                outputColumnNames: year
                Group By Operator
                  aggregations:
                        expr: count()
                  bucketGroup: false
```

```
                keys:
                      expr: year
                      type: int
                mode: hash
                outputColumnNames: _col0, _col1
                Reduce Output Operator
                  key expressions:
                        expr: _col0
                        type: int
                  sort order: +
                  Map-reduce partition columns:
                        expr: _col0
                        type: int
                  tag: -1
                  value expressions:
                        expr: _col1
                        type: bigint
      Reduce Operator Tree:
        Group By Operator
          aggregations:
                expr: count(VALUE._col0)
          bucketGroup: false
          keys:
                expr: KEY._col0
                type: int
          mode: mergepartial
          outputColumnNames: _col0, _col1
          Select Operator
            expressions:
                  expr: _col0
                  type: int
                  expr: _col1
                  type: bigint
            outputColumnNames: _col0, _col1
            Limit
              File Output Operator
                compressed: false
                GlobalTableId: 0
                table:
                    input format: org.apache.hadoop.mapred.TextInputFormat
                    output format: org.apache.hadoop.hive.ql.io.HiveIgnoreKeyTextOutputFormat

  Stage: Stage-0
    Fetch Operator
      limit: 10
```

This query contains two stages. Although it looks complicated, it is actually straightforward. Stage-1 is a standard aggregation that is implemented as a single stage MapReduce job. In the map phase, the category is used to filter rows, and the year is used to group rows. In the reduce phase, the count function aggregates values. Stage-0 simply limits the output to 10 rows.

More complex queries with multiple joins and subqueries are implemented as dependent multistage jobs. (To get an idea of what these jobs look like, you can visit `http://ysmart.cse.ohio-state.edu`.) YSmart is an SQL-to-MapReduce translator. Given a schema and an SQL query, YSmart generates Java MapReduce source code that can be compiled and executed on a Hadoop cluster.

YSmart is available as source code or as an online version. YSmart generates optimized MapReduce jobs that perform well, and the YSmart code base has been merged with Apache Hive. The YSmart home page contains a number of papers and articles describing the problem of translating SQL to MapReduce and how it is implemented within YSmart.

Figure 10-2 shows the YSmart online version.

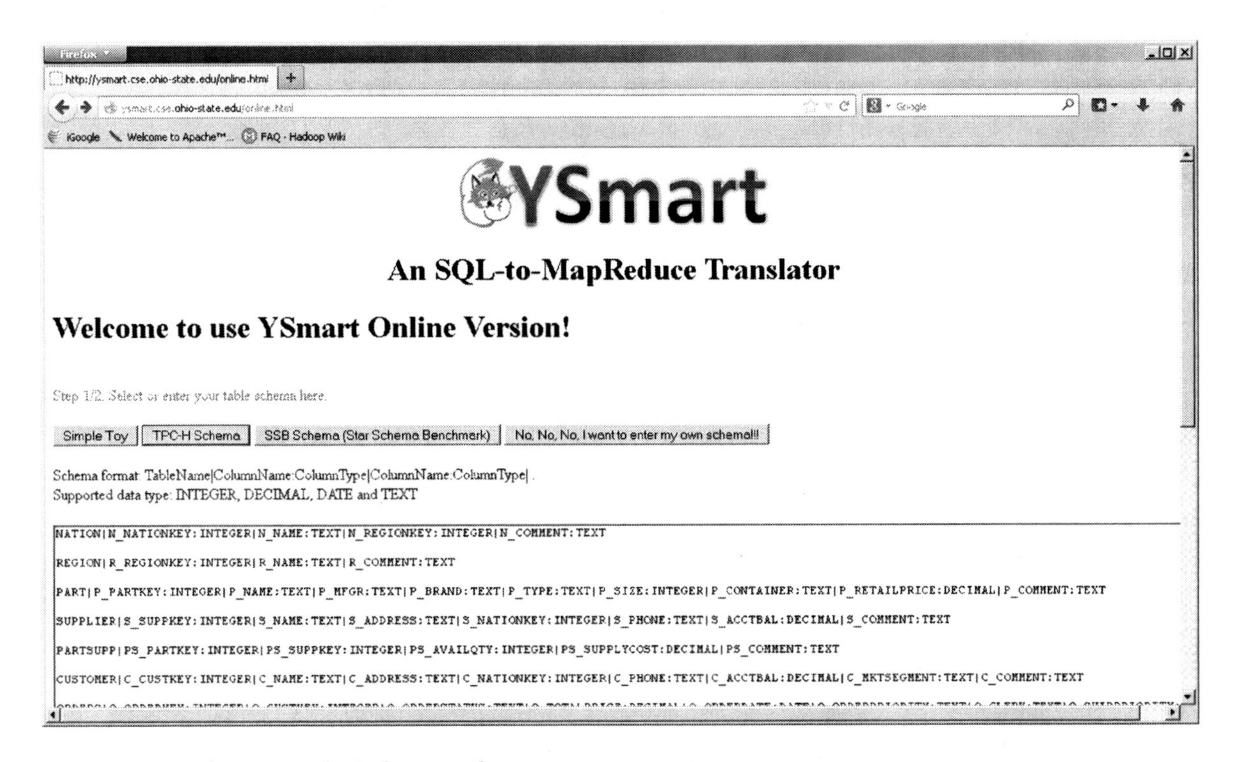


***Figure 10-2.** Only SQL-to-MapReduce translator*

## Data Definition Language

Hive supports many of the DDL found in SQL. The kinds of statements include the following:

- `CREATE DATABASE/SCHEMA, TABLE, VIEW, FUNCTION, INDEX`
- `DROP DATABASE/SCHEMA, TABLE, VIEW, INDEX`
- `TRUNCATE TABLE`
- `ALTER DATABASE/SCHEMA, TABLE, VIEW`
- `MSCK REPAIR TABLE (or ALTER TABLE RECOVER PARTITIONS)`

- SHOW DATABASES/SCHEMAS, TABLES, TBLPROPERTIES, PARTITIONS, FUNCTIONS, INDEX[ES], COLUMNS, CREATE TABLE
- DESCRIBE DATABASE, table_name, view_name

Many of these statements are self-explanatory and are not described here. These statements are well-documented in the online Hive language manual.

## Data Manipulation Language

Hive Data Manipulation Language (DML) supports many of the same operations found in SQL. Specifically, Hive DML can perform the following:

- Filter rows from a table using a `where` clause
- Select expressions using columns from tables or subqueries
- Join multiple tables using an equi-join
- Union all rows of tables or subqueries
- Evaluate aggregations on multiple "group by" columns
- Store the results of a query into another table
- Export the contents of a table to a local or HDFS directory

Hive supports both primitive and complex data types. The following primitive data types are supported:

- **Numeric Types**
  - `TINYINT` (1-byte signed integer)
  - `SMALLINT` (2-byte signed integer)
  - `INT` (4-byte signed integer)
  - `BIGINT` (8-byte signed integer)
  - `FLOAT` (4-byte single precision floating point number)
  - `DOUBLE` (8-byte double precision floating point number)
  - `DECIMAL` (user-definable precision and scale)
- **Date/Time Types**
  - `TIMESTAMP` (traditional UNIX timestamp with optional nanosecond precision)
  - `DATE` (day of year in the form *YYYY-MM-DD* without time)
- **String Types**
  - `STRING` (no length specified)
  - `VARCHAR` (1 to 65535 characters)
  - `CHAR` (1 to 255 characters, fixed-length, right padded with spaces)
- **Other Types**
  - `BOOLEAN`
  - `BINARY`

Complex types can be built from primitive types and other complex types. Table 10-1 lists the complex types that are supported.

***Table 10-1.*** *Complex Data Types*

| Type | Description |
|---|---|
| `Struct` | Collection of heterogeneous types |
| `Array` | Collection of homogeneous types |
| `Map` | Collection of name-value pairs |

## Language Limitations

Hive does not support non-equal joins, which are difficult to implement as MapReduce jobs. Hive supports `UNION ALL` exclusively, so duplicate rows are possible. This behavior differs from relational databases, which commonly support both `UNION` and `UNION ALL`.

Hive is not a transactional database, so it has no support for transactions. Single row inserts, deletes, or updates are not supported. Data can be effectively inserted using `LOAD` statements, external tables, or the result of a select query.

# External Interfaces

Hive can be accessed through several interfaces, including the CLI, Beeline, and JDBC. The CLI and Beeline are interactive user interfaces that are functionally similar, but differ in some of their syntax and implementation. JDBC is a programmatic interface very similar to relational database access.

## CLI

The Hive CLI is started using the `Hive` command from a UNIX shell prompt. The CLI supports all Hive statements terminated by a semicolon (;). A number of properties can be set, such as the number of Reducers used by the underlying MapReduce jobs. These are well-documented in the online Hive language manual. Other properties that are specific to Hive and are useful with the CLI are the following:

- `hive.cli.print.header`: When set to `true`, prints the names of the columns in the query output. By default, this value is `false`, so column names are not printed. To enable this feature, enter this command:

```
hive> set hive.cli.print.header=true;
```

- `hive.cli.print.current.db`: When set to `true`, prints the name of the current database. By default, this value is set to `false`. To enable this feature, enter this command:

```
hive> set hive.cli.print.current.db =true;
hive (default)>
```

## Beeline

Beeline is an alternative to the standard command-line interface. It connects to Hive using JDBC and is based on the open-source SQLLine project. Beeline works very much like the Hive CLI, but requires explicit connection to interact with Hive:

```
$ beeline
Beeline version 0.11.0 by Apache Hive
beeline> !connect jdbc:hive:// nouser nopassword
```

The JDBC URL for local mode is jdbc:hive://. When configured with a cluster, the JDBC URL is of the form jdbc:hive://*<hostname>*:*<port>*, where *<hostname>* is the HiveServer host name, and *<port>* is the configured port (10000 by default).

At this point, any Hive statement can be executed, just like the CLI.

## JDBC

Java clients can connect to Hive using the supplied JDBC driver. The connection steps are the same as in any JDBC-compliant database. First, load the driver and then obtain a connection. The JDBC driver name is org.apache.hadoop.hive.jdbc.HiveDriver. The JDBC URL for local mode is jdbc:hive://. When configured with a cluster, the JDBC URL is of the form jdbc:hive://*<hostname>*:*<port>*, where *<hostname>* is the HiveServer host name, and *<port>* is the configured port (10000 by default).

The following is an example that connects to Hive in local mode and issues several queries:

```
import java.sql.Connection;
import java.sql.DriverManager;
import java.sql.ResultSet;
import java.sql.ResultSetMetaData;
import java.sql.Statement;

import org.apache.log4j.Level;
import org.apache.log4j.LogManager;

public class HiveJdbcClient {
    private static String driverName = "org.apache.hadoop.hive.jdbc.HiveDriver";

    public static void main(String[] args) throws Exception {
        LogManager.getRootLogger().setLevel(Level.ERROR);
        Class.forName(driverName);
        Connection con = DriverManager.getConnection(
                "jdbc:hive://", "", "");
        Statement stmt = con.createStatement();
        stmt.executeQuery("drop table videos_ex");
        ResultSet res = stmt.executeQuery("CREATE EXTERNAL TABLE videos_ex " +
                "(producer string, title string, category string, year int) " +
                "ROW FORMAT DELIMITED FIELDS TERMINATED BY \",\" LOCATION " +
                "'/home/madhu/external/videos_ex/data'");
        // show tables
        String sql = "show tables";
        System.out.println("Running: " + sql);
        res = stmt.executeQuery(sql);
```

```
        if (res.next()) {
            System.out.println(res.getString(1));
        }
        // describe table
        sql = "describe videos_ex";
        System.out.println("Running: " + sql);
        res = stmt.executeQuery(sql);
        while (res.next()) {
            System.out.println(res.getString(1) + "\t" + res.getString(2));
        }

        // select * query
        sql = "select * from videos_ex";
        System.out.println("Running: " + sql);
        res = stmt.executeQuery(sql);
        ResultSetMetaData rsmd = res.getMetaData();
        int ncols = rsmd.getColumnCount();
        for (int i = 0; i < ncols; i++) {
            System.out.print(rsmd.getColumnLabel(i+1));
            System.out.print("\t");
        }
        System.out.println();
        while (res.next()) {
            for (int i = 0; i < ncols; i++) {
                System.out.print(res.getString(i+1));
                System.out.print("\t");
            }
            System.out.println();
        }

        // regular hive query
        sql = "select count(1) from videos_ex";
        System.out.println("Running: " + sql);
        res = stmt.executeQuery(sql);
        if (res.next()) {
            System.out.println("Number of rows: " + res.getString(1));
        }
    }
}
```

## Hive Scripts

Hive queries can be expected in sequence in a script. Queries in the script can include parameters that can be substituted with command-line values. Here is an example:

```
LOAD DATA LOCAL INPATH '${hiveconf:src}' OVERWRITE INTO TABLE videos;

hive -hiveconf src=data/videos.txt -f hiveql2.txt
```

## Performance

Similar to Java MapReduce jobs, Hive is capable of efficiently processing very large volumes of data. Although Hive scales up very well, it does not scale down as well. Even the simplest of queries against small datasets can take minutes to execute. Execution time for small datasets is largely dominated by startup time for each MapReduce job. This high latency can appear sluggish when compared with mature relational databases, such as MySQL or Oracle. Remember that Hive is designed as a batch-processing platform, not an interactive query tool:

The number of reducers used is critical when working with large datasets. The number of reducers can be set in configuration files, or set using the following command:

```
set mapred.reduce.tasks=<number>
```

When joining two or more tables, Hive tries to buffer data in the earlier tables and stream rows through the last table. For this reason, it is best to order joined tables from smallest to largest. When the smaller tables can fit in memory, Hive can improve performance considerably by performing a map-side join.

## MapReduce Integration

Hive is powerful and relatively easy to use, but it cannot solve all problems, nor can it always solve problems efficiently. In many cases, MapReduce might be required to solve complex problems or improve performance. Hive is convenient for structured tabular data. Joining large datasets is far easier in Hive. MapReduce is required when analyzing unstructured data or exploiting known regularities in a dataset.

Rather than solving a large problem using all MapReduce jobs or all Hive scripts, it is possible to use both methods together. MapReduce jobs can read Hive external tables and partitions, or write files that can be accessed in Hive. In a typical data analysis pipeline, the front end might be a MapReduce job that accepts unstructured or irregular data and converts it into structured tabular data suitable for Hive. Intermediate stages could contain Hive or MapReduce jobs, depending on the nature of the data and performance requirements. The output stage could be exposed as a Hive view. Hive views act as an abstraction layer to external systems.

### Reading from Hive External Tables

Reading from Hive external tables through MapReduce is generally straightforward. Hive tables appear as a directory of files in the HDFS. The `InputFormat` to the mapper must match the match the Hive table storage format. For text file storage, `TextInputFormat` can be used. For sequence file storage, `SequenceInputFormat` can be used.

If sequence files are used, the key is ignored by Hive. The value is used as the row data.

### Writing to Hive External Tables

Writing to Hive external tables from MapReduce is accomplished by setting the output directory of a MapReduce job to the same HDFS location as the Hive external table. The external table file format must match that of the MapReduce job. Remember that Hive uses \0001 (Control+A) as the default column delimiter, unless otherwise specified in the `CREATE TABLE` statement.

## Creating Partitions

Hive partitions are implemented as subdirectories. Each partition is of the form *<column>=<value>*. Here is an example with a table with a year as the partition column:

```
create external table videos_part (producer string, title string, category string) partitioned by
(year int) row format delimited location '/user/hadoop/videos_part';
```

The directory structure containing several partitions might look like this:

```
/user/hadoop/videos_part/year=1991/...
/user/hadoop/videos_part/year=1992/...
/user/hadoop/videos_part/year=1993/...
```

The partition subdirectories can be conveniently created in MapReduce or a Map-only job using `MultipleOutputs`. The following is a Map-only job that dynamically writes Hive-compatible partitions to the HDFS:

```
public class SplitterMapper extends Mapper<Text,VideoRecording,Text,Text> {
    private Text outputKey = new Text();
    private MultipleOutputs<Text,Text> mos;

    @Override
    protected void setup(Context context) throws IOException,
            InterruptedException {
        mos = new MultipleOutputs<>(context);
    }

    @Override
    protected void map(Text title, VideoRecording video, Context context)
            throws IOException, InterruptedException {
        for (String actor : video.getActors()) {
            outputKey.set(actor);
            String baseOutputPath = String.format("year=%d/prefix", video.getYearReleased());
            mos.write("text", outputKey, title, baseOutputPath);
        }
    }

    @Override
    protected void cleanup(Context context)
            throws IOException, InterruptedException {
        mos.close();
    }
}
```

The job can be configured like this:

```
Configuration conf = new Configuration();
Job job = new Job(conf, "splitterjob");
job.setJarByClass(SplitterJob.class);
job.setInputFormatClass(SequenceFileInputFormat.class);
job.setMapperClass(SplitterMapper.class);
```

```
FileInputFormat.addInputPath(job, new Path(args[0]));
FileOutputFormat.setOutputPath(job, new Path(args[1]));

MultipleOutputs.addNamedOutput(job, "text", TextOutputFormat.class, Text.class, Text.class);

job.setNumReduceTasks(0);
job.waitForCompletion(true);
```

Data written to a partitioned external table does not automatically appear in Hive. It is necessary to either explicitly add each partition to the Hive table using the `ALTER TABLE ADD PARTITION` statement or repair the table using the `MSCK REPAIR TABLE` statement, which adds any partitions in the HDFS that do not exist in the metastore.

## User-Defined Functions

Hive can be extended through the implementation of UDFs (in fact, most Hive functions are implemented as UDFs).

To develop a UDF, a Java class must extend the `org.apache.hadoop.hive.ql.exec.UDF` class and override the `evaluate` function. The Hive API also defines the `@Description` annotation that can include information about an UDF. This information is displayed using the `DESCRIBE` statement in Hive.

The Hive source code is an excellent reference for writing UDFs. It is often convenient to find a UDF that is functionally similar, copy it, and modify to suit your needs. The following is an example UDF that converts a string to uppercase:

```
package com.madhu.udf;

import org.apache.hadoop.hive.ql.exec.Description;
import org.apache.hadoop.hive.ql.exec.UDF;
import org.apache.hadoop.io.Text;

// add jar samplecode.jar;
// create temporary function to_upper as 'com.madhu.udf.UppercaseUDF';
@Description(
    name = "to_upper",
    value = "_FUNC_(str) - Converts a string to uppercase",
    extended = "Example:\n" +
    "  > select to_upper(producer) from videos_ex;\n" +
    "  JOHN MCTIERNAN"
    )
public class UppercaseUDF extends UDF {
    public Text evaluate(Text input) {
        Text result = new Text("");
        if (input != null) {
            result.set(input.toString().toUpperCase());
        }
        return result;
    }
}
```

To make a UDF available to Hive, it must be added to the Hive classpath and registered with a unique function name. The UDF should be packaged in a JAR file. The following commands add the JAR to the Hive classpath and register the function:

```
hive> add jar samplecode.jar;
Added samplecode.jar to class path
Added resource: samplecode.jar
hive> create temporary function to_upper as 'com.madhu.udf.UppercaseUDF';
```

The function is now available in Hive:

```
hive> describe function to_upper;
OK
to_upper(str) - Converts a string to uppercase
Time taken: 0.039 seconds, Fetched: 1 row(s)
hive> describe function extended to_upper;
OK
to_upper(str) - Converts a string to uppercase
Example:
  > select to_upper(producer) from videos_ex;
  JOHN MCTIERNAN
Time taken: 0.07 seconds, Fetched: 4 row(s)
```

The following is an example of a more complex UDF that calculates the distance between two points on the Earth's surface given latitude and longitude:

```
package com.madhu.udf;

import org.apache.hadoop.hive.ql.exec.Description;
import org.apache.hadoop.hive.ql.exec.UDF;
import org.apache.hadoop.hive.serde2.io.DoubleWritable;

@Description(
    name = "geo_distance",
    value = "_FUNC_(lat1, lon1, lat2, lon2) - Computes the distance in km over the Earth's " +
            "given two latitude and longitude coordinates in degrees",
    extended = "Example:\n" +
    "  > select id, geo_distance(38, -77, lat, lon) as dist from orders_ex;\n"
    )

public class HiveDistanceUDF extends UDF {
    public DoubleWritable evaluate(DoubleWritable lat1, DoubleWritable lon1,
            DoubleWritable lat2, DoubleWritable lon2) {

        DoubleWritable result = new DoubleWritable();

        if (lat1 != null && lon1 != null && lat2 != null && lon2 != null) {
            result.set(computeDistance(lat1.get(), lon1.get(), lat2.get(), lon2.get(), 6371.009));
        }

        return result;
    }
```

```
    /**
     * Compute distance between given latitude and longitude
     *
     * @param lat1 - Latitude of first point in degrees (positive values North of equator)
     * @param lon1 - Longitude of first point in degrees (negative values West of prime meridian)
     * @param lat2 - Latitude of second point in degrees (positive values North of equator)
     * @param lon2 - Longitude of second point in degrees (negative values West of prime meridian)
     * @param radius - Mean radius of Earth, 6371.009 km or 3958.761 mi
     * @return
     */
    private double computeDistance(double lat1, double lon1, double lat2, double lon2, double
radius) {
        double dLat = (lat2-lat1) * Math.PI / 180.0;
        double dLon = (lon2-lon1) * Math.PI / 180.0;
        lat1 = lat1 * Math.PI / 180.0;
        lat2 = lat2 * Math.PI / 180.0;

        double a = Math.sin(dLat/2) * Math.sin(dLat/2) +
                Math.sin(dLon/2) * Math.sin(dLon/2) * Math.cos(lat1) * Math.cos(lat2);
        double c = 2 * Math.atan2(Math.sqrt(a), Math.sqrt(1-a));
        return radius * c;
    }
}
```

## User-Defined Aggregate Functions and Table Functions

Functions such as `sum()`, `count()`, `min()`, `avg()`, and so on are aggregate functions. They accept multiple rows and return a scalar value. Functions such as `explode(ARRAY)` or `json_tuple()` are table functions that return one or more rows. Implementing UDAFs or user-defined table functions (UDTFs) is more complicated than UDFs because greater knowledge of the Hive API is required. The easiest way to implement a UDAF or UDTF is to find a similar function, such as `sum()`, copy the source code, and modify it for your needs.

# Impala

Impala was developed by Cloudera as an alternative to Hive for interactive queries. Hive is great at batch processing of very large datasets, but startup delays and writing intermediate results to disk make Hive appear sluggish at times. Because Hive uses MapReduce under the hood, it also suffers from some of the limitations of MapReduce.

Impala was designed to overcome some of the limitations of Hive, but not as a replacement for Hive. Impala does not rely on MapReduce, but it does maintain significant compatibility with the Hive metastore, underlying storage format, and even UDFs. Hive is more reliable for long-running batch queries. Impala queries are cancelled if a hardware failure is encountered during a long-running query.

## Impala Architecture

To avoid the limitations of MapReduce, Impala uses a distributed MPP database engine. This alternative parallel model has been used effectively with many other parallel databases.

Impala relies on a daemon called impalad installed on each node, which directly accesses the HDFS or HBase tables. Clients can access Impala through Open Database Connectivity (ODBC) or JDBC. When a client connects an impalad instance on the cluster and issues a query, that instance acts as the coordinator for that query. The query is parsed, analyzed, and distributed for execution using the remaining impalad daemons on the cluster. Each daemon returns results to the coordinator, which returns results to the client.

Interestingly, a query can be submitted to any node in a cluster, which automatically becomes the coordinator. When submitting queries interactively through the command-line interface, it is common to submit each query to the same node. Cloudera recommends that noninteractive production jobs should be submitted to different nodes in a round robin fashion to achieve load balancing.

Impala achieves fault tolerance through a statestore component that checks the health of all daemons in a cluster. If a daemon fails to respond, all other nodes are notified so that node is no longer used. The statestore is not critical to the operation of the cluster. If the statestore fails, other nodes continue to operate, although with reduced reliability. When the statestore is restarted, it automatically begins monitoring other nodes and reports health as usual. One downside to this model is that hardware failures during query execution do not complete.

Impala supports the same metastore as used with Hive. As long as Hive tables use the same column types supported by Impala, it is possible to share the metastore between Hive and Impala. Impala keeps track of HDFS blocks and caches metadata to improve performance of queries against the same table. Impala can also access HBase table through an SQL-like interface, making HBase more user friendly.

## Impala Features

Impala supports many of the same features as Hive, such as internal (managed) and external tables, DDL, DML, subqueries, insert/insert overwrites, partitions, and UDFs. As with Hive, `UPDATE` and `DELETE` statements are not supported. The Cloudera web site describes the full language syntax in detail.

Like Hive, Impala supports partitioned tables. Partitions are created as subdirectories, so the number of partitions for a table is limited by the HDFS.

Impala supports several types of joins, including inner-, outer-, semi-, and cross-joins. One advantage over Hive is that Impala supports non-equal joins. Operators such as != (not equal), < (less than), or > (greater than) can be used in join conditions.

Impala UDFs can be written in C++ or Java. Java UDFs are compatible with Hive and enjoy the advantage of reuse. Higher performance can be achieved with C++ UDFs for functions that are known to be compute-bound. Impala caches results from previous calls of a UDF, so there should be no expectation that a function will be called for all apparent invocation.

## Impala Limitations

Because Impala does not checkpoint or save intermediate results during a query, hardware failure at that time causes the entire query to be cancelled. This can be an issue for long-running queries, so Hive can often be a better alternative.

Impala does not support customer serializers and deserializers; only common native formats with built-in serializers are supported. Window functions are not supported, indexing is not supported, nor is full text search on text fields.

# Shark

Shark, another open-source alternative to Hive, is built on top of Apache Spark, which is a general-purpose cluster computing system. Shark boasts as much as 100x performance improvement over Hive. Spark achieves high performance through cyclic data flow and in memory computing. Spark supports many high-level operations to simplify parallel programming. In addition to the Shark SQL engine, Spark also supports Spark Streaming, MLib (machine learning), and GraphX (graph analytics). Spark runs on a Hadoop 2 YARN cluster, so it is part of the Hadoop ecosystem. More information about Spark can be found on the Spark home page: `spark.apache.org`.

Shark leverages the Hive front end and the metastore, so it is compatible with Hive queries, schemas, and UDFs. Unlike Impala, Shark can recover from mid-query hardware failures. For this reason, Shark can be used reliably for both short interactive queries and long-running batch queries.

## Shark/Spark Architecture

Hive/Hadoop is built on top of the MapReduce parallel processing model. This model is straightforward and resilient. Failure of the underlying map or reduce tasks is not catastrophic because intermediate data is written to disk. Although MapReduce scales well for long-running jobs, complex multistage applications and data sharing are not a good fit. Data sharing in MapReduce is slow due to replication, serialization, and disk I/O.

Shark/Spark is an analogous parallel processing model built on top of Resilient Distributed Datasets (RDDs). RDDs are distributed collections of objects that support numerous parallel operations. RDDs are designed to be fault-tolerant; they are automatically rebuilt if there is a hardware failure. RDDs support the following operations (as well as many others):

- `map`
- `reduce`
- `filter`
- `count`
- `cogroup`
- `groupBy`
- `partitionBy`
- `sort`
- `join`
- `union`
- `leftOuterJoin`
- `rightOuterJoin`

The Shark/Spark architecture at a high level is very similar to Hive (see Figure 10-3).

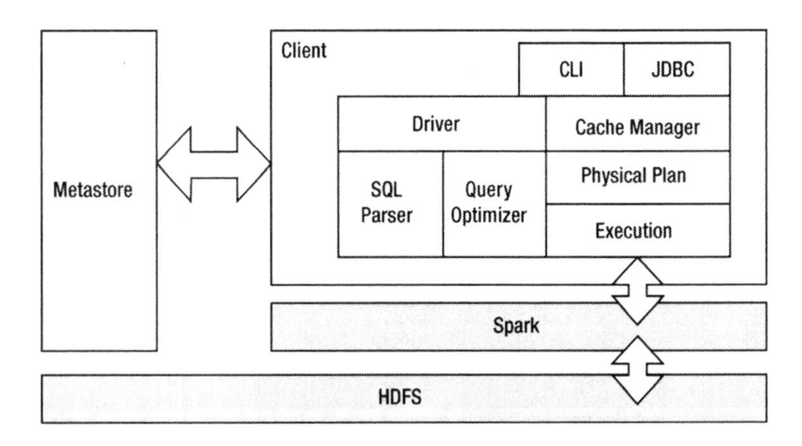


***Figure 10-3.*** *Shark architecture*

Spark shares the same front end and metastore as Hive, so it is compatible and leverages much of the heavy lifting done by the query pipeline.

# Summary

This chapter explored some of the data warehousing solutions available within the Hadoop ecosystem. Hive, Impala, and Shark support many of the familiar SQL capabilities available in conventional databases at a much larger scale.

Hive is a reliable solution built on top of MapReduce. Although it might not win awards for its responsiveness, it is highly reliable when executing long-running batch queries. Hive is the most mature and has been used successfully on many large-scale production applications.

Impala is a good tool for interactive queries when immediate response is needed. It is not as reliable as Hive for long-running queries, so it is best suited for shorter queries.

Shark might offer the best of both worlds: reliability and responsiveness. Shark uses high-speed memory when available and gracefully falls back to disk when necessary.

These systems are all capable of accessing data within the HDFS and even share the same metastore. All these solutions offer the same capability from a user perspective, but are suited to differing workloads due to their internal implementation.

CHAPTER 11

# Data Processing Using Pig

So far in this book, we have explored how to develop MapReduce programs using Java. Chapter 10 introduced Hive, the SQL engine on top of the HDFS. You learned how the Hive compiler converts high-level SQL commands into MapReduce programs, which avoids having to write low-level Java programs; you can instead focus on high-level business requirements. Hive is suitable for BI developers who want to treat the HDFS as a data warehouse.

This chapter focuses on another type of developer: the ETL developer who sees data as flowing through a complex data pipeline. SQL is a declarative language, and an ETL developer sees data following a series of transformation steps. Such a developer prefers a procedural language in which each extract/transform/load step can be seen as a part of a larger process. Pig, which was designed with such a user in mind, provides a high-level language to support development of data pipelines for non-Java developers.

This chapter also briefly discusses the Apache Crunch API, which is designed for developing data pipelines using Java.

## An Introduction to Pig

Pig is a platform for analyzing large datasets using a high-level and expressive language called Pig Latin, which enables users to describe data-processing steps. The Pig compiler converts these steps into a series of MapReduce programs that are executed in the Hadoop platform. In this respect, Pig is similar to Hive. The key difference between Pig and Hive is that Hive supports a declarative SQL-like language, whereas Pig supports a flow language that is suitable for describing data-processing steps of the type found in a typical ETL pipeline. Pig Latin allows users to describe how data will be transformed. Pig Latin scripts can produce multiple outputs, which allow users to describe a graph of complex data pipelines comprising multiple inputs, transformations, and outputs.

The high-level features of Pig Latin are as follows:

- Pig Latin provides data structures such as relations that are similar to database tables that contain rows (also called tuples, which are ordered sets of fields). It also supports standard data types such as `chararray(string)`, `int`, `long`, `float`, and so on.
- It supports various types of operations such as `LOAD`, `FILTER`, `JOIN`, `FOREACH`, `GROUP`, `STORE` and many more (discussed in subsequent sections).
- It is extensible; it supports UDFs if the existing functions do not meet user needs.

The Pig compiler also includes an optimizer that not only exploits opportunities for parallelism but also reorders steps to optimize the processing. A Pig compiler creates a logical plan for execution.

Let us illustrate how Pig works with a simple example. Assume that we have two very simple datasets: A and B. (Their respective locations in the source code accompanying this book are `a.txt` and `b.txt`, located in the `src/main/resources/input/pigsample` folder.)

Note that the files do not have a header. By default, Pig does not support headers for files; it provides a custom storage function, `CSVExcelStorage`, which supports a more comprehensive set of features, including header support. The documentation of `CSVExcelStorage` provides details of the various configurations supported. We will discuss Pig Load and Store functions in the "Loading and Storing" section.

***Listing 11-1.*** Files a.txt and b.txt

```
a.txt            b.txt
--------         --------
1,A,10           10,A,1
3,B,20           30,B,2
5,C,30           50,C,3
7,A,40           70,A,4
9,B,50           90,B,5
2,C,60           20,C,6
4,A,70           40,A,7
6,B,80           60,B,8
8,C,90           80,C,9
10,A,100         100,A,10
```

This sample Pig program is available in the following path in the book's source code:

```
src/main/resources/scripts/pig/sample.pig
```

Listing 11-2 shows this program.

***Listing 11-2.*** sample.pig

```
A = LOAD 'input/pigsample/a.txt' USING PigStorage(',') AS (a:int,b:chararray,c:int);
B = LOAD 'input/pigsample/b.txt' USING PigStorage(',') AS (d:int,e:chararray,f:int);
C = FILTER A BY (b=='A') or (b=='B');
D = JOIN C by a,B by f;
E = GROUP D by b;
F = FOREACH E GENERATE group, COUNT(D);
STORE F INTO 'output/pigsample/';
```

The Pig compiler compiles the program and optimizes certain paths to work in parallel (see Figure 11-1). Notice that `Load A` and `Filter A` execute in parallel to `Load B`. Eventually, the Pig compiler executes the program as a series of MapReduce programs. Yet Pig Latin allows users to express what they want as a series of data-processing steps using the high-level expressive language.

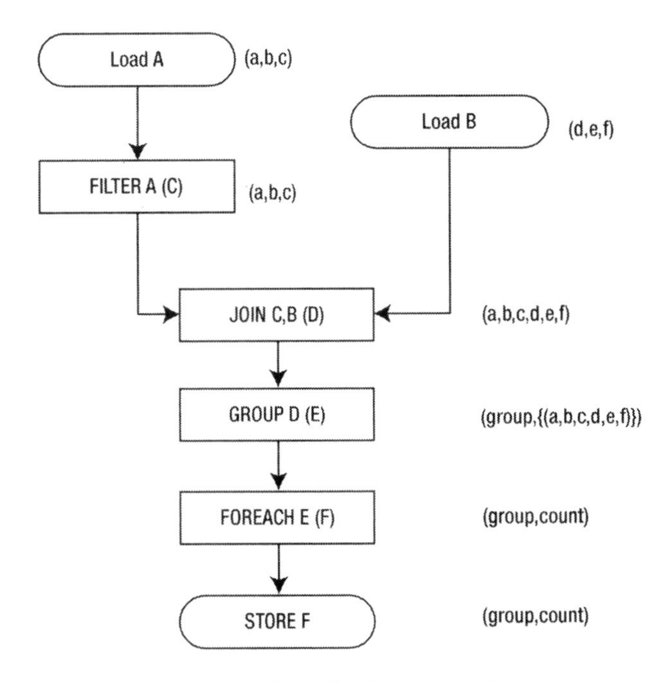


***Figure 11-1.*** *Plan developed by the Pig compiler to execute sample.pig*

The actual job gets executed only when the `STORE` command is called. The `STORE` command executes the job and stores the output to disk. Depending on the mode in which it is executed, the output is stored either to local disk or the HDFS (modes are discussed in the next section). An alternative command to `STORE` is `DUMP`, which dumps the output of the relation to the console instead of writing to a file (the syntax is `DUMP` *`<relation>`*`}`). It can be used only in development for small datasets and is useful during the development and test cycle.

# Running Pig

In this chapter, we assume that Pig is installed along with the Hadoop installation. If you are using a virtual machine, it is preinstalled as well. Pig executes in the following two modes:

- *Local*: Jobs are run locally using the local file system. In this mode, the input and output files are located on the local file system. To start Pig in local mode, type `pig -x local` at the command prompt. This process starts Grunt, which is a Pig shell in which you can execute commands interactively. For example, you can open the Grunt shell and copy the commands from `sample.pig` in the Grunt shell to run the `sample.pig` program.
- *MapReduce*: Jobs are executed on the Hadoop system using the HDFS. To start the Grunt shell in MapReduce mode, type `pig`. The Grunt shell starts in cluster mode.

Pig programs can be executed in the following three ways:

- Series of interactive commands in the Grunt shell
- Pig script executed from the command line
- Embedded mode from a Java program

## Executing in the Grunt Shell

As mentioned previously, execute commands in the Grunt shell using one of the following commands:

- `pig -x local`
- `pig`

The Grunt shell starts and looks like the following on the console:

```
grunt>
```

You can copy and paste the commands from `sample.pig` shown earlier into the Grunt shell to execute the commands interactively. Note that we hard-coded the paths in the program. The Grunt shell must be started from a working directory so that the paths referenced in the program must be relative to it.

## Executing a Pig Script

From the command line, execute one of the following commands, depending on the mode in which you want to execute the Pig script:

- `pig -x local sample.pig`
- `pig sample.pig`

Although we hard-coded the paths in `sample.pig`, we want to pass the paths as parameters because we do not want to update the Pig scripts when the paths change. We provide a parameterized version of `sample.pig` at `scripts/pig/sampleparameterized.pig`. Listing 11-3 shows this program.

***Listing 11-3.*** sampleparameterized.pig

```
A = LOAD '$A_LOC' USING PigStorage(',') AS (a:int,b:chararray,c:int);
B = LOAD '$B_LOC' USING PigStorage(',') AS (d:int,e:chararray,f:int);
C = FILTER A BY (b=='A') or (b=='B');
D = JOIN C by a,B by f;
E = GROUP D by b;
F = FOREACH E GENERATE group, COUNT(D);
STORE F INTO '$OUTPUT_LOC';
```

The program has three parameters

- `A_LOC`: Path to the input file `a.txt`
- `B_LOC`: Path to the input file `b.txt`
- `OUTPUT_LOC`: Path to the output directory

On the command line, execute one of the following commands to execute the script, depending on the mode in which you want to execute the script:

```
pig -x local -param A_LOC=input/pigsample/a.txt -param B_LOC=input/pigsample/b.txt -param
OUTPUT_LOC=sampleout2 sampleparameterized.pig
pig-param A_LOC=input/pigsample/a.txt -param B_LOC=input/pigsample/b.txt -param
OUTPUT_LOC=sampleout2 sampleparameterized.pig
```

Notice that we passed the value of the parameters using the following syntax:

```
-param <PARAM_NAME>=<PARAM_VALUE>
```

## Embedded Java Program

Now we demonstrate how to run a Pig script from an embedded Java program. First, we add the following dependencies to the pom.xml file:

```
<dependency>
   <groupId>org.apache.pig</groupId>
   <artifactId>pig</artifactId>
   <version>0.12.1</version>
</dependency>
<dependency>
   <groupId>jline</groupId>
   <artifactId>jline</artifactId>
   <version>0.9.94</version>
</dependency>
<dependency>
   <groupId>org.antlr</groupId>
   <artifactId>antlr-runtime</artifactId>
   <version> 3.2 </version>
</dependency>
<dependency>
   <groupId>joda-time</groupId>
   <artifactId>joda-time</artifactId>
   <version>2.1</version>
</dependency>
```

The Java program that executes the embedded Pig server is shown in Listing 11-4.

***Listing 11-4.*** EmbeddedPigServer.java

```
package org.apress.prohadoop.c11;

import java.util.HashMap;
import java.util.Map;
import org.apache.pig.ExecType;
import org.apache.pig.PigServer;

public class EmbeddedPigServer {

    public static void main(String[] args) throws Exception {
        boolean isLocal = Boolean.parseBoolean(args[0]);
        String pigScriptPath = args[1];
        String parameters = args[2];
        String[] params = parameters.split(",");
        Map<String, String> paramMap = new HashMap<String, String>();
        for (String p : params) {
```

```
            String[] param = p.split("=");
            paramMap.put(param[0], param[1]);
        }

        PigServer pigServer;
        if (isLocal) {
            pigServer = new PigServer(ExecType.LOCAL);
        } else {
            pigServer = new PigServer(ExecType.MAPREDUCE);
        }
        pigServer.registerScript(pigScriptPath, paramMap);
    }
}
```

sampleparameterized.pig can be executed using the following command:

```
hadoop jar prohadoop-0.0.1-SNAPSHOT.jar
org.apress.prohadoop.c11.EmbeddedPigServer \
true \
scripts/pig/sampleparameterized.pig \
A_LOC=input/pigsample/a.txt,B_LOC=input/pigsample/b.txt,OUTPUT_LOC=sampleout
```

The parameters of the program are as follows:

- The first parameter indicates whether the Pig server is running in local mode (similar to the `-x local` parameter using Pig shell). A value of `true` indicates that the job runs in local mode.
- The second parameter is the path to the script being executed.
- The third parameter is a comma-separated list of name value pairs that are passed as parameters to the Pig script.

# Pig Latin

This section discusses the various Pig commands. A Pig script is a series of commands, and each command ends with a semicolon, as shown here:

```
D = JOIN C by a,B by f;
```

From inside the Pig shell, various `Hadoop fs` commands can also be executed. Each command ends with a slash, as shown here:

```
ls /
```

## Comments in a Pig Script

Comments in Pig scripts can be provided in two ways:

- Single-line comments are on a single comment line; here's an example:

  ```
  ---This is a single comment line
  ```

- Multiline comments are provided in the same manner in which Java multiline comments are expressed:

```
/* This is a
 * multiline comment
 */
```

# Execution of Pig Statements

A Pig Latin script is executed by the Pig compiler, which performs variable binding, syntax verification, and execution plan generation based on the commands expressed in the script. The plan that is generated is executed when one of the following Pig commands is encountered:

- `DUMP`: Has the syntax `DUMP` *<relation_name>*. This command prints the relation data to the console. When this command is encountered, the logical plan is executed as a series of MapReduce jobs. Each subsequent `DUMP` command executes a new series of MapReduce jobs. `DUMP` is a debugging and diagnostic tool; it is not suitable for production.
- `STORE`: Has the syntax `STORE` *<relation_name> <output_folder>*. The `STORE` command also executes a series of MapReduce jobs. However, `STORE` is different from `DUMP`; each `STORE` does not execute a new MapReduce job. If the Pig compiler can optimize the `STORE`s into a single MapReduce job, it does so. Consider the following lines:

```
X = LOAD '$A_LOC' USING PigStorage(',')
                  AS (a:int,b:chararray,c:int);
SPLIT X INTO A IF b=='A', B IF b=='B', C IF b='C', D otherwise;
STORE A into '$OUTPUT_PATH/A/';
STORE B into '$OUTPUT_PATH/B/';
STORE C into '$OUTPUT_PATH/C/';
STORE D into '$OUTPUT_PATH/D/';
```

This program is executed as a Map-only job in which the input is split based on the value of the second attribute (`b`) and is sent to the output folders in the same Mapper using `MultiOutputFormat`. Thus, the Pig compiler intelligently optimizes the list of commands into the fewest possible MapReduce programs.

# Pig Commands

So far, we have discussed a few Pig commands, most notably the `LOAD`, `STORE`, and `DUMP` commands. Pig Latin comprises an extensive library of commands, and this section categorizes the Pig commands and describes the role of each category. A few key commands from each category are also illustrated.

The command list can be broken down into multiple categories:

- Loading and storing
- Diagnostic functions
- Relational operators
- Functions
- Macros
- Utility commands

## Loading and Storing

These commands were discussed earlier:

- The LOAD command defines a relation based on a file. The default data type for a field if none is provided is a ByteArray, and the class used to load the file into a relation is PigStorage. Its default field separator is a TAB, but it can be configured to a user-defined one. In our sample.pig, we configured it to be a COMMA. We also defined the data types for each field of the relation. It is common in large datasets to have corrupt records in the data. If the underlying field data for a certain row does not confirm to the specified data type, Pig converts it to a NULL value. It provides a warning that indicates how many such values it has replaced, but Pig does not halt the processing, which is in sharp contrast to the way databases function.
- The STORE command stores a relation to disk. Although its commands trigger the MapReduce chain of jobs, Pig tries to optimize the processing and not execute a separate MapReduce job for each STORE command.
- The DUMP command prints the relation to the screen. It is useful for debugging jobs.

## Diagnostic Functions

The various diagnostic functions provided are the following:

- DESCRIBE *<relation_name>* returns the schema for the alias used to represent the relation. For example, if we provided DESCRIBE A in the sample.pig, the following output would display onscreen:

```
A: {a: int,b: chararray,c: int}
```

- DUMP displays the results of a relation onscreen. Each DUMP program starts a new MapReduce chain of jobs.
- EXPLAIN reviews the logical, physical, and MapReduce execution plans used to execute the script.
- ILLUSTRATE *<relation_name>* reviews how the data is transformed through each preceding relation to produce the relation used as a parameter to ILLUSTRATE.

Refer to the Pig documentation for examples of how to use the ILLUSTRATE and EXPLAIN commands.

## Relational Functions

The following sets of commands, which are very familiar to SQL users, execute relational operators:

- The GROUP command is similar to the GROUP BY clause in SQL. It groups the tuples in a relation based on a field value. Our sample.pig program has an example of the GROUP command.
- Although COGROUP is similar to the GROUP command, GROUP works with a single relation, and COGROUP works with multiple relations. For example, we execute the following commands in the Grunt shell in local mode:

```
A = LOAD 'input/pigsample/a.txt' USING PigStorage(',')
                                 AS (a:int,b:chararray,c:int);
B = LOAD 'input/pigsample/b.txt' USING PigStorage(',')
                                 AS (d:int,e:chararray,f:int);
X = COGROUP A BY b, B BY e;
DUMP X;
```

The output is shown following. Notice how relations A and B, are grouped:

```
(A,{(1,A,10),(7,A,40),(4,A,70),(10,A,100)},{(10,A,1),(70,A,4),(40,A,7),(100,A,10)})
(B,{(3,B,20),(9,B,50),(6,B,80)},{(30,B,2),(90,B,5),(60,B,8)})
(C,{(5,C,30),(2,C,60),(8,C,90)},{(50,C,3),(20,C,6),(80,C,9)})
```

The form of the output is the following:

```
(GROUP_ID,{RELATION A TUPLES},{RELATION B TUPLES})
```

- The ORDER command is similar to the ORDER BY clause in SQL. The relation can be ordered based on one or several fields.
- The JOIN command, which is similar to SQL joins, allows two or more relations to be joined based on join criteria. JOIN supports both inner and outer joins, as well as multiple flavors. One example is MERGE JOIN, which sorts two relations based on criteria to be joined using a Map-only job. (You saw an example of such a join in Chapter 7 when CompositeInputFormat was discussed.) Refer to the Pig documentation to understand the rich variety of joins supported by the Pig framework.
- The FOREACH command performs data transformations based on columns of data. It is similar to the SELECT clause in SQL. Various aggregation functions can also be applied in FOREACH. (You saw an example of the COUNT function applied in FOREACH when sample.pig ran earlier in the chapter.) Following is the line reproduced from sample.pig:

  ```
  F = FOREACH E GENERATE group, COUNT(D);
  ```

- The UNION command computes the union of two relations. It does not require that both relations be of the same schema, nor does it preserve ordering. For example, we could compute the union of A and B as follows:

  ```
  A = LOAD 'input/pigsample/a.txt' USING PigStorage(',')
                                  AS (a:int,b:chararray,c:int);
  ---Note we only load two fields to have different schemas for
  ---A and B
  B = LOAD 'input/pigsample/b.txt' USING PigStorage(',')
                                      AS (d:int,e:chararray);
  X = UNION A,B;
  DUMP X;
  ```

  The output is shown here. If you execute DESCRIBE X, you are told that the schema of X is unknown. Usually in practice, both A and B have the same schema, in which case X inherits the schema.

  ```
  (10,A)
  (30,B)
  (50,C)
  (70,A)
  (90,B)
  (20,C)
  (40,A)
  (60,B)
  (80,C)
  ```

```
(100,A)
(1,A,10)
(3,B,20)
(5,C,30)
(7,A,40)
(9,B,50)
(2,C,60)
(4,A,70)
(6,B,80)
(8,C,90)
(10,A,100)
```

- The DISTINCT command takes a relation and returns distinct values. For example, consider the following set of lines:

```
A = LOAD 'input/pigsample/a.txt' USING PigStorage(',')
                          AS (a:int,b:chararray,c:int);
B = LOAD 'input/pigsample/a.txt' USING PigStorage(',')
                          AS (a:int,b:chararray,c:int);
C = FILTER A BY (b=='A') or (b=='B');
D = FILTER B BY (b=='B');
E = UNION C,D;
F = DISTINCT E;
```

If the command DUMP E is executed, the result is the following output:

```
(3,B,20)
(9,B,50)
(6,B,80)
(1,A,10)
(3,B,20)
(7,A,40)
(9,B,50)
(4,A,70)
(6,B,80)
(10,A,100)
```

If the command DUMP F is run, the result is the following output (demonstrating how the DISTINCT clause removes duplicate tuples):

```
(1,A,10)
(3,B,20)
(4,A,70)
(6,B,80)
(7,A,40)
(9,B,50)
(10,A,100)
```

- The FILTER command is similar to the WHERE clause in SQL. (You saw an example of the FILTER command when sample.pig ran.) Here is the line from sample.pig:

```
C = FILTER A BY (b=='A') or (b=='B');
```

## Functions

Pig supports the following types of functions.

- The Eval function takes a relation and produces another relation or a scalar value. There are several examples of built-in functions, including COUNT, MAX, MIN, AVG, and SUM. Some of these functions are also aggregate functions; they operate on a bag of tuples and produce a scalar value. Examples of such functions are AVG and SUM. Some Eval functions are algebraic and can operate incrementally on each value. An example of an algebraic function is COUNT, which can be computed by counting the number of elements in a subset of data and then computing the sum. In MapReduce world, this implies using a Mapper and a Combiner before computing the final results in the Reducer. In a later section, we discuss Eval functions in detail.
- The Filter function is a special type of Eval function that returns a Boolean. An example of such a function is isEmpty(). We discuss how to develop Filter functions later in this chapter.
- The Load/Store functions load and store data into a relation from/to external storage. Several types of Load/Store functions are available in the standard Pig library. These functions enable loading/storing from various formats such as Text, Json, HBase, and so on. The most commonly used function is PigStorage, which loads and stores relations using a field- and line-delimited format. By default, each line is a tuple, and each field in a tuple is tab-separated. The field separator can be configured. (We use a comma separator in our examples.) A third-party library, Elephant-Bird, was developed at Twitter includes additional Pig Load and Store functions. It is available at https://github.com/kevinweil/elephant-bird/.

---

▒ **Note** Although we do not discuss how to develop custom Load/Store functions in this chapter, you can refer to the documentation online for details on how to develop such functions. We chose to skip this discussion because you seldom need to develop such functions. If you have to do so, the online documentation provides a detailed guide on how to develop such functions.

---

Pig also provides an API to develop the custom UDFs mentioned previously. We discuss this feature in a later section of the chapter.

## Macro Functions

Pig enables you to define macro functions that can be invoked, enabling you to prevent code duplication. Consider the example shown in Listing 11-5. Imagine that we need to apply a custom filter before we use relation C. The condition for the custom filter is provided by the mycriteria parameter to the Pig script. The get_A_relation is a user-defined macro that takes a runtime parameter.

***Listing 11-5.*** samplewithmacro.pig

```
DEFINE get_A_relation(criteria1) returns C {
   A = LOAD 'input/pigsample/a.txt' USING PigStorage(',')
       AS (a:int,b:chararray,c:int);
   $C = FILTER A BY (b=='$criteria1');
};
```

```
B = LOAD 'input/pigsample/b.txt' USING PigStorage(',') AS (d:int,e:chararray,f:int);
C = get_A_relation ($mycriteria);
D = JOIN C by a,B by f;
E = GROUP D by b;
F = FOREACH E GENERATE group, COUNT(D);
STORE F INTO 'output/pigsample/';
```

## The SPLIT Function

A SPLIT function is a special type of relational operator that splits a relation into two or more relations based on the value of a field. Each split relation can then be used in its own data pipelines. For example, in our airline dataset we could define a macro to define a field that indicates whether the flight is delayed or on time. We could then split based on the return value of this macro. The delayed flight records could be handled in their own subpipeline with a different set of steps as the on-time pipeline. Listing 11-6 shows how the SPLIT function is used.

***Listing 11-6.*** samplewithsplit.pig

```
X = LOAD 'input/pigsample/a.txt' USING PigStorage(',')
                                 AS (a:int,b:chararray,c:int);
SPLIT X into A if b=='A',B if b=='B',C if b=='C',D otherwise;
STORE A INTO 'output/A/';
STORE B INTO 'output/B/';
STORE C INTO 'output/C/';
STORE D INTO 'output/D/';
```

# User-Defined Functions

Although Pig Latin provides a rich API, there are times when it is not enough, and you have to plug in custom functionality. Pig provides a way to define your own functions: using UDFs. A UDF in Pig is a class that extends the abstract class EvanFunc<T>. To implement an Eval function, the exec() method of the EvalFunc<T> class needs to be implemented. The method is shown here in the context of the EvalFunc<T> class:

```
public abstract class EvalFunc<T extends Datum>  {
   abstract public T exec(Tuple input) throws IOException;

   /*Other methods of the EvalFunc class*/
}
```

The T represents the return type of invoking function. The input tuple can be a single record or can contain multiple records inside an instance of DataBag or Map. This is the key point. EvalFunc.exec() can be invoked in the map() method of the Mapper or in the reduce() method of the Reducer. In the former case, the input tuple to the exec() method contains a single record. In the latter case, it can contain multiple records as returned by the value iterator in the reduce call.

## Eval Functions Invoked in the Mapper

Let us look at an example of an EvalFunc that would be invoked on the Mapper. Its exec() method should have a single record as input and returns a single record as output (its return value). Consider an EvalFunc that converts all text instances passed to it to uppercase. We call it ToUpper. Following is an implementation of this class:

```
public class ToUpper extends EvalFunc<Text>
{
    public Text exec(Tuple input) throws IOException {
        if (input == null || input.toString().size() == 0)
            return new Text() ;
        try{
            Text str = (Text)input.get(0);
            return new Text(str.toString().toUpperCase());
        }catch(Exception e){
           //Skipped for brevity
        }
    }
}
```

An illustration of how the preceding function can be invoked is FOREACH A GENERATE ToUpper(f1), in which the attribute f1 of relation A is converted to uppercase and returned as a new text instance.

The preceding call does not need a reduce phase and can be executed completely in the Mapper. The next section looks at an example of EvalFunc that is executed in the Reducer.

## Eval Functions Invoked in the Reducer

Eval functions are routinely invoked in the Reducer, and the most typical Reducer-based invocation is as follows:

```
A=LOAD...
B = GROUP A by a;
C = FOREACH B GENERATE group, COUNT(B);
```

The preceding set of commands is similar to the following SQL call:

```
SELECT a, COUNT(*) from A GROUP BY a
```

In Chapter 6, you learned how GROUP BY implementations in MapReduce require the Reduce phase. The COUNT(B) function in the Pig Latin script can be completed only on the Reducer node. We consider three separate implementations of COUNT from the Pig library:

- An implementation in which the entire set of values for a given key (group) must be collected in a DataBag instance. The DataBag instance is wrapped in a tuple instance that is passed as a parameter to the exec() method of the EvalFunc. A UDF must always provide a complete implementation of the UDF in the EvalFunc.exec() method. (The reasons why this is true will become clear soon.) This implementation is the most memory intensive because the value iterator from Reducer might return a large number of records that could overwhelm the memory in the reduce call.

- An implementation in which the values are batched in multiple exec() method invocations per key (group) instead of sending the entire set of values for a key (group). The results of each batch are accumulated until the final result can be computed. This implementation is feasible only when the final result can be aggregated from partial results computed from partial datasets. It is memory efficient because the entire set of values in a reduce call does not have to maintained in memory to invoke the UDF. Pig provides a specialized interface called Accumulator to support batched computations on the Reducer node.
- When batching is possible, we can often do better. For example, the COUNT function needs to know only the previous count and the current value, which is an ideal scenario for using the Combiner. (Combiners were discussed in Chapter 6, in which you learned how the use of Combiner can significantly reduce the I/O between the Mapper and Reducer nodes.) Pig provides a specialized Algebraic interface that a UDF class can implement to use the Combiner.

To summarize the previous discussion, a class must extend the EvalFunc abstract class to implement a UDF. Optionally, it can implement the Accumulator and Algebraic interfaces. A single FOREACH-GENERATE call may invoke several UDFs. All the UDFs extend the abstract class EvalFunc, but only some can implement the Aggregate and Algebraic interfaces. This can lead to some tricky situations with respect to how Pig decides which interface to use to execute the UDF. We discuss how Pig resolves this dilemma soon.

In the next three subsections, we review the source code of the COUNT function provided by Pig. First, we discuss the simple exec() method implementation, followed by a discussion of how COUNT benefits from the implementation of the Accumulator interface. Finally, we discuss how COUNT implements the Algebraic interface.

## Aggregation Functions Using the EvalFunc.exec() Method

The COUNT UDF class has the following signature:

```
public class COUNT extends EvalFunc<Long> implements Algebraic, Accumulator<Long>
```

For the purposes of this section, consider only the EvalFunc<Long> parent class of the COUNT UDF. All UDFs must implement the abstract exec() method, which is the default implementation that Pig reverts to when it cannot use the optimized implementations provided by the Accumulator and Algebraic interfaces discussed in subsequent sections.

Listing 11-7 shows the implementation of the exec() method in the COUNT class. The input tuple contains the values for the given Reducer key (group) in the Pig script. These values are included in the DataBag instance. The input tuple can contain more than one value, and the order of the values in the input tuple is the order in which the parameters were passed to the UDF invocation. For COUNT, we pass the entire relation that Pig converts to a DataBag instance per key (group). The input.get(0) call returns this instance of DataBag. The final count is obtained by iterating the DataBag instance to determine the number of elements in the DataBag instance.

***Listing 11-7.*** Implementation of the COUNT.exec() Function in Pig

```
@Override
public Long exec(Tuple input) throws IOException {
   try {
      DataBag bag = (DataBag)input.get(0);
      if(bag==null)
         return null;

      Iterator it = bag.iterator();
      long cnt = 0;
```

```
        while (it.hasNext()){
            Tuple t = (Tuple)it.next();
            if (t != null && t.size() > 0 && t.get(0) != null )
                cnt++;
        }
        return cnt;
    } catch (ExecException ee) {
        //Skipped for brevity
    } catch (Exception e) {
        //Skipped for brevity
}
```

Imagine this method invoked from the reduce() method of a Reducer. For the exec() method to return the true value of the count for a given key (group), the entire set of values for the given key (group) would have to be passed inside the input tuple (as a DataBag instance). Because this set can be huge, it can cause the underlying JVM on the Reducer node to run out of memory. The DataBag instance supports functionality to spill to the disk if the number of records contained in it gets too large. See the following note for how to take advantage of the spill-to-disk feature of DataBag.

---

■ **Note** To use the disk-spilling feature of DataBag, you should first create a DataBag instance by making the call BagFactory.newDefaultBag() and then adding entries to a DataBag instance by invoking the addTuple(Tuple T) method instead of creating a list of tuples and creating a DataBag as BagFactory.newDefaultBag(List<Tuple> listOfTuples). The addTuple() method of DataBag supports spilling to disk when the number of tuples held in the memory of the DataBag instance grows large.

---

Spilling to disk is a time-consuming operation that adversely affects the performance of the Pig script, so it should be avoided. But this requires that the number of records passed to the exec() method must be small enough. Pig allows input to the exec() method to be broken into smaller chunks. The results on the smaller chunks can be accumulated to obtain the final result. This feature is supported via the Accumulator interface discussed in the next section.

## Aggregation Functions Using the Accumulator Interface

This section focuses on the Accumulator interface. As discussed earlier, the COUNT class also implements this interface:

```
public class COUNT extends EvalFunc<Long> implements Algebraic, Accumulator<Long>
```

When a UDF implements the Accumulator interface, Pig passes the values received by the Reduce invocation in chunks to the UDF. Pig does not directly call the exec() method; it delegates the exec() method invocation to the accumulate() method, which is responsible for accumulating the results of each exec invocation. The accumulate() method is guaranteed to be called one or more times. Each accumulate invocation is passed a tuple that contains a chunk of records from the value iterator in the reduce() method, wrapped in a DataBag instance (this is similar to the exec() method described in the previous section). The chunk size is constrained by the pig.accumulative.batchsize property. The default value of this property is 20000, which means that 20,000 values are batched in each invocation to the accumulate() method. This value should be small enough to prevent disk spilling by the DataBag instance as it is being populated. You should change this value based on your knowledge of how memory intensive your custom value objects are.

Listing 11-8 shows the Accumulator interface.

***Listing 11-8.*** Accumulator Interface

```
public interface Accumulator <T> {
    public void accumulate(Tuple b) throws IOException;
    public T getValue();
    public void cleanup();
}
```

The accumulate() method is invoked multiple times for each group of values in the invocation of the FOREACH function. The partial result computed by each accumulate invocation is used to aggregate the final result. Between the multiple accumulate invocations, this partial result is maintained as a state variable in the UDF instance. The getValue function is invoked when all the tuples for a given key (group) are processed via the accumulate invocations. This method should return the final value as accumulated in the previous calls to the accumulate function. The cleanup resets the intermediate value that maintains state across the accumulate invocations and is invoked after the getValue function is invoked.

Listing 11-9 shows the implementation of the Accumulator interface in the COUNT class.

***Listing 11-9.*** Accumulator Implementation of COUNT

```
public class COUNT extends EvalFunc<Long>
                implements Algebraic, Accumulator<Long>{
   /*All the Algebraic interface code from previous section goes here*/

   /*Default exec implementation of count. Could lead to memory
    *problems
    */
    @Override
    public Long exec(Tuple input) throws IOException {
        try {
            DataBag bag = (DataBag)input.get(0);
            if(bag==null)
                return null;

            Iterator it = bag.iterator();
            long cnt = 0;
            while (it.hasNext()){
                    Tuple t = (Tuple)it.next();
                    if (t != null && t.size() > 0 &&
                        t.get(0) != null )
                            cnt++;
            }
            return cnt;
        } catch (ExecException ee) {
            throw ee;
        } catch (Exception e) {
            //Skipped for brevity
        }
    }
    /* Accumulator interface implementation */
    private long intermediateCount = 0L;
```

```
    @Override
    public void accumulate(Tuple b) throws IOException {
        try {
            DataBag bag = (DataBag)b.get(0);
            Iterator it = bag.iterator();
            while (it.hasNext()){
                Tuple t = (Tuple)it.next();
                if (t != null && t.size() > 0 && t.get(0) != null) {
                    intermediateCount += 1;
                }
            }
        } catch (ExecException ee) {
            throw ee;
        } catch (Exception e) {
            //Skipped for brevity
        }
    }

    @Override
    public Long getValue() {
        return intermediateCount;
    }

    @Override
    public void cleanup() {
        intermediateCount = 0L;
    }
```

The accumulate function maintains partial results in an intermediateCount instance attribute that is updated in each accumulate call and finally read by the getValue() method after the last call to accumulate for a given reducer key (group). The cleanup method resets the value of intermediateCount to prepare for the processing of the next key (group).

The Accumulator-based UDF is more efficient than the EvalFunc.exec()-based implementation discussed in the previous section because it avoids using large amounts of memory as well as the cost associated with disk-spills. The implementation of the Accumulator demonstrates how COUNT can be performed in smaller chunks in the Reducer. The results of the smaller chunks are aggregated into a final value. In this case, the order of the values in the individual chunks is not relevant to the calculation. Remember the section on Secondary Sorting in Chapter 6, in which the values received by the reduce call were also sorted based on user-defined criteria. When the total order of values (Secondary Sort) received in the Reducer matters, the Accumulator is the only possible optimization over the simple EvalFunc.

Because the order of values is inconsequential to the calculation of the COUNT function, we can delegate part of this calculation to the Mapper node in the Combiner component. (Combiners are discussed in Chapter 6.) They significantly reduce the I/O between the Mapper and Reducer nodes. Pig supports the use of Combiners through the Algebraic interface we discuss in the next section.

## Aggregation Functions Using the Algebraic Interface

This section discusses the Algebraic interface in the context of its implementation in COUNT UDF from the Pig library:

```
public class COUNT extends EvalFunc<Long> implements Algebraic, Accumulator<Long>
```

Listing 11-10 shows the Algebraic interface that can be used to further optimize UDFs.

***Listing 11-10.*** Algebraic Interface

```
public interface Algebraic{
    public String getInitial();
    public String getIntermed();
    public String getFinal();
}
```

The return value of each of the preceding methods is a string that represents a fully qualified name of a class that extends the EvalFunc class. The contract for an Algebraic function is as follows:

- exec(Tuple t) of the class represented by getInitial(), which we refer to as the Initial class, is called once for each input tuple. In MapReduce terms, it involves invoking the exec function on the Initial class for each input record in the map() method of the Mapper. The output contains partial results.
- exec(Tuple t) of the class represented by getIntermed(), which we refer to as the Intermed class, is called zero or more times on the partial results of the Initial.exec() invocation. In MapReduce terms, it involves invoking the exec function of the Intermed class in the reduce() method of the Combiner. The input to the Intermed.exec() method is an instance of DataBag (wrapped in a Tuple) that contains records returned by the Initial.exec() invocations.
- exec(Tuple T) of the class represented by getFinal(), which we refer to as the Final class if called once per key (group) on the results returned by the corresponding invocation of Intermed.exec(). In MapReduce terms, it involves invoking the Final.exec() once in the reduce() method of the Reducer per key (group).

Listing 11-11 shows the implementation of the COUNT function. Notice that the Intermed and Final classes invoke the sum() method, which performs the aggregation (invoked in the Combiner and Reducer, respectively) function; and the Initial class invokes the count() method, which computes a partial result (invoked in the Mapper). The Intermed invocation produces a Tuple because it needs to feed its partial result to the Reducer, but the Final invocation produces a scalar value.

***Listing 11-11.*** COUNT Implementation with the Algebraic Interface

```
public class COUNT extends EvalFunc<Long> implements Algebraic, Accumulator<Long>{
    public Long exec(Tuple input) throws IOException {
       return count(input);
    }
    public String getInitial() {
        return Initial.class.getName();
    }
    public String getIntermed() {
        return Intermed.class.getName();
    }
    public String getFinal() {
        return Final.class.getName();
    }
    static public class Initial extends EvalFunc<Tuple> {
        public Tuple exec(Tuple input) throws IOException {
```

```
            return TupleFactory.getInstance()
                   .newTuple(count(input));
        }
    }
    static public class Intermed extends EvalFunc<Tuple> {
        public Tuple exec(Tuple input) throws IOException {
           return TupleFactory.getInstance().newTuple(sum(input));
        }
    }
    static public class Final extends EvalFunc<Long> {
        public Tuple exec(Tuple input) throws IOException {
           return sum(input);
        }
    }
    //The Initial class invokes this method. Called from map()
    static protected Long count(Tuple input)
                                        throws ExecException {
        Object values = input.get(0);
        if (values instanceof DataBag)
            return ((DataBag)values).size();
        else if (values instanceof Map)
            return new Long(((Map)values).size());
    }
    //The Intermed and Final classes invoke this method. Called from
    //reduce() call of the Combiner as well as the Reducer
    static protected Long sum(Tuple input)
                        throws ExecException,NumberFormatException {
        DataBag values = (DataBag)input.get(0);
        long sum = 0;
        for (Iterator (Tuple) it = values.iterator(); it.hasNext();)
        {
            Tuple t = it.next();
            sum += (Long)t.get(0);
        }
        return sum;
    }
}
```

## How Does Pig Decide Which Interface to Use?

The last three sections demonstrated how the COUNT function is implemented in Pig. We noted the following key points about UDFs in the process:

- A UDF must extend the EvalFunc class and provide a default and complete implementation for the exec() method, which must assume that it will be called only once. It should not assume that the methods of the Accumulator and Algebraic interfaces will be invoked, even if those interfaces are implemented by the UDF.
- The Accumulator interface is more memory efficient on the Reducer side. It is ideal when computations can be batched and the final result can be computed by accumulating the partial results. It typically should be used when the batching needs to depend on the total ordering of the values for a key in the Reducer.

- The Algebraic interface is the most memory- and I/O efficient because it delegates part of its functionality to the Mapper nodes via the Combiner. This feature can be used when the partial computations can be performed in the Mapper node.

Similar to COUNT UDF, you will encounter several UDFs that support all three interfaces, which is required if several UDFs are invoked in a single call to the FOREACH-GENERATE invocation in Pig Latin. Remember that even though several UDFs are being invoked, there is a *single MapReduce program* invoking those UDFs. Pig decides which interface to use based on the following set of rules:

- If all the UDFs implement the Algebraic interface, Pig uses the Algebraic interface. This is the most efficient method; the amount of I/O between the Mapper and Reducer is significantly reduced.
- If even one of the UDFs does not implement the Algebraic interface, but they all implement the Accumulator interface, Pig uses the Accumulator interface. In this scenario, all the values have to be sent from the Mapper to the Reducer to support the UDF not implementing the Algebraic interface. There is no advantage to using the Algebraic interface for the UDFs that support it.
- If even one of the UDFs does not support the Algebraic or Accumulator interface, Pig defaults to using EvalFunc.exec() on all the UDFs. The entire set of values for a Reducer key has to be accumulated in a DataBag instance for the UDF not supporting the Algebraic or Accumulator interface. Hence, there is no benefit to use the Accumulator or Algebraic interface on the UDFs implementing them. This is the reason why Pig requires that all UDFs correctly implement the EvalFunc.exec() method. In a FOREACH GENERATE invocation comprising multiple UDF invocations, if there is even a single UDF that does not implement the Algebraic or Accumulator interface, Pig defaults to using the EvalFunc.exec() method.

## Writing and Using a Custom FilterFunc

This section demonstrates how to write and use a custom UDF. We create a simple filter function to emulate the FILTER function of the sample.pig class. We extend the FilterFunc class, which is an implementation of the EvalFunc class that returns a Boolean value.

Listing 11-12 shows the implementation of the custom UDF that we call CustomIf. It checks whether the value contained in the input tuple contains the string "A" or "B". Note the implementation of the Boolean method isTrue(Tuple input). The if condition checks whether the input tuple contains a DataBag or (else part) contains a single record. The former is required when the UDF is invoked from the Reducer, and the latter (else part) is used when the UDF is invoked from the Mapper.

***Listing 11-12.*** CustomIf.java

```
package org.apress.prohadoop.c11;
import java.io.IOException;
import org.apache.hadoop.io.Text;
import org.apache.pig.FilterFunc;
import org.apache.pig.backend.executionengine.ExecException;
import org.apache.pig.data.Tuple;


public class CustomIf extends FilterFunc {
   @Override
    public Boolean exec(Tuple input) throws IOException {
        returnisTrue(input);
    }
```

```
    private static Boolean isTrue(Tuple input) throws IOException {
        try {
            Object o = input.get(0);
            if (o instanceof DataBag) {
                DataBag db = (DataBag) o;
                Iterator it = db.iterator();
                while(it.hasNext()){
                    Object s = it.next();
                    if(isTrue(s.toString())){
                        return true;
                    }
                }
            }
            else{
                return CustomIfOptimized.isTrue(o.toString());
            }
        } catch (ExecException ee) {
            throw ee;
        }
        return false;
    }

    private static Boolean isTrue(String s){
        if(s.equals("A") || s.equals("B")){
            return true;
        }
        return false;
    }
}
```

This function is used as follows:

1. Perform a Maven build. You should obtain the JAR file `prohadoop-0.0.1-SNAPSHOT.jar`.
2. Register the JAR file in the Pig script with this call:

    ```
    REGISTER prohadoop-0.0.1-SNAPSHOT.jar;
    ```

3. Provide a more user-friendly name for the function. Although this step is optional, it makes for more readable code.

    ```
    DEFINE MY_IF org.apress.prohadoop.c11.CustomIf;
    ```

4. Invoke `MY_IF` as follows:

    ```
    C = FILTER A BY MY_IF(b);
    ```

The modified Pig script called `samplewithcustomif.pig` is shown in Listing 11-13. The listing assumes that the JAR file is placed in the same folder as the Pig script.

***Listing 11-13.*** samplewithcustomif.pig

```
REGISTER prohadoop-0.0.1-SNAPSHOT.jar;
DEFINE MY_IF org.apress.prohadoop.c11.CustomIf;

A = LOAD 'input/pigsample/a.txt' USING PigStorage(',')
                            AS (a:int,b:chararray,c:int);
B = LOAD 'input/pigsample/b.txt' USING PigStorage(',')
                            AS (d:int,e:chararray,f:int);
C = FILTER A BY MY_IF(b);
D = JOIN C by a,B by f;
E = GROUP D by b;
F = FOREACH E GENERATE group, COUNT(D);
STORE F INTO 'output/pigcustomif/';
```

This code is included in the book source code as `org.apress.prohadoop.c11.CustomIfOptimized`. You should read the source code of this class for details on this implementation. A script that invokes `CustomIfOptimized` is as follows:

```
REGISTER prohadoop-0.0.1-SNAPSHOT.jar;
DEFINE OP_MY_IF org.apress.prohadoop.c11.CustomIfOptimized;

A = LOAD 'input/pigsample/a.txt' USING PigStorage(',')
                            AS (a:int,b:chararray,c:int);
B = LOAD 'input/pigsample/b.txt' USING PigStorage(',')
                            AS (d:int,e:chararray,f:int);

C = COGROUP A BY a, B BY b
D = FILTER C BY OP_MY_IF(A);
E = FILTER D BY OP_MY_IF(B);
```

In the custom UDF example, we did not specify the output types because Pig can usually discover the output types using reflection (this is true when Pig returns an instance of a scalar). In more complex scenarios, you need to provide this information to Pig by defining the output schema by overriding the method `public Schema outputSchema(Schema input)` in the `EvalFunc` class. Refer to the Pig documentation on UDFs for details on how to provide output schema definitions in your UDF implementation.

# Comparison of PIG versus Hive

The most commonly asked question by Hadoop developers is this: why do we need separate APIs in the form of PIG and Hive? Why not just use Hive to develop data-processing pipelines? An excellent blog entry by Alan Gates provides a very detailed answer. The entry can be accessed here: `https://developer.yahoo.com/blogs/hadoop/comparing-pig-latin-sql-constructing-data-processing-pipelines-444.html`.

Here are some of the important points from Alan Gates's blog from the point of view of this chapter:

- Pig Latin is procedural, whereas HiveQL is declarative. SQL always forces the developer to develop pipelines inside out; for example, the `LOAD` portion in SQL is encapsulated in the `FROM` clause, whereas the `STORE` portion is encapsulated in the `SELECT` clause. The `FILTER`s are applied in the `WHERE` clause; and the aggregation queries are split between the `SELECT`, `GROUP BY`, and `HAVING` clauses.
- Pig Latin is more suitable for data pipeline development because it specifies how the data flows explicitly. The `LOAD` portion precedes `FILTER` (as well as other transformations), aggregation, and `JOIN` commands. `STORE` is the last command.
- Unlike HiveQL, Pig allows users to checkpoint data at an intermediate step. If a particular process is likely to take a long time to execute, you can simply `STORE` intermediate results and have a next step in the pipeline `LOAD` the intermediate data and `PROCESS` it. So if a subsequent step fails due to a reason unrelated to the results produced by an earlier step in the process, the results of the earlier step can be reused.
- Regarding splitting pipelines, SQL is designed to produce one output. However, ETL pipelines can take a dataset and split it based on a column value. Using our airline dataset as an example, we might want separate pipelines applied for on-time flights and delayed flights. Although doing this with HiveQL is not feasible, Pig Latin provides a natural `SPLIT` function to develop such complex and highly customized data pipelines.

# Crunch API

This section provides a high-level overview of another API used to develop data pipelines: Apache Crunch. This API, which is rapidly gaining in popularity, is a Java library for writing, testing, and running MapReduce pipelines. The goal of the Crunch API is to simplify the task of developing data pipelines composed of Java-based UDFs.

## How Crunch Differs from Pig

Why should you learn another API if it does exactly what Pig provides? Following is a list of some of the benefits of using the Crunch API instead of Pig:

- *Developer-focused*: Apache Hive and Apache Pig made MapReduce accessible to data analysts who are not experts in Java. Crunch is an API developed for Java developers. Developing data pipelines using MapReduce is a tedious task, even for an expert Java developer, and Crunch provides a higher-level Java API on top of MapReduce to simplify this task.
- *High performance with access to the full MapReduce API*: The Crunch Java API is only a thin layer on top of MapReduce, which makes it very comparable in performance to custom developed raw MapReduce pipelines. Crunch API also has full access to the underlying MapReduce API in case the developer needs to access them.
- *Flexible data model*: Pig and Hive support a tuple-centric view of the data, which is adequate when data is represented like rows in a database table. Due to its nature, Crunch naturally supports richer formats such as Avro, not just tuple-centric ones.

## Sample Crunch Pipeline

This section illustrates using the Crunch pipeline to perform WordCount (refer to the documentation on the Crunch web site for installation instructions for Crunch software).

A Crunch job begins with a `Pipeline` instance, which manages the execution lifecycle of the data pipeline. Version 0.90 release has three pipelines:

- `MapReducePipeline`: Executes as a series of MapReduce jobs on a Hadoop cluster.
- `MemPipeline`: Executes completely in memory. It is suitable for unit testing in the memory of the client machine.
- `SparkPipeline`: Executes as a series of Spark jobs locally or on the Hadoop cluster. Spark is an advanced DAG execution engine for developing high-performance data pipelines using in-memory computing techniques. Spark can run up to 100 times faster than MapReduce jobs. (Refer to Chapter 10 and Chapter 15 for a brief introduction to Spark.)

The skeleton of our sample Crunch pipeline is shown in Listing 11-14. It shows all the package declarations required to write a simple WordCount application in Crunch.

***Listing 11-14.*** Skeleton of SampleCrunchPipeline.java

```
package org.apress.prohadoop.c11;
import java.io.Serializable;
import org.apache.crunch.DoFn;
import org.apache.crunch.Emitter;
import org.apache.crunch.PCollection;
import org.apache.crunch.PTable;
import org.apache.crunch.Pipeline;
import org.apache.crunch.PipelineResult;
import org.apache.crunch.impl.mem.MemPipeline;
import org.apache.crunch.impl.mr.MRPipeline;
import org.apache.crunch.types.writable.Writables;
import org.apache.hadoop.conf.Configuration;
import org.apache.hadoop.conf.Configured;
import org.apache.hadoop.util.GenericOptionsParser;
import org.apache.hadoop.util.Tool;
import org.apache.hadoop.util.ToolRunner;

public class SampleCrunchPipeline extends Configured implements Tool, Serializable{
  public  int run(String[] allArgs) throws Exception {
    String[] args = new GenericOptionsParser(getConf(),
                                allArgs).getRemainingArgs();
    boolean inMemory = Boolean.parseBoolean(args[0]);
    String inputPath = args[1];
    String outputPath = args[2];
    Pipeline pipeline = null;
    if(inMemory){
        pipeline = MemPipeline.getInstance();
    }
```

```
    else{
      pipeline = new MRPipeline(SampleCrunchPipeline.class,
                               getConf());
    }

    /*Rest of the code which performs Word Count*/
  }

  public static void main(String[] args) throws Exception {
    Configuration conf = new Configuration();
    ToolRunner.run(new SampleCrunchPipeline(), args);
  }
}
```

Notice that the class must implement `Serializable`, which is a runtime requirement of Crunch when running a `MapReducePipeline`. It is not required to run a `MemPipeline`.

The class also extends `Configured` and implements `Tool`, which allows the `ToolRunner` class to parse the command-line arguments for the MapReduce jobs and make them available via the `getConf()` method, which is inherited from the `Configured` class. The key purpose for adding the `Configured` and `Tool` interface is that it provides a simple way to override Hadoop parameters via command-line arguments.

The Crunch section of the code arrives in the `run()` method after the application parameter supplies are extracted via the command line. The command line takes three parameters:

- Is the pipeline an in-memory pipeline? If this value is `true`, we create a `MemPipeline` or a `MapReducePipeline`.
- Input directory where the text files are maintained.
- Output directory where the word count will be written to.

## Consuming Input Files in Crunch

After the pipeline is defined, we need to identify the data source. In Crunch parlance, a *source* is the location and format of the data that the pipeline processes. We do not specify `InputFormat` classes as we did with MapReduce jobs. Our Crunch pipeline looks considerably cleaner without excessive plumbing code.

```
PCollection<String> lines = pipeline.readTextFile(inputPath);
```

The core data abstraction for Crunch is `PCollection<T>`, which is similar to Pig's relation or Hive's table and represents a distributed immutable collection of records of type `T`. In our example, the lines of text in the input files are represented as a collection of `String` objects.

### Supporting Various InputFormat Types

In the preceding example, we use the `readTextFile()` helper method to consume a text file. It is similar to using `TextInputFormat`, but we are not limited to just the text format; we can consume virtually any Hadoop format.

To read from a `SequenceFile`, use the following method call, where the key is an instance of `LongWritable`, and `value` is an instance of `MyCustomWritable`. Notice the use of the `From` class, which provides static factory methods for creating common source types. Because a `SequenceFile` comprises a key/value pair, we use `PTable`, a subinterface of `PCollection`, to represent an immutable distributed multimap of key/value pairs. A *multimap* is a map with keys that can hold multiple values; it is unlike a `Map` interface, in which one key can hold only one value. A `SequenceFile` can hold repeated keys, so a multimap interface is necessary to represent a `SequenceFile`.

```
// Reference entries from a sequence file where the key is a //LongWritable and the value is a
custom Writable class.
PTable table = pipeline.read(From.sequenceFile( "/path/to/seqfiles", LongWritable.class,
MyCustomWritable.class));
```

Similar to `read()` and `AvroObject`, use the following line (`MyAvroObject` implements Avro's `SpecificRecord` interface):

```
PCollection myObjects = pipeline.read(From.avroFile("/path/to/avrofiles",
                        MyAvroObject.class));
```

To read from a custom `InputFormat`, use the following call:

```
PTable custom = pipeline.read(From.formattedFile(
                              "/path/to/customformat/",
                               MyFileInputFormat.class,
                               KeyWritable.class,
                               ValueWritable.class));
```

## Tokenizing String Instances

Next, we convert the collection of strings into a collection of words. We will write the first data-processing function that overrides the process method of the `DFn<S,T>` function. The `S` is for the source data type, and the `T` is for the target data type. Our goal is to convert the lines of text into a list of words. The source code for this implementation is shown here:

```
PCollection<String> words = lines.parallelDo(new DoFn<String,
                                          String>() {
    @Override
    public void process(String line, Emitter<String> emitter) {
    for (String word : line.split("\\s+")) {
      emitter.emit(word);
    }
  }
}, Writables.strings()); // Indicates the serialization format
```

The first parameter of `parallelDo` is the `DFn<S,T>` function instance. The `parallelDo` call executes the `DFn<S,T>.process()` function within the underlying Mapper's `map()` method in parallel with each line of the input `PCollection` and returns a new `PCollection`. (Remember that `PCollection` is immutable.) The second parameter of the `parallelDo` function defines the `PType` of the returned collection.

Following is the description of `PType` from the Crunch web page, which provides a general overview of how `PType` fits within the MapReduce context and the various alternatives available within `PType`:

> *The* PType<T> *interface is a description of how to serialize the records in a PCollection, and is used by the Crunch runtime whenever it need to checkpoint or shuffle the data in a* PCollection. *Crunch provides two different serialization frameworks with a number of convenience methods for defining* PTypes; *one is based on Hadoop's* Writable *interface, and the second is based on the Apache Avro serialization project.*

Finally, we count the words as we did in the Reducer in earlier chapters:

```
PTable<String, Long> counts = words.count();
```

The `count()` method applies a series of Crunch methods and generates a count by word; it does not know anything about the nature of data stored in the underlying `PCollection`. Internally, it combines a part of the Mapper and Reducer functionalities. In the underlying Mapper, it executes the following line:

```
context.write(word,1)
```

The actual count by word is performed in the Reducer. Listing 11-15 shows the implementation of the `Aggregate.count()` method from the Crunch source code. `PCollection.count` delegates the count functionality to `Aggregate.Count()`.

***Listing 11-15.*** Implementation of count() in Crunch code base

```
public static <S> PTable<S, Long> count(PCollection<S> collect,
                                          int numPartitions) {
    PTypeFamily tf = collect.getTypeFamily();
    return collect.parallelDo("Aggregate.count",
                             new MapFn<S, Pair<S, Long>>() {
      public Pair<S, Long> map(S input) {
        return Pair.of(input, 1L);
      }
    }, tf.tableOf(collect.getPType(), tf.longs()))
        .groupByKey(numPartitions)
        .combineValues(Aggregators.SUM_LONGS());
  }
```

# Writing to the Output Folder

Next, we tell the pipeline where to write the output. The call is as follows:

```
pipeline.writeTextFile(counts,outputPath);
```

This call writes to `TextOutputFormat`. Similar to the `From` class you saw before, there is a corresponding `To` class that allows you to write to a variety of `OutputFormats`.

## Supporting Various OutputFormat Types

To write to a text file, use the following lines (similar to the helper method call `writeTextFile` in the `PCollection` instance:

```
pipeline.write(words, To.textFile("/put/my/words/here"));
```

To write `PTable` to the sequence file, the lines of code are these:

```
// Write a PTable to a sequence file
PTable textToText = ...;
textToText.write(To.sequenceFile("/words/to/words"));
```

To write PCollection to an Avro file, the lines of code are these:

```
// Write a PCollection to an Avro data file:
PCollection objects = ...; objects.write(To.avroFile("/my/avro/files"));
```

Finally, to write to a custom output format, the lines of code are these:

```
// Write a PTable to a custom FileOutputFormat: PTable custom = ...;
pipeline.write(custom, To.formattedFile("/custom",
                            MyCustomFileOutputFormat.class));
```

## Executing the Pipeline

In the preceding steps, we defined our pipeline. None of the steps above actually executes any data processing on the cluster (or in memory). The pipeline is executed, and the run status is obtained using the following lines of code.

```
PipelineResult result = pipeline.done();

return result.succeeded() ? 0 : 1;
```

The complete source code of the run() method is shown in Listing 11-16.

***Listing 11-16.*** Complete Implementation of the run() Method

```
public  int run(String[] allArgs) throws Exception {
    String[] args = new GenericOptionsParser(getConf(), allArgs).getRemainingArgs();
    boolean inMemory = Boolean.parseBoolean(args[0]);
    String inputPath = args[1];
    String outputPath = args[2];
    Pipeline pipeline = null;
    if(inMemory){
        pipeline = MemPipeline.getInstance();
    }
    else{
        pipeline = new MRPipeline(SampleCrunchPipeline.class,
                                getConf());;
    }
    PCollection<String> lines = pipeline.readTextFile(inputPath);
    PCollection<String> words = lines.parallelDo(new DoFn<String,
                                                String>() {
        @Override
        public void process(String line, Emitter<String> emitter) {
        for (String word : line.split("\\s+")) {
          emitter.emit(word);
        }
      }
```

```
        }, Writables.strings());
        PTable<String, Long> counts = words.count();
        pipeline.writeTextFile(counts,outputPath);
        PipelineResult result = pipeline.done();
        return result.succeeded() ? 0 : 1;
    }
```

This concludes our brief introduction to Crunch. Note that we rarely used any MapReduce APIs while developing our WordCount application. Yet we have the flexibility to configure the underlying MapReduce framework in which the Crunch pipeline executes because the `SampleCrunchPipeline` class implements the `Configured` and `Tool` interfaces. We developed our pipeline using only Java. The Crunch way of developing data pipelines is very useful for Java developers who are looking for an API to assist with MapReduce development without having to constantly work with the low-level classes of the MapReduce API.

Crunch is an extensive API, so we merely scratched the surface in this chapter. You should read the Crunch documentation on its web site to explore the full scale of what Crunch can do.

# Summary

This chapter explored how data pipelines can be created on the Hadoop platform using a high-level language called Pig Latin. This language allows a user who is not a Java developer to perform data-processing operations on the data stored in Hadoop. Users familiar with data-processing languages such as SAS find Pig to be a useful abstraction to work with Hadoop data.

We also explored how to develop custom UDFs using the Java API. This feature allows Pig to be extended to support custom use-cases that are not covered by the functions that are part of the basic Pig API.

Finally, we considered the Crunch API, which is a Java API that operates slightly higher than the MapReduce API. This API is suitable for a Java developer who wants to develop data-processing pipelines using a level of abstraction above MapReduce, yet needs access to the MapReduce API if necessary.

CHAPTER 12

# HCatalog and Hadoop in the Enterprise

Previous chapters of this book explored various features of Hadoop. We started with the batch-oriented MapReduce programming model and discussed how this model can support data warehousing with Hive and data pipeline development with Pig.

One of the major obstacles of adopting Hadoop in the Enterprise is that Hadoop implementations still require a low-level understanding of the system, including working with files in the distributed file system. Users of databases and ETL systems are used to working with abstractions such as databases and tables, and Hive supports this abstraction. Hive is not suitable for ETL, however; Pig and MapReduce programs are better suited for ETL offloading tasks. Although Pig and MapReduce still work with files in the file system, ETL users and vendors that develop tools prefer to work with the familiar abstraction of databases and tables even when using Pig and MapReduce programs.

HCatalog, the new API that provides this abstraction, brings familiar abstractions to Pig and MapReduce. It is designed to support ETL tool vendors to communicate with the HDFS using abstractions familiar to database users. HCatalog is designed to facilitate the use of Hadoop in a mature Enterprise.

HCatalog is a table and storage management layer that abstracts HDFS files into a familiar database and tables relational view, which enables users of various Hadoop APIs such as Pig, Hive, and MapReduce to be insulated from the low-level details of HDFS-based data-storage. Instead of having to know which directories and files contain the data, the users of these APIs can simply provide the name of the database and table to access the data.

## HCatalog and Enterprise Data Warehouse Users

Figure 12-1 shows how HCatalog supports a typical BI or ETL user. Both of these users use SQL-based tools to support day-to-day tasks. HCatalog exposes files in the HDFS as databases and tables in the databases. HCatalog also provides a RESTful API to enable these tools to query HCatalog for a list of databases and tables contained in these databases.

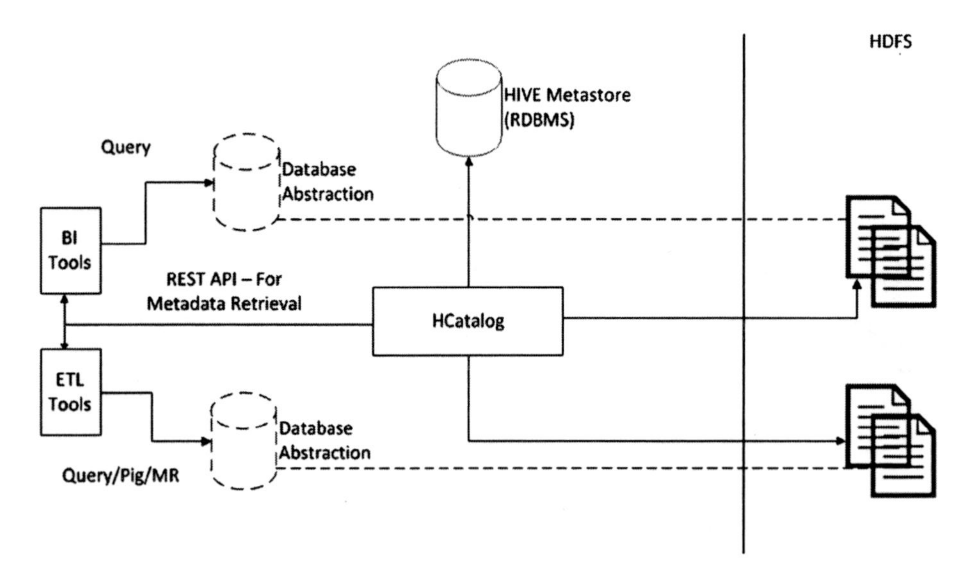


***Figure 12-1.*** *HCatalog for a SQL tools user*

Using HCatalog, Hive, Pig and MapReduce can collectively use the same metadata. A table can be generated via tools such as Pig and MapReduce and queried by using Hive. An ETL platform can now use Hive, Pig, and MapReduce in the same pipeline and interface using logical abstractions such as "databases" and "tables" instead of the physical abstractions such as HDFS folders and files, which enables ETL pipelines to become arbitrarily complex and consequently more effective. This is not unlike a user plugging a Java program into a mostly SQL query-based DataStage (IBM product) pipeline to execute a complex array of tasks in one step. Similar to this example, users can use Hive for simple tasks, plug in Pig scripts for moderately complex tasks, and use MapReduce for extremely complex components. All these components can be part of the same logical pipeline. We show how HCatalog enables this integration and simplifies the use of Hadoop or a typical Enterprise user.

# HCatalog: A Brief Technical Background

Having discussed the role of HCatalog from the point of view of a typical Enterprise user, let us look at some of the underlying technical details of HCatalog in this section. You will understand how HCatalog works internally and become prepared for upcoming sections on how HCatalog works with MapReduce and Pig frameworks.

HCatalog provides a relational view for HDFS files that have a SerDe defined for them. A SerDe, as discussed in Chapter 10, is a serializer/deserializer component that allows Hive to read data into a table and write data back to the HDFS in any format. Out-of-the-box HCatalog supports the following types of files:

- The Record Columnar File (RCFile) format is used to support column-based storage in the HDFS.
- The Optimized Row Columnar (ORC) format is a more efficient version of RCFile format (this efficiency is with respect to storage and data retrieval). Refer to this link for more details: `https://cwiki.apache.org/confluence/display/Hive/LanguageManual+ORC`
- TextFile is a typical character delimited file format. The delimiter can be custom and does not always need to be a comma.

- Javascript Object Notation (JSON) is a DOM-based format that is similar to XML except that it is more lightweight.
- SequenceFile is a native Hadoop-based binary format we discussed in Chapter 7. It allows storage of binary data as key-value pairs where keys do not need to be unique, and the keys and values can be complex writable types. It supports very efficient storage and retrieval mechanisms.

A custom format can also be supported if a SerDe is provided for it (for example, if a user is using Python or C++, and the data is stored as Avro instances). As discussed in Chapter 7, Avro is a language-neutral binary format. A custom SerDe for the specialized Avro schema allows users to read/write instances of their custom Avro schema using Hadoop streaming (a utility that supports the use of the Hadoop framework using languages other than Java). A high-level architecture for HCatalog is shown in Figure 12-2.

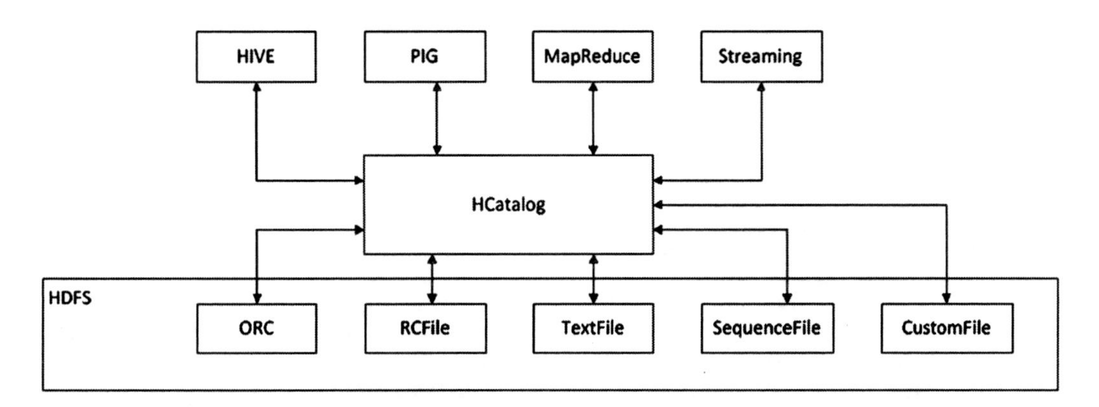


***Figure 12-2.*** *HCatalog-supported interfaces*

Figure 12-3 shows the high-level architecture for HCatalog. The key features are as follows:

- The Hive metastore client is used by HCatalog classes to retrieve and update schema definitions for the databases, tables, and partitions of tables stored as files in the HDFS.
- Similar to Hive, HCatalog uses a SerDe to read and write a record to a table.
- HCatalog uses specialized I/O format classes (`HCatalogInputFormat` and `HCatalogOutputFormat`) to allow MapReduce programs to work with common abstractions such as tables. These I/O formats abstract away the low-level details of working with the actual files in the file system. These classes use the SerDes defined for the underlying file system to read and write records to the tables. The information about appropriate SerDes for the file is retrieved from the Hive metastore.
- HCatalog also provides specialized loadand store classes (`HCatLoader` and `HCatStorer`) to allow Pig to interact with the tables defined in the HCatalog. These classes in turn use the I/O format classes mentioned earlier.

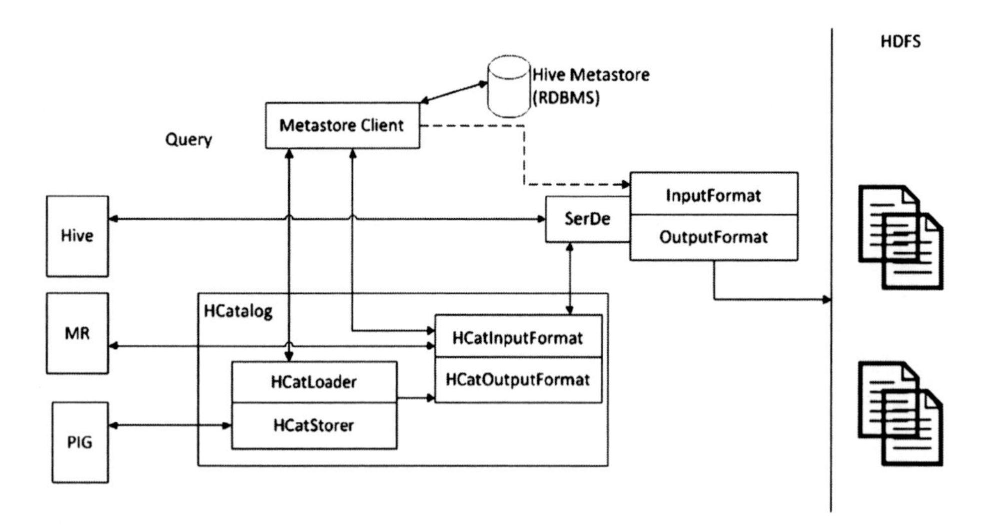


***Figure 12-3.** High-level HCatalog interfaces*

# HCatalog Command-Line Interface

HCatalog is built on top of the Hive metastore and supports Hive DDL commands. HCatalog provides a CLI to manage databases, tables, and partitions. Tables can be partitioned (refer to Chapter 10). Data partitioning makes the Hadoop programs run faster because partitions are stored in separate files. The programs need to read the files only for the partition requested. For example, assume that the Sales data file is partitioned by a year-month combination. If data is requested for 2001-01 (January 2001), only the files in the folder for 2001-01 need to be read.

HCatalog CLI supports creating, altering, and dropping databases and tables. A full listing for supported CLI functions can be found at `https://cwiki.apache.org/confluence/display/Hive/HCatalog+CLI`.

# WebHCat

`WebHCat` is a REST API for HCatalog. Using WebHCat and its underlying security schemes, programs can securely connect and perform operations on HCatalog through a general REST-based API.

- REST-based API calls are available to manage databases, tables, partitions, columns, and table properties.
- PUT calls are used to create/update; GET calls are used to describe or get listings; and DELETE calls are used to drop databases, tables, partitions, and columns.

The operations that WebHCat allows through a REST interface are the following:

- DDL operations
    - Execute a DDL command
    - List and describe databases and tables in databases
    - Create/drop/alter databases and tables
    - Manage table partitions, including operations such as creating and dropping partitions
    - Manage table properties

- Job management
    - Remotely start and manage MapReduce jobs
    - Remotely start and manage Pig jobs
    - Execute Hive queries and commands
    - Remotely obtain job status and manage job execution

## HCatalog Interface for MapReduce

MapReduce is the most flexible framework available. When using Hadoop for perfoming a typical Extract/Transform/Load (ETL) process, it is possible to encounter a complex requirement for which a MapReduce program will be the most efficient way to solve the problem. HCatalog supports development of MapReduce programs through custom implementations of `InputFormat` and `OutputFormat` classes which interface with HCatalog services. `HCatInputFormat` is the custom `InputFormat` class and `HCatOutputFormat` is the custom `OutputFormat` class which supports the use of HCatalog in custom MapReduce programs. The corresponding RecordReader and RecordWriter implementations utilize the metadata from HCatalog to identify the appropriate `SerDe` implementation class(The `SerDe` interface is discussed in Chapter 10) to use to interpret raw bytes into HCatRecord instances or to convert HCatRecord instances into raw bytes. These custom `InputFormat` and OutputFormat classes also allow developers to utilize the flexibility of writing raw MapReduce programs while still using higher level abstractions like tables and schema definitions. Remo

Listing 12-1 shows how MapReduce programs utilize HCatalog to read/write from/to a table with a well-defined schema. In this example a MapReduce program utilizes HCatalog through specialized `InputFormat` and `OutputFormat` implementations to read from a partitioned table (we discussed Partitioned tables in Chapter 10) to to perform a `GroupBy` operation. It demonstrates how the filight records in the partition for the year 1987 are read by a MapReduce program and aggregated. We calculate two aggregation values: total number of flights with arrival delays and the total number of flights with departure delays. The results of the aggregation are written into an existing output table called `FlightSummary` which has been defined with the following three columns:

- Year
- TotalArrivalDelays
- TotalDepartureDelays

***Listing 12-1.*** The run( ) method of MR program using HCatalog

```
public int run(String[] args) throws Exception {
  Configuration conf = getConf();
  args = new GenericOptionsParser(conf, args)
                              .getRemainingArgs();

  String dbName = "MyDb";
  String inputTblName = "FlightDetail";
  String outputTblName = "FlightSummary";

  /*
   * Specify the following
   * -database name
   * -table name
   * -filter criteria
   */
```

```
  Job job = new Job(conf, "MyJob");
  HCatInputFormat.setInput(job,
                           InputJobInfo.create(dbName,
                           inputTblName,
                           "year=1987"));
  // initialize HCatOutputFormat

  job.setInputFormatClass(HCatInputFormat.class);
  job.setJarByClass(MyMRJob.class);
  job.setMapperClass(MultiGroupByMapper.class);
  job.setReducerClass(MultiGroupByMapper.class);

  job.setMapOutputKeyClass(Text.class);
  job.setMapOutputValueClass(IntWritable.class);
  job.setOutputKeyClass(IntWritable.class);
  job.setOutputValueClass(DefaultHCatRecord.class);

  //Specify which partition to write to
  Map<String, String> partitions = new HashMap<String, String>();
  partitions.put("year", "1987");

  HCatOutputFormat.setOutput(job,
                             OutputJobInfo.create(dbName,
                             outputTableName, partitions));
  HCatSchema s = HCatOutputFormat.getTableSchema(job);
  HCatOutputFormat.setSchema(job, s);
  job.setOutputFormatClass(HCatOutputFormat.class);
  return (job.waitForCompletion(true) ? 0 : 1);
}
```

HCatInputFormat uses a setInput method to specify the following:

- Database
- Table
- Partition filter

HCatalog uses the metadata definition about the table stored in the Hive Metastore client to determine which SerDe implementation class to utilize. This class is then used by the underlying RecordReader (see Chapter 7 for details on RecordReader) to convert raw data into an HCatRecord instance that is fed to the Mapper class. Recall from Chapter 10 that the SerDe definition is part of the CREATE TABLE invocation. Table metadata provided in the CREATE TABLE call is stored in the Hive Metastore. Note the configuration for HCatalogOutputFormat. The following are configured:

- Database
- Table
- Partition

In this case, it does not make sense to specify a partition (it is provided only as an illustration). For writing to multiple partitions, multiple MapReduce jobs are needed.

The code for the `MultiGroupByMapper` is shown in Listing 12-2. The skeleton code demonstrates how the output of the mapper is a key comprising the year and an `IntWritable` instance that takes the value `1` if there is an arrival delay and `2` if there is a departure delay. Note from the highlighted line how the values are extracted from the `HCatRecord` instance. A large number of aggregations can be performed using this pattern.

***Listing 12-2.*** MultiGroupByMapper

```
public static class MultiGroupByMapper extends
        Mapper<WritableComparable, HCatRecord,
               IntWritable, IntWritable> {
  @Override
  protected void map(
           WritableComparable key,HCatRecord record,Context ctx)
           throws IOException, InterruptedException {
           //Perform Mapper actions
           int year = (Integer) record.get(1)
           //getXXXDelay() are custom functions
           int arrivalDelay = getArrivalDelay(record);
            int depDelay = getDepartureDelay(record);
            if(arrivalDelay>0)
              context.write(new IntWritable(year),
                            new IntWritable(1));
             if(depDelay >0)
              context.write(new IntWritable (year),
                            new IntWritable(2));
        }
    }
```

Listing 12-3 shows the `MultiGroupByReducer` listing, which counts the total number of arrival and departure delays and writes out the value. As a result, a record gets created in the underlying table partition.

***Listing 12-3.*** MultiGroupByReducer

```
public static class MultiGroupByReducer extends
        Reducer< IntWritable,Iterable<IntWritable> values,
                 IntWritable, HCatRecord> {
  @Override
  protected void reduce(
             IntWritable key,
             Iterable<IntWritable> values,
             Context context)
             throws IOException, InterruptedException {
    int year = key.getInt();
    Iterator<IntWritable> iter = values.iterator();
    int sumArrivalDelay = 0;
    int sumDepartureDelay = 0;
    while (iter.hasNext()) {
       IntWritable val = iter.next();
       if(val.get()==1) sumArrivalDelay++;
       if(val.get()==1) sumDepartureDelay ++;
    }
```

```
    HCatRecord record = new DefaultHCatRecord(3);
    record.set(0, year);
    record.set(1, sumArrivalDelay);
    record.set(1, sumDepartureDelay);

    context.write(null, record);
  }
}
```

The HCatalogOutput format has been configured with the output table name. The output table name is used to query the Hive Metastore utilized by HCatalog to obtain the SerDe implementation configured for the table. . This SerDe class is used to serialize data into raw bytes through the `context.write()` invocation in the Reducer class.

Note that the program makes no reference to the I/O paths; it simply uses high-level abstractions such as "Database" and "Tables". The programmer no longer has to be aware of the exact paths of data storage in the HDFS.

We have only skimmed the surface of the ways in which HCatalog can be used with MapReduce. Refer to the full documentation here for more information: `https://cwiki.apache.org/confluence/display/Hive/HCatalog+InputOutput`.

## HCatalog Interface for Pig

In Chapter 11 we discussed Pig which is a platform for analyzing large data sets using a high level, declarative style data flow language called Pig Latin. Pig allows users to describe data pipelines which are compiled by the Pig engine into MapReduce programs. Thus Pig allows users to focus on data flows and data transformations instead of having to worry about low level details of writing MapReduce programs. The HCatalog interface supports Pig. The `HCatLoader` and `HCatStorer` classes are used for loading and storing tables using HCatalog. Internally these classes depend on `HCatInputFormat` and `HCatOutputFormat`, respectively for reading and writing data.

In the previous section we demonstrated how HCatalog can be used with MapReduce programs. In this section we will use the same example to demonstrate how HCatalog simplifies the use of Pig.

Before we discuss Pig in the context of HCatalog let us quickly review how Pig can be used without HCatalog. First, we need to load the data into a variable. If the data is partitioned by year, the `LOAD` statement is this:

```
A = LOAD '/user/hdfs/airline/1987/' as (year:int,......)
```

Not only do we need to know the exact path in the HDFS file system but we also have to specify the schema during `LOAD`.

Next, we perform various operations and finally store them back to a file. The Pig command to do that is shown here. `C` is the name of the variable that contains the summary results (we skipped showing the lines that calculate the summaries).

```
STORE C into '/user/hdfs/airlinesummary/1987/'
```

Now let us review how HCatalog simplies the use of Pig. The same Pig script using HCatalog reads as follows:

```
A = LOAD 'FlightDetails' USING org.apache.hcatalog.pig.HCatLoader();
//To apply a parition, call filter immediately after load
B = filter A by year == 1987

//Remaining PIG script including UDF invocations to perform
//the same calculations we did in the MR programs
Store C into 'FlightSummary' using HCatStorer()
```

Note the key difference: we are working with tables, not files. The HCatLoader and HCatStorer will use the table names provided to them to query the metastore to determine the schema and the SerDe implementation classes. There is no need to explicitly specify the schema. Not only is the code much easier to read but it is also much easier to integrate with other components of the ETL pipeline because they can all share the same metadata. If there are any metadata changes, such as the schema being extended or the data location changing, all components continue to operate because the I/O formats transparently incorporate these changes through the metastore integration. Using HCatalog, we have decoupled the administration of the system from the use of the system. This feature makes HCatalog an ideal integration interface for the Hadoop APIs.

In this section, we have only skimmed the surface of using HCatalog with Pig. For a detailed description of the data types supported and how to work with table partitions, refer to the HCatalog documentation at `https://cwiki.apache.org/confluence/display/Hive/HCatalog+LoadStore`.

## HCatalog Notification Interface

HCatalog supports a notification system. Currently, notifications are supported for two types of events:

- When a new partition is added to a table.
- When a set of partitions is added to a table. These notifications are useful when you want to perform a follow-up action only after all the required partitions are added.

These push-based notifications are delivered via the open Java Message Service (JMS) interfaces. Workflow tools such as Oozie respond to these notifications by executing an ETL workflow in response.

From an Enterprise perspective, this feature is valuable for several reasons:

- Without the notification feature, programs that use a freshly loaded table or partition of the table have to keep polling the HIVE Metastore via the Metastore client to determine if the table or partition is available.
- With the notification feature, a workflow can be launched based on the notification.

Detailed documentation on how to create topics, and subscribe and respond to a message sent to a topic can be obtained here: `https://cwiki.apache.org/confluence/display/Hive/HCatalog+Notification`.

# Security and Authorization in HCatalog

HCatalog makes Hive easier to use, and both authentication models are identical. But because HCatalog also supports Pig and MapReduce, the Hive authorization model is inadequate for HCatalog.

The default Hive authorization model uses familiar database concepts such as users and roles that are granted permissions to perform DDL and DML operations on databases and tables. However, unlike databases, Hive does not have full control over the data stored in its tables; these tables are stored in the HDFS file system as files. The table abstraction is simply an interface. Users can access the data by going to the files in the HDFS, even if their rights to a table based on those files are revoked. Conversely, users might not be able to access the table, even if they have access rights to the files underlying the table in the HDFS.

HCatalog does not use this default model; instead it uses storage-based authorization. Users' privileges for a database and table are inherited from the privileges they have on the files in the file system, which is more secure because it is not possible to subvert the security of the system by changing the abstraction levels. Although it might be more limited than traditional database style security in the sense that fine-grained, column-level permissions are not feasible, it is more appropriate because HCatalog supports Pig and MapReduce in addition to Hive.

# Bringing It All Together

At a high level, a typical Enterprise-level ETL system comprises of periodic batch jobs which perform three steps:

1. Fetch daily transaction data from an operations database
2. Perform aggregations on the transaction data
3. Load the aggregations back into an Enterprise Data Warehouse (EDW). The EDW is queried by the Business Intelligence applications in the Enterprise.

HCatalog has enough features to enable such an Enterprise-level ETL system. The following workflow is suitable for an ETL system:

1. The HCatalog administrator defines databases and tables in the databases.
2. Data is exported from the operations databases to files
    a. Transaction sata is bulk exported into files from the tables of the operations database.
    b. Master data is bulk exported into files from the master data management system.
3. A user or a metadata management system adds these files to the HDFS in preconfigured directories. Examples of commands used to perform this operation are `distcp` and `hdfs fs -copyFromLocal`. If possible steps 2 and 3 should be combined into one step. It can be operationally time consuming to first copy tables into a local file system and then copy these local files into HDFS. I have utilized the UNIX Named Pipe feature to stream data into HDFS as it is being exported from the transaction tables. A named pipe is used to transfer data from one application (table export process) to another application (the Hadoop `fs -copyFromLocal` command) without the use of an intermediate file.
4. A user uses the CLI, or a metadata management system uses the REST interface provided by WebHCat, to load or create partitions for the tables defined in HCatalog, using the files copied to HDFS in the previous step.
5. In response to the data load into the HCatalog based table partitions in the previous step, the HCatalog system sends an event notification (via the HCatalog notification interface) to the subscribers of this event. Such a subscriber could respond to this event notification by triggering an HCatalog based Pig or Hive script or a MapReduce program. This program will perform various aggregations on the raw data loaded into the HCatalog tables in the previous step.
6. The previous step produces output files which are mapped to tables and table partitions transparently using HCatalog.
7. In response to the generation of output partitions in the previous step, the HCatalog system sends notification events to downstream processes in the ETL workflow to trigger a set of jobs to bulk load these output files into the Fact and Dimension tables of an EDW. Again HCatalog supports these processes by providing them with higher level abstractions like Databases and Tables on top of raw HDFS files.

# Summary

From an Enterprise perspective, Hadoop solves the same problems that databases have solved for decades, except that it does this when the scale of data is so large that it cannot be done using databases without investing in expensive MPP-based database systems. Yet enterprises have been cautious about adopting Hadoop and making it an integral part of their operations systems. The reasons for this caution have been Hadoop's lack of tooling and its low-level interfaces.

HCatalog is the step in the direction of abstracting away these complexities for high-level BI and ETL users. It is a promising API that allows high-level tools to be developed. These tools not only bring Hadoop to the traditional enterprise user but also boost productivity by allowing users to work with familiar abstractions such as databases and tables instead of files, `InputFormats` and `OutputFormats`.

This chapter explored how HCatalog is the metadata glue that binds Hive, Pig, and MapReduce together. You saw how an implementation of custom SerDes can enable an adaptation of highly customized formats to familiar abstractions such as a relational table.

The HCatalog REST-based endpoints for administration and development enable tools developers to develop user-friendly ETL tools on top of Hadoop. The notification mechanism allows push-based ETL workflows. ETL workflows are usually executed during nonworking hours, and a notification mechanism enables steps in a workflow to be coordinated without requiring human interaction.

CHAPTER 13

# Log Analysis Using Hadoop

The explosive growth of the Web toward the end of the 20th century led to web scale data, particularly log files. Suddenly, everyone who had a web site generated lots and lots of web access logs that were initially used as to debug problems with a web site. Eventually, organizations realized that web access logs were a rich source of information about their customers and potential customers. Click stream analysis offered insights into customer behavior within a web site, and search query analysis offered examples of products and services most important to customers.

Log files are not limited to web servers. In fact, almost every application generates log files of some kind. This information is of great value in identifying root causes of failures to improve software quality and improve customer experience.

Hadoop's capability to analyze large-scale unstructured data makes it an excellent platform for scalable log analysis. This chapter explores some applications of log file analysis and architectures that can be used to analyze log files.

## Log File Analysis Applications

This section examines some common applications that involve log file analysis. These examples help facilitate understanding of the kinds of problems that can be handled using log file analysis.

### Web Analytics

Web analytics started as a way to measure web traffic. In the beginning of the Web, web counters (see the following image) were common on web sites.

These counters increased every time the web page was accessed. The click count indicated the popularity of a site or particular pages within a site. These simple counters offered a clue about how many times a particular page was accessed, but did not determine who was accessing the page. In some cases, many of the accesses came from automatic web crawlers for search engines.

Over time, web analytics matured considerably. With the exponential growth of web sites and web traffic, web analytics became an efficient tool for market research. The effectiveness of product launches and advertising campaigns could be measured in real time using web analytics.

Every access to a web page automatically creates a record on a web server. Each record contains useful information such as time and date, complete URL accessed by the client, client IP address, and so on. Analyzing these logs at a particular time can provide insight into the most popular pages within a site, the referrer (or how a client got to the site), and a general idea of clients' geographic locations.

Open-source tools such as Webalizer can analyze logs from modest-sized web sites and provide user-friendly charts and graphs to analyze demographics of the clients visiting the web site. For example, a Webalizer chart can show the distribution of users based on geographic locations. More advanced click analytics or customer lifecycle analytics can determine the percentage of vistors that become paying customers or identify reasons why they choose not to buy.

Although a snapshot in time is useful, trend analysis over time can scientifically measure the efficacy of product launches or marketing campaigns. Although brief snapshots of logs can be analyzed on single machines, trend analysis of long-term data can grow to levels that require a scalable solution such as Hadoop. Hive, Pig, MapReduce, and higher-level tools such as RHadoop can also be used for complex, domain-specific analysis.

## Security Compliance and Forensics

Network security has become as critical, if not more critical, than physical security for any organization with Internet connectivity. High-profile security failures have led to a significant loss of customer confidence and, in some cases, to resignations of chief executives of large organizations.

Although it is difficult (if not impossible) to eliminate all security breaches, log file analytics can identify vulnerabilities, enforce security compliance, and collect forensic evidence in the event of a security breach.

Similar to web logs, network equipment such as routers and firewalls can log access or attempted access to irregular addresses and ports. Analyzing these logs can identify the kinds of attacks a perpetrator is attempting. For example, port scans from certain IP addresses can be used to proactively block specific IP addresses and reduce the potential for a successful attack.

In many cases, vulnerabilities originate from inside an organization. Employees might access web sites or services with known risks. This access might be inadvertent or intentional, depending on the nature and regularity of the conduct. Given the number of employees and services constantly accessing the Internet in most organizations, the volume of firewall logs requires a scalable solution for analysis.

In the event of a breach, logs can be used to reconstruct how security measures failed and how they can be improved to avoid future attacks. Similarly, legal compliance with the Health Insurance Portability and Accountability Act (HIPAA) and Sarbanes-Oxley requires regular security audits. Analyses of server and firewall logs are methods of measuring compliance.

## Monitoring and Alerts

Data collection and monitoring, once done by hand, is now increasingly automated. When aggregated, modern systems can generate large volumes of structured and unstructured data. Here are some examples:

- Utility readings, such as electrical energy, gas, and water usage
- Security monitors, including open/close status, motion detectors, and video cameras
- Industrial control systems; physical plants such as heating, ventilation, and air conditioning (HVAC); and Supervisory Control and Data Acquisition (SCADA) systems
- Transportation systems, including traffic monitors, automated toll collection, subways, trains, fleet management, elevators, escalators, and many others
- Network equipment, switches, routers, terminals, and set top boxes

Data from these systems can be collected through wired or wireless connections. Data used to be used only once and discarded as soon as an adequate data retention period had elapsed. Organizations are increasingly finding that collecting more of this data automatically offers unique business opportunities. For example, collecting electrical power consumption every minute rather than after a month can more efficiently manage electrical power generation and transmission at the local level.

## Internet of Things

In the 1990s, many industry experts predicted the Internet of Things (IoT), referring to a broad range of devices, large and small, connected to the Internet. Figure 13-1 illustrates one vision of technology progression as it relates to the IoT.

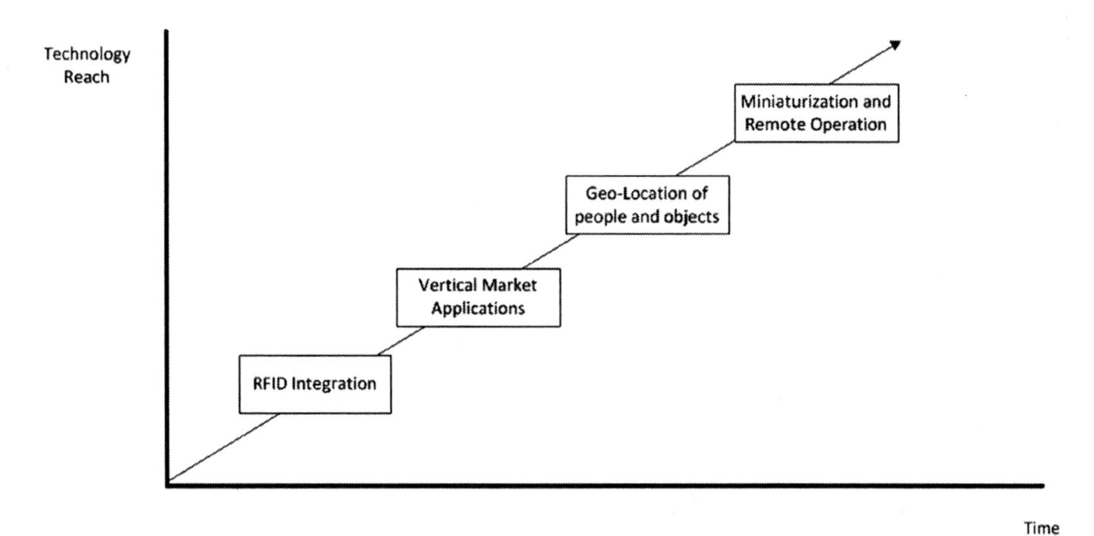


***Figure 13-1.*** *Internet of Things*

The early 2000s saw radio frequency identification (RFID) integration, which was beneficial for the supply chain management industry. It supported applications such as seamless inventory management.

In the first decade of this century, the Web had a major impact on software applications, which led to a preponderance of custom applications with a ubiquitous web interface. Not only the interface but also the underlying technology became ubiquitous. This development resulted in nearly standard methodologies for log generation.

In late 2007, mobile devices became increasingly smarter; by the end of the decade, they had become pervasive, which took geo-location to an entirely new level. Now applications could discover your geo-location and recommend places of interest. Shoppers can not only search for deals online but also receive deals sorted by the nearest locations. Applications can dynamically discover traffic patterns based on how fast certain types of devices are changing their geo-locations and recommend driving directions that are relatively free of heavy traffic. However, this development led to an explosion of the amount of useful data collected based on user behavior, which resulted in log analysis transitioning into a Big Data problem.

The future will lead to further explosion of this data collection and distant objects will not be able to communicate. Imagine a refrigerator communicating with a grocery store via bar-coded items stored in it to request an automatic replenishment delivered to your doorstep. Or an intelligent temperature controller that communicates with your cell phone to decide when to turn it on or off. All these changes will also lead to increased data collection.

It is estimated that the number of devices connected either directly or indirectly to the Internet could reach 30 billion by 2020. As with other automated systems, IoT will regularly generate very large volumes of log data, so Big Data solutions such as Hadoop will be critical to analyzing this data efficiently.

# Analysis Steps

Log file analysis follows a basic three-step approach: loading, refining, and visualizing, regardless of the data source or type of log information.

## Load

In the load step, raw source data is collected from one or more sources and placed in a data store accessible by a Hadoop cluster. The data is usually stored in the HDFS, but it can also be stored in HBase or MongoDB, for example. The choice of data store depends on the structure and nature of the data.

Many tools can be used to load the data. Common tools include Apache Sqoop and Flume. Sqoop (SQL to Hadoop) is specifically designed to transfer data between relational databases and HDFS. Sqoop can also populate Hive or HBase tables directly. Flume is described later in this chapter.

## Refine

Refining raw data can be done with any number of Hadoop platform components: Hive, Pig, MapReduce, Spark, Shark, and so on. Component selection depends on the nature of the data, the kinds of data reduction required, and developer skillsets. There is no one-size-fits-all solution for refinement and analytics. Business value in a particular log file is business-specific, which can be difficult to generalize.

For structured data, Hive, Pig, and Shark work well. For unstructured data, which is common with log files, a MapReduce ETL job might be required. The job could accept raw data from the HDFS and add sufficient structure and regularity for use with higher-level tools such as Hive, Pig, or Shark. For example, when the ETL job encounters a Java exception stack trace, it could extract important information, such as the exception class, message, and list of source file names/line numbers. MapReduce can be used exclusively for analysis and refinement, but doing so can be tedious and lack interactive capabilities. For these reasons, higher-level query tools are commonly used.

The initial stages of refinement usually include filtering to reduce the amount of data processed in subsequent stages. Aggregations by time, geographic location, or business-specific domain provide additional insight. Later versions of Hive support sophisticated window and analytic functions that can be helpful for more complex analysis. Time series and trend analysis can be useful, as illustrated by the query in Google trends (see Figure 13-2).

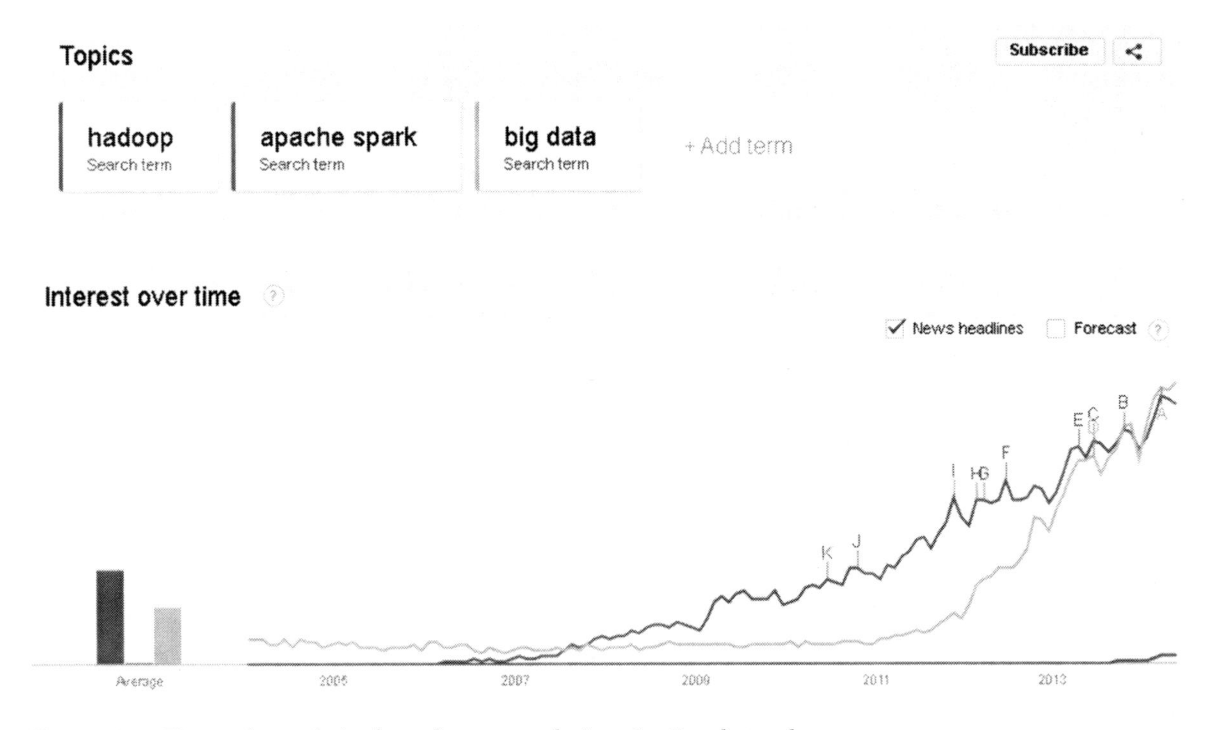


*Figure 13-2. Time series analysis of search term popularity using Google trends*

## Visualize

After data is refined, it can be exported into small manageable files containing the critical information. Many tools exist for visualization, from the most common, such as Microsoft Excel, to more sophisticated tools such as R. These tools work well for graphs and charts. For automated report generation, Java reporting libraries such as JasperReports work well. For visualization of graphs, GraphViz is a powerful and easy-to-use tool.

# Apache Flume

Apache Flume is an open-source project that can be used to reliably perform the load step in log file analysis. Flume can collect data from multiple streaming sources and place them on the HDFS for refinement and analysis. Flume supports failover and recovery to deliver data reliably.

Flume is designed to collect event messages from a number of streaming sources. For example, Flume can collect social media feeds, network packet feeds, and e-mail messages. Flume supports a flexible data source model, which can be easily customized for any type of data source.

## Core Concepts

Flume provides a distributed, highly available, and reliable system for efficiently collecting, aggregating, and moving large amounts of discrete log data from a variety of sources to a centralized data store for analysis. Some of the concepts underlying Flume are as follows:

- *Event*: A binary payload with optional header information that represents the unit of data that is transported from origin to destination.
- *Client*: Responsible for sending data to a Flume agent for processing. In most cases, it is a simple Java class integrated into the application generating the events. Examples of such an application are a web application or the backend backbone for a mobile application suite.
- *Agent*: A high-level construct for various types of Flume-based processes such as sources, channels, and sinks. It is capable of receiving, storing, and forwarding events. Flume supports a number of standard input sources. The following is a list of some of the sources built into Flume:
    - Avro
    - Thrift
    - Java Message Service (JMS)
    - Spooling directory (listen for new files in a directory)
    - NetCat (listen on a specified TCP port)
    - Syslog
    - HTTP
- *Source*: A class that implements specific interfaces to allow it to operate as a source. The role of this component is to receive event messages and forward them to the channel components. For example, an Avro source is an implementation used to receive Avro events from clients or other agents in the data flow.
- *Channel*: A temporary store for events that are delivered to it by various agents. Events stay in the channel until they are consumed by the connected sinks that remove them for processing or further transport.
- *Sink*: A class that implements specific interfaces. Its role is to remove events from the channels and transmit them to the next agent in the flow or to the event's final destination. A sink that transmits its event to the final destination is called a terminal sink. An example of such a sink is the HDFS sink.

Figure 13-3 shows the previous concepts in the context of the overall Flume Architecture.

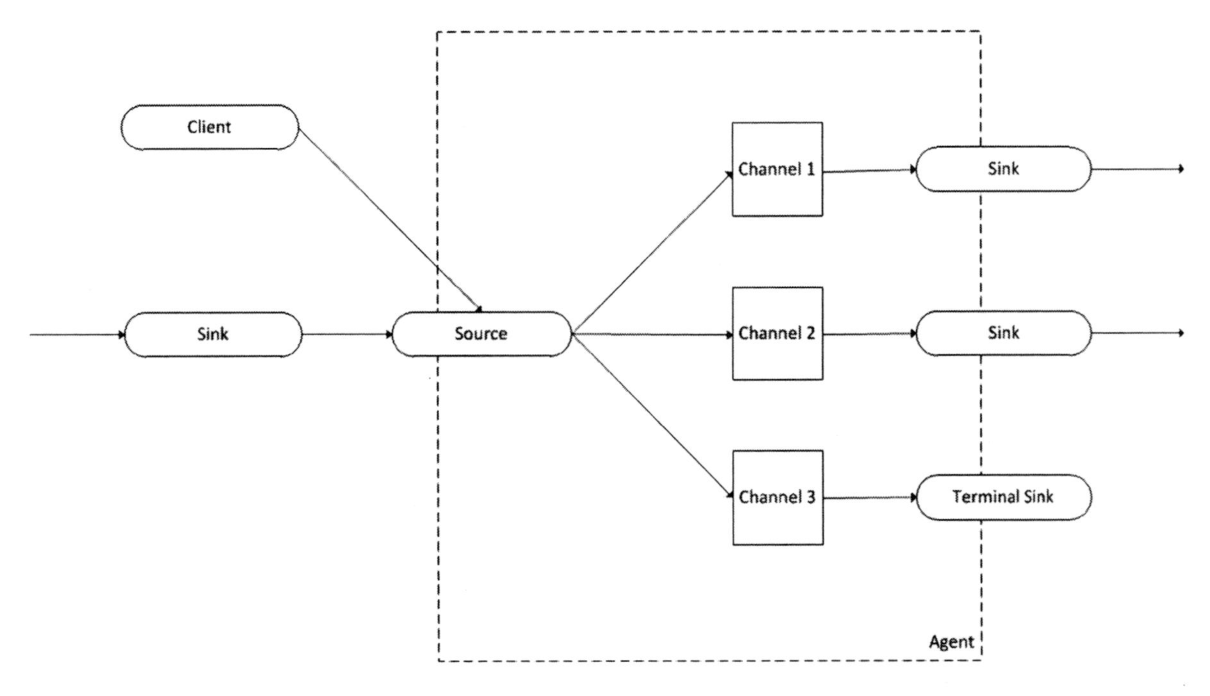


***Figure 13-3.** Flume architecture*

This simple example is a flow with one source, one channel, and one sink. More-complex flows containing multiple sources and sinks are also possible. It is all configured in a file syntactically similar to a Java properties file. At a high level, the configuration looks something like this:

```
# properties for sources
<Agent>.sources.<Source>.<someProperty> = <someValue>

# properties for channels
<Agent>.channel.<Channel>.<someProperty> = <someValue>

# properties for sinks
<Agent>.sources.<Sink>.<someProperty> = <someValue>
```

The details of the configuration are described in the Apache Flume user guide.

Flume performs transactions between sources and sinks to achieve reliability. Events are removed from a channel only after it has been successfully stored on the next channel of the next agent. If an agent fails, events are not lost. Events stored in persistent channels, such as a file channel, can be recovered. In-memory channels are faster than file channels, but memory channel contents are lost if its agent fails.

The Flume user guide provides excellent examples on how to create a Flume-based application: `https://flume.apache.org/FlumeUserGuide.html`.

# Netflix Suro

Netflix Suro is another open-source project with elements that are similar to Apache Flume. Figure 13-4 illustrates the important components of Suro.

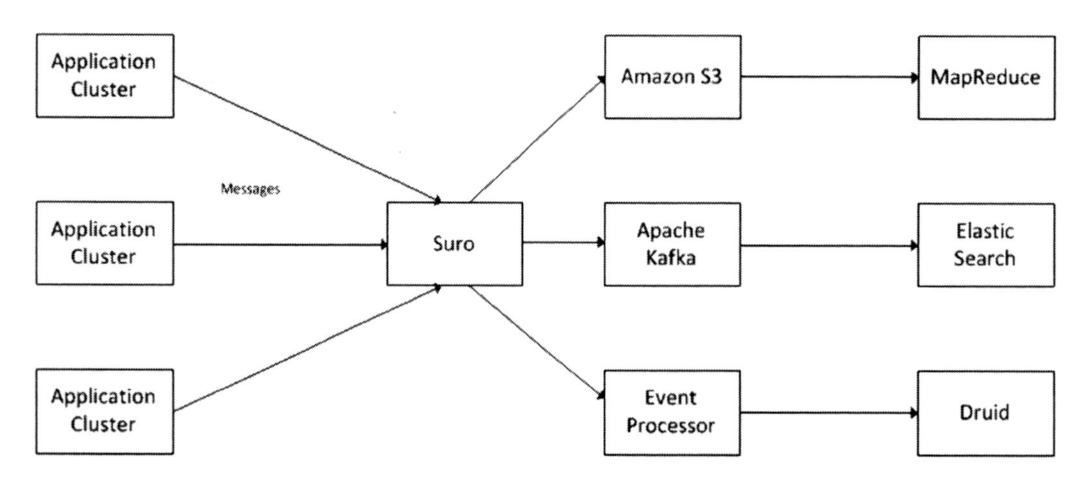


***Figure 13-4.*** *Suro architecture*

Suro, which originated at Netflix, supports batch- and real-time processing of discrete log messages. Netflix collects a lot of user data: about 1.5 million events per second during peak hours and more than 80 billion events per day. Data can be user-activity data or system-generated operational log data. To make effective use of this data, Netflix has to analyze it in a timely fashion. Suro enables this business requirement.

Data analysis is critical for Netflix's business execution. For the scale of data generated, Netflix needed a software infrastructure to be highly scalable and available with a capability to deliver arbitrary events with minimal latency. Suro provides this infrastructure.

Suro was designed to be pluggable from the ground up and it has support for various types of consumers. Out of the box, it supports Amazon Simple Storage Service (S3)-backed MapReduce, Kafka-backed Elastic Search, and Druid. Kafka, which is a publish-subscribe messaging system that originated at LinkedIn, supports messaging on a Big-Data scale using its unique pull-based (instead of the traditional push-based) architecture.

Elastic Search, which is similar to Apache Solr, is a faceted search engine that allows users to rapidly search massive amounts of unstructured and semistructured data tagged with low-dimensional facet information. For example, a shopping web site typically tags its products with category and subcategory metadata. These tags act as facets, and a typical user can navigate them on the left bar of most shopping web sites. Extend this concept to log messages by tagging them with date and hour quarter information and originating application information, and it becomes a powerful cross-dimensional search and navigation tool. Druid is a structured data analysis tool that allows real-time ingestion and analysis of structured time-series data.

You can design your own custom consumers to work with Suro. For example, if your organization is an avid user of Apache Solr instead of Elastic Search, you can create such a consumer. Similarly, if you want to perform analysis using Shark or Impala (refer to Chapter 10), you can also create such a consumer.

A typical data flow into Suro is as follows:

1. Various systems generate events that are sent to Suro along with their desired routing details.

2. Suro routes these events to various consumers.

    a. Events designed for analysis with MapReduce are routed to Amazon S3, where a scheduled MapReduce job picks them up.

    b. Events designed for faceted search and time-series analysis are routed to a Kafka queue. Various subscribers to the Kafka queue then receive the messages based on the type of content they are interested in, index it in Elastic Search, and ingest it into a time-series analysis store such as Druid.

    c. Custom consumers handle the events in custom ways.

3. MapReduce job then generates analysis; business users analyze logs using Elastic Search and Druid.

In a nutshell, Suro is a Big Data orchestration technology for events generated and required to be processed using varying levels of service level agreements typical of batch or real-time processing.

# Cloud Solutions

Amazon Kinesis is a service for streaming large volumes of data into an Elastic MapReduce (EMR) cluster. The data can include logs from multiple sources. EMR is a Hadoop distribution, which is capable of analyzing data using tools such as Hive, Pig, MapReduce, and so on. The processed data can then be written into Amazon S3 or DynamoDB.

Using Amazon Kinesis requires three steps:

1. Create a Kinesis data stream with a name and capacity. The capacity is limited by the number of shards; each shard is capable of 1 MB write capacity and 2 MB read capacity.

2. Write data to a stream. Data blobs are written to the stream through an HTTP `POST` request. At a minimum, a stream name, partition key, and data blob are required. Amazon provides APIs to simplify client interaction with Kinesis streams.

3. Consume data from a stream. Stream data can be consumed by Kinesis applications through a client library. Kinesis applications can write data to Amazon S3, EMR HDFS, or other streams.

When Kinesis is integrated with Amazon EMR, additional analysis is possible. Given that Hadoop is a batch-processing system, a common model for stream processing is to launch analysis jobs at regular intervals. For example, web logs can be analyzed every 10 minutes to identify the most popular search queries or content.

# Summary

Log processing is a special case of the more general class of problems called complex and streaming events processing. Log files contain useful business features, such as using web analytics, identifying security vulnerabilities, conducting audits, and monitoring the emerging Internet of Things. Hadoop, along with other open-source technologies, can be used to efficiently analyze log files and extract business value. This chapter examined the context behind log processing.

We then discussed various technologies available for log processing in the Big Data space. First, we explored Apache Flume, which is integrated with Hadoop. Next, we discussed Apache Suro at a high level. Suro, which originated at Netflix, provides an excellent way to perform not just log processing but also integrated analytics based on tools such as Elastic Search and Druid. Finally, we explored Amazon Kinesis, a cloud-based option for events processing.

CHAPTER 14

# Building Real-Time Systems Using HBase

So far in this book, you have studied MapReduce and its derivative APIs. You learned about batch processes that emphasize throughput (the amount of work done per unit of time) over latency (the response time). Hadoop jobs can take hours to run, but the amount of work done per unit time is phenomenal. Yet there are use-cases for which response time is important. When a Facebook user posts a comment, it is sent as a notification to all the friends. This function needs to happen in near-real time, and it would not be ideal to have to wait until the end of the day for a batch job to execute to get these notifications.

This chapter examines HBase, a high-performance distributed database that is ideal for high-response time type use-cases. HBase is an example of a NoSQL database, which is modeled differently from the traditional tabular structure of an RDBMS. It can store and retrieve both structured and semistructured data for use-cases for which response time is critical. HBase is suitable for random reads and writes in which the throughput supported is thousands of transactions per node. HBase can also support MapReduce programs in which input to the MR program is a set of HBase tables.

## What Is HBase?

In very basic terms, HBase is a key/value store. In Java, a HashMap is a key/value store. HBase is also multidimensional, so the keys are multidimensional. For example, the key for a HashMap in Java is an object, and a key in HBase is a combination of the following:

- Row key
- Column key

This is overly simplistic, but we will start unsimplifying soon. From a logical perspective, you provide a combination of row key and column key, and HBase returns a value for that combination. Or you provide a row key, and HBase provides all the columns and values for those columns.

When you imagine HBase this way, you can easily appreciate why it is sparse from a columns perspective. In a typical database, every table has a fixed number of columns, and a row has all columns, regardless of whether they are populated. Although the number of columns for a table is limited to a few hundred, given the way HBase stores rows and columns, you can add columns dynamically. Each row can have millions of columns if necessary, and they can all be different.

Consider an equivalent implementation using HashMap: the key is equal to a row key, and the value is a HashMap whose key is the column key. Thus, the value for the inner HashMap is the combination of the row key and column key. (Bear in mind that this is only an example; it is not the way HBase is actually implemented.)

HBase also stores versions of the values, so when you overwrite a value it does not replace its value; it creates a new version for it. Thus, going back to the HashMap example, the internal HashMap no longer has a single value; it has a queue of values, with the most recently inserted value at the top of the queue. Figure 14-1 shows this intuition behind HBase.

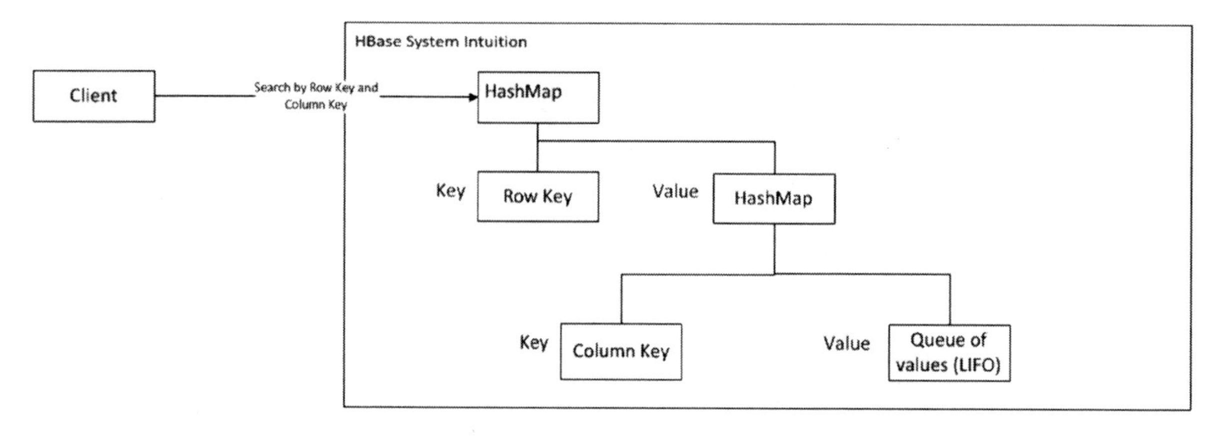


***Figure 14-1.*** *Client-side intuition behind HBase*

Despite the overly simplistic nature of this intuition, it should be easy to understand the following characteristics of HBase:

- It is multidimensional.
- Unlike a typical database, it can have a different set of columns for each row.
- It can have millions of columns for each row.
- Nulls are free. If a column exists in one row but not in the other, no extra space is needed to hold the data or metadata of the empty column in the second row.

# Typical HBase Use-Case Scenarios

HBase is used in scenarios where you need thousands of random read and writes per second on massive amounts of data using predictable access patterns. In this section we will look at some example use-cases for HBase.

A social networking site such as Facebook has terabytes of data. If a user wants to retrieve the Friends list of a user, the response time is expected to be consistent with what users of an interactive system have come to expect.

Imagine a document management system that contains millions of documents, and each document contains pages. For example, the United States Patent and Trademark Office (USPTO) patent documents can be accessed here: `https://www.google.com/googlebooks/uspto-patents.html`.

A sample USPTO document can be downloaded from here: `https://www.google.com/googlebooks/uspto-patents.html`.

Each document is a set of pages, and each page is small (a few megabytes). Imagine an access pattern in which users can access documents by providing page numbers. Also imagine that some documents can have thousands of pages, making it impractical to just load the document in memory, and support for page-by-page access is essential.

The HDFS works best with large files, but not so well with small files (refer to Chapter 2) because the NameNode can get overwhelmed with large numbers of small files. In such a case, HBase tables can be large file containers for the smaller files. HBase is a multidimensional key/value store, and its multidimensionality is provided by columns. At a high level, a value in HBase is stored with respect to a key that is combination of row keys and column keys. An HBase table is a collection of files similar to sequence files.

Thus, each document can be maintained as a key comprising a document identifier and column key as the page identifier. The value is the byte array of the page data. If the metadata about the page can be obtained either from a separate metadata table or the cell, millions of documents become accessible in real time.

The beauty of the preceding design is that if the document is retrieved by document ID, all the pages can be retrieved in one call. If the document is retrieved by document ID and a set of column identifiers, only the requested pages can be retrieved. The granularity of the data returned is controlled by the client, but the results are obtained in near real time (or interactive transaction time because it is not real time in the strict sense of the term).

HBase should only be used when the access patterns for data stored in HBase are predictable. Unlike a RDBMS where normalized tables can serve a variety of access patterns due to the rich nature of SQL, HBase table designs are centered on their access patterns. A table design that is very fast for a certain type of access pattern will be abysmally slow for another type of access pattern. Revisit this discussion when you have read the sections on Table and Row Key design.

# HBase Data Model

The following sections discuss HBase data models: the logical data model and the physical data model.

## HBase Logical or Client-Side View

This section describes how HBase appears to an HBase client. The key aspects of HBase are these:

- Table
- Column family
- Column qualifier
- Row key
- Cell
- Version number

Figure 14-2 demonstrates all these concepts except the version number. It is sample table that encapsulates the entire USPTO use-case described in the previous section.

- The table is called `PATENT`.
- It has two column families: `PATENT_DETAILS` and `IMAGES`.
- The row key is the patent ID that is used to retrieve the entire patent metadata from the `PATENT_DETAILS` column family as well as from the `IMAGES` column family. The column family can also be specified; each column family is stored in its own separate file. For multiple column families, significant I/O can be saved by electing to read a specific column family. Rows are sorted by row key in a table.
- The column and the values (cell) are described inside the column family (for example, `Inventor_1:'James...`').

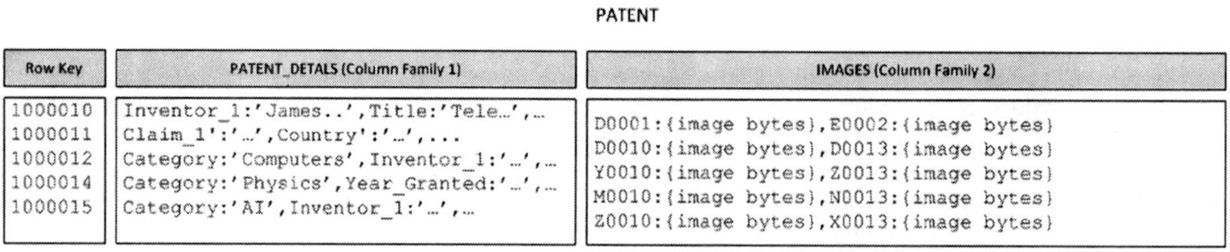

PATENT

| Row Key | PATENT_DETALS (Column Family 1) | IMAGES (Column Family 2) |
|---|---|---|
| 1000010 | Inventor_1:'James..',Title:'Tele...',... | D0001:{image bytes},E0002:{image bytes} |
| 1000011 | Claim_1':'...',Country':'...',... | D0010:{image bytes},D0013:{image bytes} |
| 1000012 | Category:'Computers',Inventor_1:'...',... | Y0010:{image bytes},Z0013:{image bytes} |
| 1000014 | Category:'Physics',Year_Granted:'...',... | M0010:{image bytes},N0013:{image bytes} |
| 1000015 | Category:'AI',Inventor_1:'...',... | Z0010:{image bytes},X0013:{image bytes} |

***Figure 14-2.*** *HBase table: logical view*

HBase table and row key design is discussed in subsequent sections, but the key takeaway from the preceding example is that HBase tables are highly denormalized. Also, for those coming from a database background, the multiple column families are like a join query in which the join criterion is the row key. The nature of HBase design ensures that very few tables and column families are needed. This is in stark contrast to RDBMSs, which typically have a highly normalized structure and consequently significantly larger number of tables as compared with the corresponding HBase design. This feature is not unique to HBase; it is typical of all NoSQL databases such as Cassandra or MongoDB.

# Differences Between HBase and RDBMSs

We now have the context to discuss the key differences between HBase and RDBMSs. This discussion is not limited to HBase; it can be applied to any NoSQL database such as MongoDB or Cassandra.

## RDBMSs

An RDBMS has the following characteristics

- *Schema-oriented*: A RDBMS table has a fixed schema, and the type of data stored in the table is defined in the schema. The schema is defined at table creation time and cannot be easily changed. Adding elements to the schema usually takes the form of additional tables with a one-to-one relationship between the new table and the original table. This limitation makes RDBMSs suitable for highly structured use-cases such as financial data storage.
- *Normalized data*: Although RDBMSs typically store highly normalized data, it is not true when an RDBMS operates in data warehouse mode in which data can be very denormalized. But data warehouses are designed for reporting use-cases. A normalized data store is appropriate for a transactional use-case.
- *Thin tables*: An RDBMS typically has only a small number of columns with the maximum limit in hundreds of columns. This causes an RDBMS to have multiple tables with a variety of relationships between them: one-to-one, one-to-many, and many-to-many.

## HBase (or NoSQL Database)

HBase has the following characteristics:

- *Schema-less*: As discussed, an HBase is effectively a map of a map, so it has no schema. The user can define the columns at runtime, and each row can have its own columns. The responsibility is on the application to interpret the values stored in HBase. This makes HBase very suitable for applications in which the schema is flexible. For example, the earlier example shows that the schema for the patent application has to be flexible; the document needs to be accessed in one retrieval to improve I/O performance. Yet each document can have a variable number of pages, and each page has its own ID. Unlike a database, in which a separate metadata table is needed to describe the page identifiers, each column key describes itself in HBase. And we can accommodate cases in which certain documents can have a very large number of pages. The other documents do not need null values in empty columns to accommodate such documents.
- *Denormalized data*: The previous example about USPTO documents emphasized the denormalized nature of an HBase table. Denormalization improves performance for row retrieval. Each row retrieval from an HBase table should return a maximum amount of information from the use-case perspective to reduce server round trips and support a large number of simultaneous requests. Thus, entire reports can be retrieved from a single HBase row. In the USPTO example, a single retrieval can fetch the entire document with all its metadata and pages.

## HBase Tables

The HBase table is the logical unit of storage of records in HBase. It comprises rows, and a row comprises columns in a column family. The key features about rows and columns in an HBase table are these:

- The rows are sorted lexicographically by row key.
- The columns are sorted lexicographically by column key within the column family. A column is also referred to as a column qualifier.

## HBase Cells

The intersection of a row and a column is called a cell and it is versioned. If a new value is provided for a cell, the older value is not overwritten. Instead, a new cell is generated with a higher version number. The original cell is retained, and the cells are retained in the descending order of their version number, with the most recent version at the top.

A cell is stored as an array of bytes, so HBase can store any digital data. The responsibility for interpreting the bytes is left to the user program. It can be simple string, a serialized integer, or even an Avro object.

## HBase Column Family

As described in the previous section, an HBase table is composed of rows and columns. Rows are identified by a row key; and an entire HBase row can be queried by the row key.

Columns are grouped by column family. Some of the characteristics of a column family are as follows:

- A table usually has a single column family and it should not have more than five column families.
- Column families must be declared at table creation time.

- The column families of the table should never change.
- A column family can have an unlimited number of columns.
- All columns of a column family are stored together.
- Columns are sorted within the column family.
- Columns can be created at runtime.
- Columns exist only when inserted. Null values are not stored, so they are free.

Lookups by the same row key talk to the same physical node. If the table has two column families, files belonging to the two column families reside on the same node. Thus, there is a one-to-one relationship between a row key and node.

Each column family is stored in its own set of files. Retrieving columns from a column family is sequential I/O. Reading from two column families implies reading two separate files and blocks in the HDFS. All storage settings must be specified at the column family level. Examples of these settings include the following:

- *Compression scheme*: The default compression scheme is NONE. Values applicable are GZ, LZO, and SNAPPY.
- *No. of versions*: By default, HBase defines three versions.
- *Time to Live (TTL) length in seconds*: Applies only to elements of the row for the column family. After the expiration time is reached, HBase deletes the rows.
- *Block size*: HBase adopts a default block size of 64 KB, which is different from the default HDFS block size of 64 MB because HBase is for random access. Once a candidate block for a row key is discovered, it is serially accessed to identify the cell requested. It is faster to scan a 64 KB block than a 64 MB block.
- *In-memory*: The default value is `false`. If this value is set to `true`, HBase attempts to keep the entire column family in memory. It is also persisted to disk for durability reasons. However, during runtime HBase attempts to keep the entire table in memory.
- *Block cache*: The default is `true`. Block cache is an in-memory store that HBase uses to keep the most recently used blocks in memory. Blocks are dropped from the block cache using the least recently used (LRU) criteria.
- *Bloom filter*: As discussed in a subsequent section, a Bloom filter is a space-efficient probabilistic data structure that enables the client to test whether the element (a row key or a combination of row keys and column qualifiers) is *definitely not* in the table or *may be* present in the table. The default value is `NONE`. But the other values are `ROW`, which means that the Bloom filter can be used at the row key level; and `ROWCOL`, which means that the Bloom filter applies at the row key and column qualifier level.

# HBase Commands and APIs

This section describes the HBase shell. In your OS command prompt, type the following command to open the HBase shell:

```
hbase shell
```

This command opens the HBase shell (see Figure 14-3).

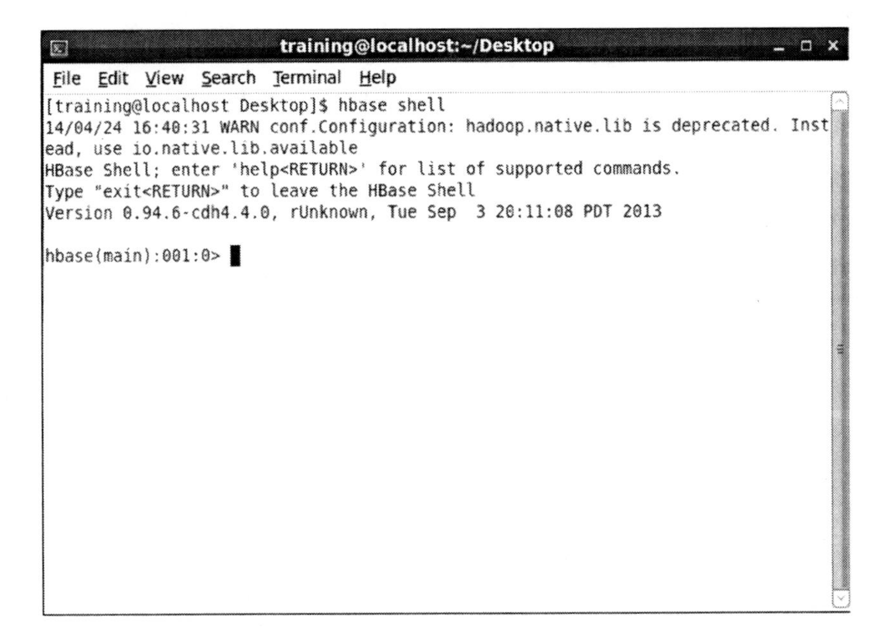


***Figure 14-3.*** *HBase command shell*

We can now execute the DDL and DML commands in the shell.

## Getting a Command List: help Command

Use the `help` command to get a list of commands categorized by type. The command groups can be categorized as follows (for a full list of commands under each category, refer to the HBase documentation):

- *General*: These commands discover general information at the cluster level. Examples include `status`, `whoami`, and `version`.
- *DDL*: These commands create, alter, and drop HBase tables. Examples include `alter`, `create`, `describe`, `drop`, `enable`, `disable`, and `list`.
- *DML*: These commands add, modify, and delete data from HBase tables. Examples include `get`, `put`, `delete`, `deleteall`, `get_counter`, `incr`, `put`, `scan`, and `truncate`.
- *Tools*: These commands perform maintenance of the HBase cluster. Examples include `assign`, `close_region`, `compact`, `flush`, `hlog_roll`, and `major_compact`.
- *Replication*: These commands support adding and removing nodes in the cluster. Examples include `add_peer`, `enable_peer`, `disable_peer`, `list_peers`, `start_replication`, and `stop_replication`.
- *Snapshot*: These commands support taking snapshots of the HBase cluster for backup and recovery. Examples include `snapshot`, `restore_snapshot`, and `list_snapshots`.
- *Security*: These commands assign security in HBase. Examples include `grant` and `revoke`.

## Creating a Table: create Command

Use the create command to create a table from the HBase shell. The syntax of this command requires the following:

- Table name
- Dictionary of specification for each column family
- Optional dictionary of table configurations

An example of this command is shown here. We are creating a table 'tbl1' with two column families. The first one is 'cf1', whose cells will have a maximum of 5 versions, a TTL of 1 day, and block cache turned on. The second column family is 'cf2', which uses the default settings for a column family. The table also has table-level configurations defined. This table will have a maximum of 15 regions and will be presplit using the HexStringSplit algorithm (table splitting and presplits are discussed in a later section).

```
create 'tbl1¢, {NAME => 'cf1¢, VERSIONS => 5, TTL => 86400, BLOCKCACHE => true},{NAME => 'cf2¢},
{NUMREGIONS=> 10, SPLITALGO => 'HexStringSplit'}
```

## Adding Rows to a Table: put Command

The put command is used to insert row into an HBase table. The form of this command is the following:

```
put '<TBL_NAME>', '<ROW_KEY>', '<COLUMN_FAMILY>:<COLUMN>','<VALUE' [,<TIMESTAMP>]
```

An example of the put command is the following:

```
put 'tbl1', 'r1', 'cf1:c1','1'
put 'tbl1', 'r1', 'cf2:c1','1',1234567890123
```

## Retrieving Rows from the Table: get Command

The get command retrieves data from HBase. The form of this command is this:

```
get '<TABLE_NAME>', '<ROW_KEY>' [,<OPTIONS>]
```

By default, the get command retrieves all the columns from each column family. An example of a get command is shown here: it retrieves the key 'r1' from 'tb1'. The columns retrieved are 'c1' from the column family 'cf1', and the last two versions are retrieved.

```
get 'tbl1', 'r1', {COLUMN => 'cf1:c1', VERSIONS => 2}
```

## Reading Multiple Rows: scan Command

The scan command allows you to retrieve multiple rows of the table. By default, the command form is this:

```
scan <TABLE_NAME>
```

The command retrieves all the rows in the table in order, but you can provide options to limit the retrieved records. Some of the options are these:

- Columns to be retrieved
- Start row ID (inclusive)
- Stop row ID (exclusive)
- Limit count that will limit the number of rows returned

An example of the `scan` command is this:

```
scan 'tbl1', {COLUMNS => 'cf1:c1}, {COLUMNS => 'cf2:'},{STARTROW=>'r1', STOPROW=>'r200', LIMIT=>75}
```

The `get` command retrieves only the column `'c1'` from the column family `'cf1'` and all the columns from the column family `'cf2'`. The starting row for the scan is `'r1'`, and the ending row is `'r200'`. The maximum rows returned are 75.

## Counting the Rows in the Table: count Command

The `count` command allows the user to get the count of rows in a given table in HBase. Be aware that for large tables, this command takes a very long time to run.

An example of this command is this:

```
count  '<TABLE_NAME>'[,interval]
```

If an interval is specified, it shows progress every `'interval'` number of rows. If it is not specified, the count is shown every 1,000 rows.

## Deleting Rows: delete Command

The `delete` command comes in several flavors:

- To delete an entire row:

  ```
  deleteall '<TABLE_NAME>', '<ROW_KEY>'
  ```

- To delete columns from a row:

  ```
  delete '<TABLE_NAME>', '<ROW_KEY>', '<COLUMN_FAMILY>:<COLUMN_NAME>'
  ```

Deleting a row does not physically delete the record from HBase; it only creates a tombstone for the deleted entity. The actual physical delete happens during the process of compaction (discussed later in the chapter).

## Truncating a Table: truncate Command

Truncating a table effectively performs the following set of operations:

- Disable a table
- Drop a table
- Create a table

The command format for the `truncate` command is this:

```
truncate '<TABLE_NAME>'
```

## Dropping a Table: drop Command

Dropping a table removes the table, but it must be disabled before it is dropped. Here is the sequence of steps:

1. Disable a table
2. Drop a table

The following group of commands is executed to drop a table:

```
disable '<TABLE_NAME>'
drop '<TABLE_NAME>'
```

## Altering a Table: alter Command

The column family options for a table can be altered. Note that only the options can be changed; the name of the column family cannot be altered. The table must first be disabled with this command:

```
disable '<TABLE_NAME>'
```

The column family options can be added or altered with the following command:

```
alter '<TABLE_NAME>', {NAME => '<COLUMN_FAMILY>', [,<OPTIONS>]}
```

Finally, the table is re-enabled with this command:

```
enable '<TABLE_NAME>'
```

# HBase Architecture

This chapter describes the architecture of HBase in the context of the Hadoop cluster. We will delve into how the various components are laid out across the cluster and what their roles are in the client-to-HBase interaction. The various components of the HBase architecture are shown in Figure 14-4.

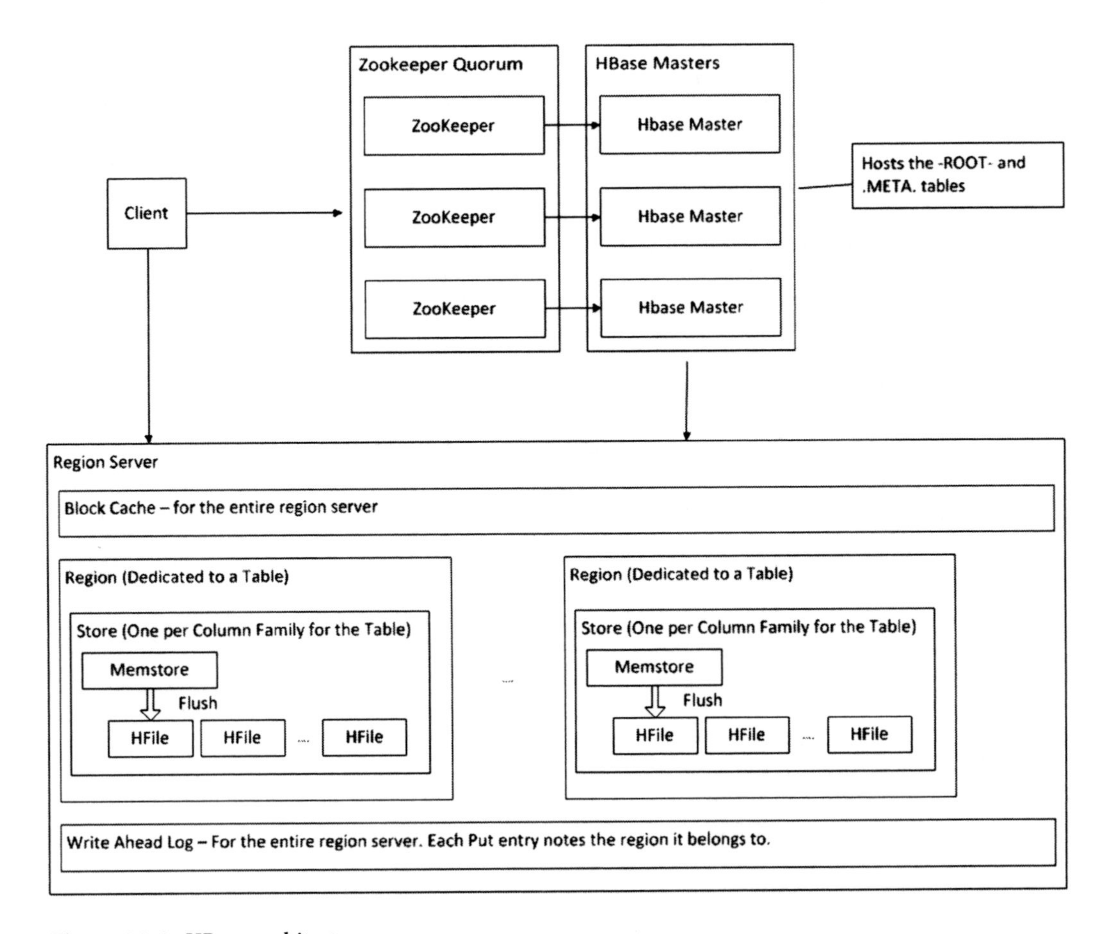


***Figure 14-4.*** *HBase architecture*

# HBase Components

The following are the major components of the HBase architecture:

- ZooKeeper
- Master that hosts the catalog tables
- Region server
    - Memstore
    - WAL
    - Block cache
- Region
- Zero or more files; one set for each column family of a table represented by the region
- Bloom filter

HBase can cache aggressively if the block cache is enabled, and it caches a whole block at a time. Reading a record by key caches the entire record set stored in the same block from which the record is retrieved. Subsequent retrievals can be from the cache.

## ZooKeeper

Apache ZooKeeper is the Hadoop distributed coordination service that simplifies writing distributed applications by owning the responsibility to coordinate across components distributed across the cluster. In any distributed application, the components coordinate with each other about the state of their health using a mechanism called heartbeats. They also communicate via messages, but sometimes these messages can get lost. And the sender has to determine whether the lost message is due to the network failure, receiver failure, or just a general delay in the network. ZooKeeper handles the task of managing such partial failures.

The ZooKeeper documentation explains the various features of ZooKeeper in detail. For our purposes, just remember that ZooKeeper is simple, distributed; and fast-naming, as well as a coordination service built from the ground up to support loosely coupled messaging between remote components. ZooKeeper also provides a detailed API that enables developers to develop distributed applications that use ZooKeeper to perform coordination between components. HBase uses ZooKeeper.

## HBase Master

There is more than one HBase Master, as shown in Figure 14-4, but only one is playing the role of the Master at any one point. The rest exist to take over if the current Master fails. When the cluster starts up, all the Master nodes compete to become the true Master. If the Master loses its lease with the ZooKeeper, the remaining Masters once again compete, and only one succeeds.

The role of the Master is to perform administrative operations, keep track of which data is maintained on which node (determined by the row key), and control load balancing and failover.

## Region Server

In a Hadoop cluster, a region server runs on a data node, and its role is to host multiple regions. The data is actually stored in the region. Each region server has a Write Ahead Log (WAL) file and block cache for the entire region server that is shared by all regions hosted on the region server.

### Write Ahead Log (WAL)

The purpose of the WAL is to ensure that data is not lost if there is a failure and recovery. Every update operation is written to the WAL first. The WAL contains all the edits made to the Region Server that is not yet flushed to HFiles (discussed soon). HBase edits are written to an in-memory data structure called MemStore, an in-memory data store which is flushed to disk (HFile) when it reaches a certain size. MemStore is not persistent memory and if the RegionServer crashes all updates to MemStore are lost. WAL allows recovering from a Region Server crash by replaying the unsaved edits from the WAL. WAL can be disabled but when you do that you risk losing data if a Region Server crashes. Disabling the WAL can improve the performance of HBase but risks losing data in the event of a Region Server crash. It is only appropriate in situations where you have a process writing to HBase which is notified about Region Server failure and can restart data loads to HBase in response to this notification. Also data loads process needs to be idempotent which means that loading the same data multiple times results leads to the same end result in HBase.

### Block Cache

A block cache is an in-memory data structure that follows the eviction semantics of least recently used (LRU). By default, it is enabled for the table. Any block read from HBase during a `SCAN` and `GET` operation is placed in the block cache (assuming that it is turned on), which enables repeat reads close in time to be read directly from the memory instead of performing a disk I/O. Starting with the 0.92 release HBase began supporting OffHeap memory storage for the Block Cache. Prior to this release HBase only supported `LruBlockCache` which stores its contents entirely in the JVM heap memory which is managed entirely by the Garbage Collector. HBase 0.92 release provided `SlabCache` and 0.96 release supported `BucketCache`. These classes support usage of memory outside the JVM heap memory by utilizing the Java NIO API. This significantly reduces the GC pressure experienced by the region server process.

## Region

A region server hosts several regions, and each region is pinned to a table and manages rows for a key range of that table. A region contains one store per column family for a table to which it is assigned.

A table can have many regions across region servers. Each region server will have zero or more regions for a table. A region contains a separate set of files (HFiles) for each column family of the table. Each region manages a contiguous range of row keys. The range of row keys managed by regions in a single region server need not be continuous. A region managing a key range 1-10000 may reside on one region server with another region managing a key range 20001-30000 while the region managing the key range 10001-20000 may reside on a separate region server. A region gets dedicated memory and resources, so there can be only so many active regions in a cluster. This decision becomes critical when presplitting a table.

One of the common design patterns in HBase is to use a salted key, which is a key that is prefixed with a small set of bytes. This process is known as salting, and the resulting key is called a salted key. The salt is usually a small number left-padded with zeroes to ensure constant length for the salt. For example, an application may decide to have a maximum of 5,000 salts (range 0000 to 4999). This salt is determined by a simple process: obtain the hash code of the key and perform modulo 5000 to get the salt. The salt is then zero-padded to ensure a 4-character salt that ranges from 0000 to 4999 and prepended to the row key. A table can be defined to have 50 regions at the outset. This is known as presplitting a table, and each region gets 100 salts. Region 0 gets salts 0000-0099, Region 1 gets salts 0100-0199, and so on until Region 49 gets salts 4950-4999. If we define several such tables, we could easily run out of regions, which would cause performance problems in our cluster.

The default behavior (no presplits) of HBase is to first allocate a region on a region server to a table. All `Puts` go to the same region. When the size of the data stored in the region grows to a certain point (as defined in the configuration files in the `hbase.hregion.max.filesize` property; the default size is 10 GB), the region is split. At this point, there are two regions usually homed in separate region servers. This process involves some data migration. (We will discuss the process of splits later in this chapter.)

A region has two main subcomponents, discussed in the following sections:

- Memstore
- HFile

### MemStore

When an update is made to the data in HBase (a `Put` or a `Delete`), it is written to the WAL and then to the Memstore, which is an in-memory structure that stores all the writes in memory. From a logical perspective a MemStore is essentially as sorted map per Region hosted by the Region Server containing all the recently written data. When the size of the MemStore reaches a threshold defined by the `hbase.hregion.memstore.flush.size` parameter, the MemStore is flushed to a file known as an HFile. Each flush produces a new HFile per column family in the region. Any failure in the HBase cluster between writing to the MemStore and flushing to the HFile can be recovered through using the entries in the WAL. This is the role of the WAL.

When data is being read from HBase, the read first goes to the MemStore, which then reads the HFile if the data is not read from the HFile (assuming that the data was not found in the block cache).

---

■ **Note** HBase can be optimized for reads or writes by tilting more memory toward the block cache or MemStore. The higher the block cache memory allocation, the faster the frequent re-reads. The more the value of the MemStore flush size, the less frequent are the MemStore flushes to file and the faster is the write performance. Remember that a MemStore flush is a blocking operation.

---

Early implementations of MemStore lead to heap fragmentation resulting in the frequent invocations of full Garbage Collection which causes long pauses. One of the key concepts in MemStore implementation is the MemStore-Local Allocation Buffer (MSLAB). MSLAB is a component added to HBase which dramatically reduces long Garbage Collection pauses by allocating contiguous chunks of memory. Using MSLAB, memory is allocated in larger chunks (default 2MB) as byte arrays. The `KeyValue` instances written to the MemStore are stored in these byte arrays. This allows larger chunks of memory (2MB by default) to be allocated and garbage collected. This reduces heap memory fragmentation which in turn lowers the occurrence of full Garbage Collection and the resulting long pauses.

MSLAB concept is described well in the HBase documentation, and Cloudera has some excellent blog entries about it as well. You should read these blogs as understanding the MSLAB gives insights into some very complex engineering decisions involved in building a product such as HBase.

Another excellent blog entry about MemStore configuration is `http://blog.sematext.com/2012/07/16/hbase-memstore-what-you-should-know/`. It delves into details on how to optimally configure MemStores.

## HFile

The HFile is the file in which HBase rows are finally maintained. An HFile belongs to a table and column family, and the rows are sorted lexicographically by row key, column. They are sorted in descending order by column version numbers. Each cell in HBase is stored in HFile, as shown in Figure 14-5.

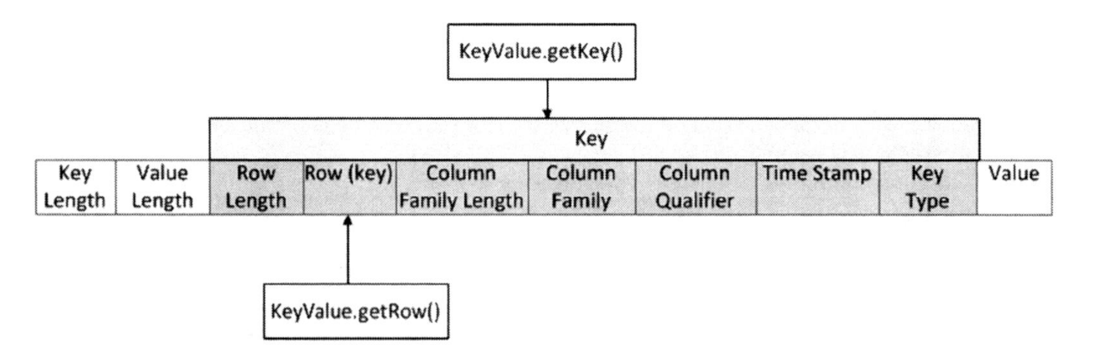


***Figure 14-5.*** *Physical structure of an HBase cell*

The key-value storage in `HFile` starts with storage of two numbers, each of which has a fixed length. These numbers indicate the length of the key and value, respectively, which allows the client to get the byte offset to jump directly to the value portion.

Figure 14-5 shows an important class in the HBase API: the `KeyValue` class. This diversion is essential to understand the difference between a key and a row from an HBase perspective. The `getKey()` method returns the entire key, as denoted by the shaded portion; the `getRow() method` returns only the row key.

Other important features regarding the key include these:

- The row length is a fixed-length number that indicates the length of the row that follows. It provides an offset to start reading the column family length.
- The column family length is a fixed-length number that enables reading just the column family and provides a way to jump to the byte offset of the column qualifier.
- The Time Stamp is the version number for the cell stored as a long value.
- The key type is stored in a single byte. Key type indicates if the given row is a Put or a Delete. Additional types are supported. Refer to the documentation of the `org.apache.hadooop.hbase.KeyValue.Type` for more details.
- The column qualifier length can be inferred from the key length, column qualifier length, number of bytes to store a long value which stores the timestamp and the number of bytes needed to store the key type (single byte). This is the reason why the length of the column qualifier is not stored in the row key.

In the context of this structure, note that rows in HBase column families are sorted first in ascending order by row key, column family and column qualifier, and then in descending order by timestamp.

Remember that an HFile is created per table and column family managed by a region on each Memstore flush, which is how entries are sorted correctly in a single HFile. They are not sorted across multiple HFiles that are created on different Memstore flushes. HFiles are regularly merged through a process of minor and major compaction. When the HFiles are merged to a single HFile, a merge sort re-sorts the entries in the destination HFile.

Minor compaction runs regularly during the day. Its role is to merge HFiles into a single HFile when their number for a region crosses a threshold. A major compaction is an expensive process that runs at a specified time in the day (usually a time of low activity). Its role is to merge all the HFiles for a region into a single HFile. We will discuss compaction in a subsequent section.

## Catalog Tables in the HBase Master

Now you have the background to learn about the role of catalog tables in the HBase Master. How does a client know which region server is hosting the region containing a specified key? The answer lies in catalog tables, which are periodically downloaded by the client when the connection with HBase is initiated.

Catalog tables are maintained as HBase tables. There are two types of catalog tables:

- The `-ROOT-` table keeps track of the `.META.` table. This table is just a reference to the location of the `.META.` table in the cluster. It notes the server node in which this table is hosted.
- The `.META.` table maintains metadata about every region in the cluster. The key class here is the `HRegionInfo` class, which maintains the table and the start key (inclusive) and end key (exclusive).

When the client initiates a connection to the HBase cluster, it follows these steps:

1. The client connects to the ZooKeeper ensemble. This information needs to be provided to the client (either provided directly or can be read from the HBase configuration file). The client inquires about the location of the `-ROOT-` table from the ZooKeeper. This is the first query made by the client.
2. The client connects to the `-ROOT-` table and gets the information about the location of the `.META.` table. This is the second query made by the client.
3. The client connects to the `.META.` table to download a list of all regions and their locations. This is the third query made by the client.

4. The client uses the information downloaded from the `.META.` table to connect directly to the region servers and the regions in them to operate on the data. These are a list of queries performed by the client.

5. The client caches the information in the first three queries. If the client discovers that the cache is stale, it refreshes the cache. This happens if the client is directed to a region server but discovers that the region server is no longer managing the key range provided by the `.META.` table.

## Bloom Filters

Bloom filters are a probabilistic data structure which allows you to test if an object satisfies one of the following criteria:

- It is definitely *not* in the data structure.
- It was *probably* added to the data structure.

In other words, when you test for an object in the Bloom filter, you get one of the two answers: No or Maybe. Figure 14-6 shows how Bloom filters work; a description of the process follows.

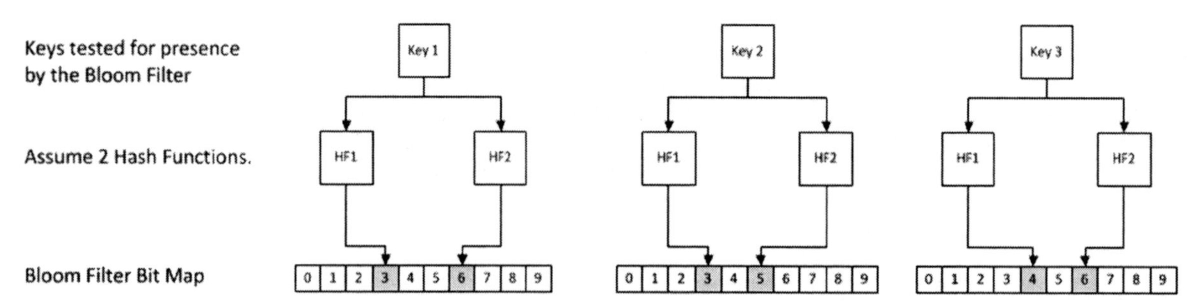


***Figure 14-6.*** *A simplified Bloom filter illustration*

1. In this simplified example, imagine that there is a byte array. The size of this byte array is determined by two factors: the maximum number of objects likely to be added to the dataset and the percentage of false positives (Maybes that are actually No) that are tolerable. Remember that a Maybe requires the actual data structure to be scanned, so too many false positives would be expensive. The size of the byte array responsible for maintaining the Bloom filter grows with the maximum number of entries that will be inserted into the data structure and decreases as we increase our tolerance for false positives.

2. Imagine a simplistic Bloom filter as a byte array comprising as many bytes as the maximum number of elements, and each byte index represents the object. We simply have to set the appropriate byte when the object is inserted. Such an implementation is ideal but impractical because we do not know the exact number of keys in advance and the hash code of the key is rarely a unique number in the range of the number of keys.

3. Next, imagine that we have two hash functions that return a number in the range between 0 and the size of `(byte array above)-1`.
4. When a new object arrives, simply compute the hash values using the preceding two functions and set the appropriate bits.
5. When an object's existence needs to be tested, the two hash functions are executed on it, and the byte locations are computed. If one of the two bytes is not set, the object is not contained in the dataset. If both of the bits are set, the object may exist in the dataset. Figure 14-6 shows exactly why this is a Maybe instead of a Yes.

Bloom filters are used to maintain a lightweight index of the keys in the HFile. They come in two flavors that can be set using the `HColumnDescriptor` class:

```
HColumnDescriptor.setBloomFilterType(NONE | ROW | ROWCOL)
```

- `ROW FILTER` makes a determination only on the basis of the row key. Use this type of filter when your query pattern is typically row key–only.
- `ROWCOL FILTER` makes a determination on the basis of the row and column qualifier. The column family is not included because Bloom filters operate on an HFile, and an HFile contains values from a single column family. Use this type of Bloom filter when your query pattern uses both the row key and column qualifier to search for values.

The previous discussion might give the impression that Bloom filters are always a good thing. However, like most things in Hadoop, discretion needs to be applied when electing to use the Bloom filter.

For example, in the best case a Bloom filter adds one byte of storage for each new record that can be added. We have discussed that a smaller byte array leads to more false positives that invalidate the purpose of using the Bloom filter. What if the key-value pairs are small in size? For example, if the key-value pairs take only a total of 20 bytes, the Bloom filter is 1/20 the size of the entire HFile. For a 5 GB file, this results in a 250 MB Bloom filter data structure in memory, which is quite large and strains the JVM from a garbage collection perspective. Also bear in mind that this is per HFile, and a region server contains many regions and hence many HFiles. As you can see, the situation can quickly get out of hand.

But if the key-value pair sizes are large, say 1,000 bytes, the Bloom filter is only 1/1000 the size of the data in the HFile. For a 5 GB file, there is a 5 KB Bloom filter that is very reasonable. Put this discussion in the context of deciding between a `ROW` and a `ROWCOL` filter, and it is easy to see why `ROWCOL` filters can get much more expensive. It is possible for a `ROW` to be large, but it is quite unlikely that an individual cell will be small.

A detailed discussion on this trade-off can be found here: `http://www.quora.com/How-are-bloom-filters-used-in-HBase`.

# Compaction and Splits in HBase

In this section we will discuss how HBase manages its regions. We will first discuss how the HBase splits regions to manage the size of each region. Next we will discuss compaction.

## Region Splits

When a table starts getting populated, there is a single region. When the size of the region crosses a certain threshold defined by the property (default value 10GB) - `hbase.hregion.max.filesize`.

HBase will split the region. Initially the split is logical. HBase only maintains the key where the split happens. The existing file continues to grow. During the process known as major compaction (discussed soon), the original data file is split into separate files into the new region's directory in HDFS. The original region is physically split into two regions and each region may belong to separate region servers.

The process of splitting is described is as follows:

1. The clients of a table keep making `Puts` on a new table.
2. The region server's `CompactSplitThread` monitors the size of the region. When it crosses the limit specified by `hbase.hregion.max.filesize`, it initiates the following set of actions:
    a. The region server temporarily closes the region to prevent changes during the split.
    b. The region is split into two logical daughter regions. The split is not done physically at this point to make the splitting process fast. The split is maintained as a split key, which is a key in the middle of the key range maintained for the region.
    c. The region server updates the `.META.` table. Note that both split regions are maintained in the same region server.
    d. The region server once again opens the newly created regions for updates.
3. The major compaction creates two separate files for the daughter regions created. This process may assign the two regions to separate region servers.

A very detailed description of this process can be found at this Hortonworks blog entry: `http://hortonworks.com/blog/apache-hbase-region-splitting-and-merging/`.

The region size impacts performance; some key points linked to region size include these:

- Too many regions can impact performance because resources are allocated per region. For example, the MSLAB data structure described in the section on MemStore requires 2MB per region per column family (2MB is default and can be configured to a lower value which can lead to GC related consequences). Having 2000 regions will soak up 4GB of heap memory (MSLAB pre-allocates to manage heap fragmentation) before it even stores any data.
- Too few regions can impact performance because parallel performance is limited.
- The region count must be managed per region server. It should be optimal, which often translates to a few 10s to 100s per region server. The exact range depends on the amount of resources available per region server.

# Compaction

Each region in a region server collects edits in the MemStore. When a MemStore reaches a threshold size it flushes to disk and produces a file known as an HFile. Recall that HBase stores cell values sorted by row key, column qualifier and version number for a column family. Having multiple HFiles for a region requires Scan and Get processes to read multiple HFiles simultaneously to ensure that cell values are returned in the order of row key, column key and version no.

Compaction is the process of combining HFiles in a region. Having fewer HFile improves the performance of the HBase system since the Scan and Get operation needs to read fewer files. Compaction process compacts multiple HFiles in to a single Hfile Compaction processes come in two flavors: minor and major. An ideal HBase store is one HFile per region, which translates to one HFile per table and column family in a region. However, this is not possible under a high transaction load. As the MemStore gets filled up and flushed, multiple HFiles are created. Each HFile is sorted by row key and column key, but the sorting is not total across HFiles for a region. This leads to multiple disk reads to find a row key requested by the client. Compaction tries to reduce the HFiles per store (a combination of table and column family in a region) to reduce unnecessary I/O.

## HBase Minor Compaction

Minor compaction combines several HFiles into a single HFile. It is specified by the hbase.hstore.compactionThreshold parameter (the default value is 3).

When the number of HFiles crosses this threshold in a region (a table and column family combination), the minor compaction process starts. During this process, the clients can continue making Puts to the Memstore, but the Memstore cannot flush when minor compaction is running. If the Memstore fills up, the clients to that region server simply block until the minor compaction process completes.

## HBase Major Compaction

Major compaction takes all the HFiles for a region and merges them to a single HFile. It is a heavyweight operation that should be run when the load is low. During this process, all the Deletes and cells with expired versions are removed. The major compaction runs once per day by default and is controlled by the hbase.hregion.majorcompaction property.

This property controls the amount of time between two major compactions. Its default value is 604,800,000 milliseconds, or 7 days. Setting it to 0 disables auto major compactions. The administrator can start the major compaction from the HBase shell.

# HBase Configuration: An Overview

This book assumes that HBase is already installed. It is necessary to have a basic understanding of the HBase configuration files to operate HBase or to write programs for HBase. The goal of this section is to provide this basic understanding. The following three configuration directories are important:

- $HBASE_HOME is the directory in which HBase is installed.
- $HBASE_HOME/conf is the configuration directory for HBase.
- $HBASE_HOME/bin contains programs to start and stop HBase.

## hbase-default.xml and hbase-site.xml

Similar to the way Hadoop configuration files are defined, there is an hbase-default.xml file and a custom hbase-site.xml file. The latter is used to override the properties of the hbase-default.xml file.

HBase uses the default setup properties in the hdfs-site.xml file for HDFS-related properties. For example, if the default replication factor (the hdfs-default.xml setting) is set to 3, the files created by HBase use 3 as the default replication factor by default. If you change this property to 5 in the hdfs-site.xml file, HBase does not see it unless you add $HADOOP_CONF_DIR to the $HBASE_CLASSPATH configured in the $HBASE_HOME/bin/hbase-env.sh file.

Alternatively, if you want only a new replication factor (or other HDFS-related properties) to be used by HBase alone without affecting the rest of the cluster, do one of the following:

- Copy the hdfs-site.xml file in the $HBASE_HOME/conf directory and update these properties.
- Update the hbase-site.xml file with the properties from hdfs-site.xml that you want to override.

The key property in the hbase-site.xml file needed to operate HBase is the following:

```
<property>
   <name>hbase.zookeeper.quorum</name>
   <value>
      zk_server1:zk_port,zk_server2:zk_port,zk_server3:zk_port
   </value>
</property>
```

As noted in the earlier section, the client uses these files to connect to the ZooKeeper, which is used by the client to get access to the two HBase Master tables:

- The -ROOT table contains the entries that identify where the .META. table is stored. This table is never split.
- The .META. table contains the metadata for each table stored in HBase, and it lists regions and their locations based on the HBase row key. Thus, when a client requests a row from a table for a given key, this table informs the client which region server to connect to and what region to search for to find the record. Likewise, when the client attempts to add or update a column for a key in a table, this .META. table informs the client which region server and region to add the update to.

The client accesses HBase as follows:

1. The client queries the ZooKeeper to get the location of the -ROOT- table. The location of the ZooKeeper is obtained in the ZooKeeper quorum in hbase-site.xml.
2. The client then connects to the server in which the -ROOT- table is maintained and gets the location of the .META. table.
3. Finally, the client connects to the server in which the .META. table is stored and downloads the entries for the .META. table.

The regions for these queries are cached locally by the client. Then, based on the action requested, the client connects to the region server in which the region for the data is held.

# HBase Application Design

This section discusses issues related to the design of HBase tables.

## Tall vs. Wide vs. Narrow Table Design

Some of the design decisions required for HBase table design are as follows:

- More rows or more versions? The latter uses the intrinsic versioning capability of HBase. However, you have to define the maximum number of versions for a column family, which can become tricky. It is best to use multiple rows for each version, which requires having the timestamp portion added to the key. Patterns of data retrieval define how and where the timestamp portion is defined in the row key.

- More rows or more columns per row? HBase can support millions of columns for a single row, and each row can have its own qualifiers. In the USPTO use-cases earlier in the chapter, you saw how each page identifier formed its own column qualifier in the IMAGES column family. The decision is similar to the earlier one: add the column qualifier to some part of the row key and store one image per line or multiple images in a single row? This decision is based on the application's data retrieval pattern.
- Rows as column design? This is a middle ground between many columns and many rows. One example is a logs database that stores the entire machine-generated log per system. You can define each row as a combination of system identifier (the system producing the logs) and a range of time (for example, in 15-minute intervals) and each column to be a discrete log event. This method is more complex from an application development perspective. See section on Row Key Design next for an illustration on how the row key for such as system can be designed.

## Row Key Design

HBase access is by row key. Scan filters can enable specifying a large number of keys at once. However, HBase recognizes only a row based on its row key. Row key design is essential to ensure good HBase performance. Some of the key design decisions are as follows:

- Row keys are stamped on every cell of HBase, and the longer the key, the more I/O is associated with storing a cell. Sometimes, a substitution key is used in the form of a MD5 hash of a fixed length.
- With a composite row key, the order of each component is determined by the access pattern. With a logs database that stores log events by host name and event type, the key could be in any one of these formats:
    - [*hostname*][*event-type*][*timestamp*]: Suitable when the access patterns require the logs to be accessed by host name and event type types.
    - [*event-type*][*timestamp*][*hostname*]: Suitable when the access patterns require the logs to be accessed by event type and timestamp.
    - [*event-type*][*reversed timestamp*][*hostname*]: A reversed timestamp is a value equal to `Long.MAX_VALUE-timestamp`, which ensures that data is stored in the order of more recent events first. This is suitable when data needs to be processed in the order of more recent events first.
- In the previous section we considered tables which use "rows as columns" design. A sample row key design for such a scheme is as follows `[event type][hostname][ [yyyymmdd][hh][interval]` where `[yyyymmdd]` refers to the date, `[hh]` refers to hour of the day (24 hour clock scheme) and the `[interval]` will take values from `0-3`, where `0` refers to interval `00-14` minutes, `1` refers to `15-29` minutes, `2` refers to `30-44` minutes and `3` refers to `45-59` minutes. Individual log entries will be separate columns in the row for the appropriate interval bucket. The column qualifier could be a reverse timestamp or just the timestamp depending on whether you want to store the log entries in ascending or descending order of timestamp. This method is more complex from an application development perspective. If the user scans the table based on start row key constructed using only the event type and hostname alone, HBase will return rows corresponding to all logs from for specific event type and hostname. If the user scans the table based on a start row key comprised of event type, hostname, date and hour all the logs for that event type, hostname, date and hour are returned.

Row key design is a complex topic, and we have only scratched the surface. But the previous discussion should provide the reader with some background in the nature of decision making that goes with row key design.

# HBase Operations Using Java API

We now describe the Java API for HBase. The process of using it is straightforward: put the HBase JAR files in the host name and start using it. The API does need the path to the ZooKeeper ensemble, which can either be provided directly or read from the `hbase-site.xml` file. This section is only an overview of the API. You should complement this section with other resources such as the HBase documentation.

## HBase Treats Everything as Bytes

HBase treats everything as bytes, including not just the cell values but also the names of the tables, column families, and columns (which are also referred in the API as column qualifiers).

The HBase helper class to work with bytes is this:

```
org.apache.hadoop.hbase.util.Bytes
```

This class has helper methods to work directly with bytes, even though you are using instances of various classes and primitive types. For example, the call to convert a `String` instance to bytes is this:

```
byte[] byteArray = Bytes.toBytes("columnName")
```

And the call to convert a byte array back to a `String` is this:

```
String columnName = Bytes.toString(byteArray)
```

Similar operations can be performed on other data types. For example, the conversion to bytes can be performed as follows:

```
byte[] intArray = Bytes.toBytes(123);
byte[] dblArray = Bytes.toBytes(123.1D);
byte[] boolArray = Bytes.toBytes(true);
```

A reconversion from bytes to the appropriate data type can be performed using the following calls:

```
int i = Bytes.toInt(intArray);
double d = Bytes.toDouble(dblArray);
boolean b = Bytes.toDouble(boolArray);
```

## Create an HBase Table

The key classes involved in creating the HBase Table are as follows:

- `org.apache.hadoop.hbase.HBaseConfiguration`: Instances of this class obtain the reference to the HBase `Configuration` object. The ZooKeeper ensemble can be maintained in this configuration read by an instance of this class.
- `org.apache.hadoop.hbase.client.HBaseAdmin`: A client utility that can be used to run the HBase administrative and DDL commands.
- `org.apache.hadoop.hbase.HTableDescriptor`: Maintains the metadata for the table being created or referenced. Table-level properties can be set using this class.
- `org.apache.hadoop.hbase.HColumnDescriptor`: Used to configure the column family.

Listing 14-1 shows the source code to create an HBase table using the Java API.

***Listing 14-1.*** Creating an HBase Table

```
String hbaseZKQuorum = "server1,server2,server3";
String hbaseZKClientPort = 2180;

Configuration conf = new Configuration();
Configuration hConf = HBaseConfiguration.create(conf);
hConf.set(Constants.HBASE_CONFIGURATION_ZOOKEEPER_QUORUM,
          hbaseZKQuorum);
hConf.set(Constants.HBASE_CONFIGURATION_ZOOKEEPER_CLIENTPORT,
          hbaseZKClientPort);
HBaseAdmin admin = new HBaseAdmin(hConf);
HTableDescriptor tblDesc =
                new HTableDescriptor(Bytes.toBytes("myTable"));
HColumnDescriptor colFamDesc =
                 new HColumnDescriptor(Bytes.toBytes("myColFam1"));
tblDesc.addFamily(colFamDesc);
admin.createTable(tblDesc);
admin.close();
```

## Administrative Functions Using HBaseAdmin

Various administrative functions can be performed using the `HBaseAdmin` class. The listing of all the tables in HBase can be obtained by making the following call:

```
HTableDescriptor[] descriptors = admin.listTables();
```

This call returns an array of `HTableDescriptors`.

A description of a specific table can be obtained with the following call:

```
HTableDescriptor descriptor =
             admin.getTableDescriptor(Bytes.toBytes("myTbl"));
```

Disabling of a table can be performed by using this call:

```
admin.disableTable("myTable")
```

The table can be re-enabled by making the following call:

```
admin.enableTable("myTable")
```

## Accessing Data Using the Java API

Data is accessed using the `HTableInterface` class. This class is not thread-safe, and each thread should use its own instance. Each instance connects to a single table in HBase. It must be closed using the `close()` call after all operations are completed.

```
HTableInterface myTbl = new HTable("myTbl");
...
myTbl.close();
```

## Get

The `Get` class is used to retrieve a specific set of cells from HBase. Listing 14-2 shows how a `Get` call can be used to retrieve a specific cell from HBase.

***Listing 14-2.*** Get Calls on an HBase Table

```
String hbaseZookeeperQuorum = "server1,server2,server3";
String hbaseZookeeperClientPort = 2180;

//First - create a Configuration Object
Configuration conf = new Configuration();
Configuration hConf = HBaseConfiguration.create(conf);
hConf.set(Constants.HBASE_CONFIGURATION_ZOOKEEPER_QUORUM,
          hbaseZookeeperQuorum);
hConf.set(Constants.HBASE_CONFIGURATION_ZOOKEEPER_CLIENTPORT,
          hbaseZookeeperClientPort);

//Second - Connect to a table using the Table Name
HTableInterface table = new HTable(conf, "myTbl");

//Third - Create a Get instance using the row key
Get g = new Get(Bytes.toBytes("rKey1"));

//Fourth - Retrieve a result from HTableInterface using Get instance
Result r = table.get(g);

//Fifth - Extract components of the result
String rowKey = Bytes.toString(r.getRow()); //Get Row Key
//Get the Cell Value
byte[] byteArray = r.getValue(Bytes.toBytes("colFam1"),
                              Bytes.toBytes("col1"));
```

```
//Convert the column value to a Java Object from bytes
String value = Bytes.toString(byteArray);

//Close the HTableInterface instance
table.close();
```

Note the Result class in the previous listing. The Get instance returns a Result instance. The actual values are returned from the Result instance, which takes two parameters:

- Column family
- Column descriptor

Also note that the value is returned as a bytes array that needs to be reconverted to an appropriate Java class. We emphasized early in the chapter that HBase is schema-less, which is one meaning of that phrase. HBase only understands bytes; it is up to the client to provide meaning to those bytes.

We noted in earlier sections that HBase can store versions. By default, the Get instance returns the most recent version for the combination of column family and column qualifier. It can be made to return multiple versions by making this call:

```
/*Get which returns upto 10 version
 *The multiple versions can be returned using the call
 */
get.setMaxVersion(10);
NavigableMap<byte[],NavigableMap<byte[],NavigableMap<Long,byte[]>>> map = get.getMap();
```

The preceding return value can be interpreted as follows:

- The outermost map returns a map of columns by column family.
- The second inner map is a map of values by column qualifier.
- The innermost map stores the values by version number.

## Put

The Put class is used to insert cells into HBase. Listing 14-3 shows how a Put call can be used to insert a record in HBase.

***Listing 14-3.*** An HBase Put

```
String hbaseZookeeperQuorum = "server1,server2,server3";
String hbaseZookeeperClientPort = 2180;

//First - create a Configuration Object
Configuration conf = new Configuration();
Configuration hConf = HBaseConfiguration.create(conf);
hConf.set(Constants.HBASE_CONFIGURATION_ZOOKEEPER_QUORUM,
          hbaseZookeeperQuorum);
hConf.set(Constants.HBASE_CONFIGURATION_ZOOKEEPER_CLIENTPORT,
          hbaseZookeeperClientPort);
```

```
//Second - Connect to a table using the Table Name
HTableInterface table = new HTable(conf, "myTbl");

//Third - Configure a Put instance using the row key
Put p = new Put(Bytes.toBytes("rKey2"));
p.add(Bytes.toBytes("colFam1"),
      Bytes.toBytes("cq10"),
      Bytes.toBytes("my value"));

//Fourth - Insert the Put instance using the HTableInterface
table.put(p)

//Close the HTableInterface instance
table.close();
```

HBase has no separate concepts for Insert and Update; each update is an insert. When a Put is made for an existing row key, column qualifier, and column name, a new version of the cell is made in HBase. The original version continues to exist. The maximum number of versions for a column family can be configured at creation time. The process of compaction removes the additional versions for a cell.

## Delete

HBase records can be deleted by a row key. In such cases, the entire row is deleted, including all the column families and columns in that column family. Specific cells that are combinations of a column family and column can also be deleted. Listing 14-4 shows the HBase Delete using the Java API.

***Listing 14-4.*** An HBase Delete

```
String hbaseZookeeperQuorum = "server1,server2,server3";
String hbaseZookeeperClientPort = 2180;

//First - create a Configuration Object
Configuration conf = new Configuration();
Configuration hConf = HBaseConfiguration.create(conf);
hConf.set(Constants.HBASE_CONFIGURATION_ZOOKEEPER_QUORUM,
        hbaseZookeeperQuorum);
hConf.set(Constants.HBASE_CONFIGURATION_ZOOKEEPER_CLIENTPORT,
        hbaseZookeeperClientPort);

//Second - Connect to a table using the Table Name
HTableInterface table = new HTable(conf, "myTbl");

//Third - Configure a Put instance using the row key
Delete row = new Delete(Bytes.toBytes("rKey10"));

//Fourth - Execute the Delete using the HTableInterface
table.delete(d)

//Close the HTableInterface instance
table.close();
```

If a specific column has to be deleted, make the following call on the `Delete` instance before executing the `table.delete(d)` call:

```
deleteColumns(byte[] family,byte[] colQualifier);
```

Other similar calls are as explained here:

- To delete columns with a timestamp less than or equal to a specific timestamp, replace the earlier mentioned call by the following one:

  ```
  deleteColumns(byte[] family, byte[] qualifier, long timestamp)
  ```

- To delete all columns of a column family, the relevant call is this:

  ```
  deleteFamily(byte[] family)
  ```

- To delete all columns of a column family with a timestamp before or equal to a specific timestamp value, the relevant call is this:

  ```
  deleteFamily(byte[] family,long timestamp)
  ```

## Scan

In HBase, a Scan is used when you want to access more than one row at a time. The characteristics of an HBase Scan are the following:

- A Scan can be for a group of rows matching a user defined criteria. Such a Scan requires additional filters configured. Refer to the HBase documentation for details on configuring scan filters.
- Alternatively, a Scan can be limited by a start row key (inclusive) and a stop row key (exclusive).
- A Scan can be on an entire table. Such a Scan is configured without a filter or start and stop keys.

  A Scan can be configured to return only specific column families and columns for those column families. It is often done to limit the I/O.

Listing 14-5 shows a simple Scan on an entire table.

***Listing 14-5.*** An HBase Scan

```
String hbaseZookeeperQuorum = "server1,server2,server3";
String hbaseZookeeperClientPort = 2180;

//First - create a Configuration Object
Configuration conf = new Configuration();
Configuration hConf = HBaseConfiguration.create(conf);
hConf.set(Constants.HBASE_CONFIGURATION_ZOOKEEPER_QUORUM,
          hbaseZookeeperQuorum);
hConf.set(Constants.HBASE_CONFIGURATION_ZOOKEEPER_CLIENTPORT,
          hbaseZookeeperClientPort);

//Second - Connect to a table using the Table Name
HTableInterface table = new HTable(conf, "myTbl");
```

```
//Third - Configure a full table scan
Scan s = new Scan();

//Fourth - Get a reference to the scanner
ResultScanner rs = table.getScanner(s);

//Fifth - Start Scanning
for (Result r : rs) {
  String rKey = Bytes.toString(r.getRow());
  byte[] valBytes = r.getValue(Bytes.toBytes("colFam1"),
                               Bytes.toBytes("colName1"));
   String val = Bytes.toString(valBytes);
}
//Close the HTableInterface instance
table.close();
```

By default, the Scan object returns one row at a time, which is a slow process. It can be improved by making the following call on the Scan object that will significantly improve performance (be wary of using a very high value because it increases the memory usage):

```
setCaching(10000); //Reads 10000 results in memory
```

This section skimmed through the various capabilities of Scan. We left out discussions about scan filters, which are very important to using the Scan functionality in practice. Refer to the HBase documentation for details on scan filters.

# HBase MapReduce Integration

In this section we will review how HBase can be used directly with MapReduce. MapReduce requires an input file which is stored on the HDFS in the cluster mode. MapReduce can also utilize HBase tables as input to the MapReduce job via specialized `InputFormat` implementation. Likewise HBase provides specialized `OutputFormat` implementation to support generation of HFiles which can be bulk loaded into HBase

# A MapReduce Job to Read an HBase Table

First, we consider a very simple MapReduce job that reads an HBase table. Listing 14-6 shows the listing for such a job.

***Listing 14-6.*** Reading an HBase Table from a MapReduce Job

```
Configuration config = HBaseConfiguration.create();
Job job = new Job(config, "Single Table Read");
job.setJarByClass(MyJob.class);

Scan scan = new Scan();
scan.setCaching(10000);
scan.setCacheBlocks(false);
scan.addFamily(columnFamily.getBytes());
// set other scan attrs
...
```

```
String tableName = "myTbl";
TableMapReduceUtil.initTableMapperJob(
  tableName,          // input HBase table name
  scan,// Scan instance to control CF and attribute selection
  ReadMapper.class,   // mapper
  Text.class,         // mapper output key
  Text.class,         // mapper output value
  job);
job.setOutputFormatClass(NullOutputFormat.class
boolean b = job.waitForCompletion(true);
```

The job internally uses a specialized InputFormat called TableInputFormat that uses the Scan instance template provided to read the table. A Mapper is created for each region of the table. The MapReduce job processes the table in parallel where the unit of parallelism is a region. It is important that the table have several regions. When tables are presplit, as discussed in an earlier section, ensure that the tables used for MapReduce have a sufficient number of splits to ensure that they are processed fast enough through a MapReduce job.

A key feature to note in Listing 14-6 is the Scan object. Notice that we specified the column family. For a table with multiple column families, a region will host files from all the column families. In our example the MapReduce job will only operate on the specified column family. If the column family was not specified on the Scan object, all of the column families would be processed simultaneously for a region in a Mapper instance. You also configure row caching by invoking the setCaching is a reasonable number (100 or 1000 or even 10000 for narrow tables), depending on table characteristics. This parameter defines how many rows are returned by the scanner based on the Scan instance for each fetch operation. As mentioned in the previous section the default value is 1 which will cause your MapReduce jobs to run very slow as fetching each row of the table requires a network call. You should also consider invoking scan.setBatching(noOfColumns) if you have really wide rows. For example, if you have a row that can have a million columns, the entire column list will be fetched in memory without the setBatching call, which can easily lead to an OutOfMemoryException. The invocation of the setBatching method ensures that only the noOfColumns specified as the parameter to setBatching is returned per call. If the value of noOfColumns is 1,000, and the row has 10,000 columns, the same row is returned 10 times with a different set of 1,000 columns consecutively in the same Mapper instance. Your program needs to take this factor into account by adding specialized implementation to handle this scenario.

Listing 14-7 shows the Mapper class. Note the key/value classes for the Mapper. ImmutableBytesWritable represents the HBase row key, and the result represents the HBase row returned as a list of key-value pairs where key is the column qualifier and the value is the column value. Note the Mapper output key-value classes and the Mapper signature in Listing 14-7. Also note the various method calls to fetch the row key, column qualifier and column (cell) value.

***Listing 14-7.*** ReadMapper Class

```
public static class ReadMapper extends TableMapper<Text, Text> {

  public void map(ImmutableBytesWritable row,
               Result value, Context context)
              throws InterruptedException, IOException {
    byte[] rowKey = row.get();        //Get byte array for the row key
    for(KeyValue kv:result.raw()){
       byte[] rowKey2 = kv.getRow(); //Alternative way to get byte array for row key
       byte[] fullKey = kv.getKey(); //Full Key
       byte[] columnQualifier = kv.getQualifier();
       byte[] columnValue = kv.getValue();
    }
}
```

It is possible to add a Reducer to the mix by adding the following call after the call to TableMapReduceUtil.initTableMapperJob:

```
TableMapReduceUtil.initTableReducerJob(
            targetTable,        // output table
            WriteReducer.class, // reducer class
            job);
```

The signature of the WriteReducer.class (a custom class) is as follows:

```
public static class MyTableReducer extends
  TableReducer<Text, Text, ImmutableBytesWritable>
```

Note that the input key-value classes match the Mapper output key-value classes. However, we define only the key for the output that should match the format for the row key. The value class is implied and it is of type Mutation. The Put and Delete classes extend the Mutation class, and it is expected that the Reducer only produces Put or Delete as the value classes along with the row key as the output key as an instance of ImmutableBytesWritable. The table used for writing the output is specified in the call to initTableReducerJob.

It is possible to use multiple tables as input to the job. The appropriate type of initTableMapperJob method to call in such cases is this:

```
initTableMapperJob(
               List<Scan> scans,
               Class<? extends TableMapper> mapper,
               Class<? ExtendsWritableComparable> outputKeyClass,
               Class<? extends Writable> outputValueClass, Job job)
```

We have to pass a list of Scan instances, and each Scan instance must be configured with its own table name. To do that, augment each Scan instance with the following call:

```
Scan.setAttribute(Scan.SCAN_ATTRIBUTES_TABLE_NAME,
                  tableName.getBytes());
```

Each Scan instance can also specify its caching and batching attributes. The underlying job uses MultiTableInputFormat.class. This class abstracts away the details of using the multiple Scan instances provided to read the multiple tables for the MapReduce job.

Inside the Mapper, if you want to determine which table is being processed, invoke the following call in the Mapper setup method:

```
@Override
 public void setup(Context context)
            throws IOException,InterruptedException {
     TableSplit split = (TableSplit) context.getInputSplit();
     this.tableName=new String(split.getTableName());
}
```

A TableSplit processed belongs to a region, and a region belongs to a table and column family combination. Hence, we need to identify the table name only once in the setup method.

This section demonstrated using MapReduce with HBase tables. We have only scratched the surface. We used the `TableMapReduceUtil` class to hide some details for the purpose of illustration. However, we can write a MapReduce job from scratch using various Hadoop I/O formats to handle HBase tables. For example, we can use `HFileOutputFormat` to directly produce HFile files in the HDFS that can be bulk loaded into HBase. In fact, the `TableMapReduceUtil` class that uses the Reducer in the example does just that.

Read the HBase documentation for details on how to achieve this. The goal of this section was to show how HBase and MapReduce work together. The key takeaway is that there are `InputFormat` and `OutputFormat` classes provided by HBase that allow programmers to read in an HBase table and produce HBase output files that are bulk loaded.

# HBase and MapReduce Clusters

A common question that newcomers to HBase often ask is - Should my HBase and MapReduce cluster be the same or should I run them on separate clusters?

There is no easy answer. However we will consider scenarios which will provide some guidance.

## Scenario I: Frequent MapReduce Jobs Against HBase Tables

If your use-cases include frequent execution of MapReduce jobs on HBase tables, your MapReduce jobs should execute on the same cluster as HBase. The reason is data locality. When the Mapper processes the region, it is usually located on the same node as the data for your region. This configuration enables your MapReduce jobs to execute much more quickly.

## Scenario II: HBase and MapReduce have Independent SLAs

Assume that you are a social media site that heavily uses HBase, and your users are constantly querying your HBase cluster. You also run regular MapReduce jobs to perform analytics. In this scenario, you have independent service level agreements (SLAs) for HBase and MapReduce.

HBase and MapReduce access patterns for cluster resources are markedly different. HBase is CPU and memory-intensive while sporadically and randomly accessing the disk. MapReduce is much more I/O-intensive and depends on sequential disk access. MapReduce also depends on having full control of the disk for better performance. However, MapReduce accesses memory and CPU only sporadically. In other words, HBase is CPU- and memory-bound, and MapReduce is I/O-bound. Combine them and the result is unpredictable performance due to resource contention conflicts. (If you combine the two, you should also allocate fewer Map and Reduce slots than you would allocate if you had a dedicated MapReduce cluster.)

This discussion might make it seem as if it is not advisable to run both MapReduce and HBase on the same cluster. In practice, it is not an easy decision. There are MapReduce loads that need to run on HBase tables in a typical HBase shop, and it is sometimes beneficial to run them on the same cluster. If you are using the cloud (see Chapter 16), you might have the option of adding more nodes to your cluster when you need to run the nightly MapReduce job that can alleviate some of the concerns illustrated in this section. Like everything else in this book, consider your unique requirements (business and technical) to configure the cluster that achieves the best trade-offs.

# Summary

In this chapter we explored HBase the transaction database based on the Hadoop File System. HBase is a column oriented schema less NoSQL database which enables high volume transactional use-cases. HBase supports structured and semi-structured data. We explored how HBase is different from RDBMS and the nature of use-cases it supports. We discussed the logical and physical architecture of HBase as well as explored the command line and Java API for HBase.

CHAPTER 15

# Data Science with Hadoop

"Data science" is a broad term that describes the study of extracting knowledge from data. Data science is interdisciplinary in that it usually requires expertise in a variety of fields. Some examples include these:

- Business domain expertise
- Mathematics and statistics
- Scientific method
- Computer science and data engineering
- Visualization

The specific business domain and nature of the data strongly influence the techniques needed to solve specific problems. There is no one-size-fits-all solution for data science. Some problems of interest include these:

- Text mining and natural language processing
- Machine learning
- Signal and image processing
- Statistics and analytics

The applications of data science continue to grow as new sources of data and the techniques to analyze them become available. For economic reasons, much of the data produced historically in organizations was not collected or (in many cases) deleted. As more and more organizations learn that significant business value can be derived from this "exhaust data," data science will grow to fill this need.

The data of interest can be big or small; all can present unique challenges. This chapter focuses on Big Data because Hadoop is designed to solve this class of problems.

The broad topic of data science can easily fill a book, so this chapter focuses on technologies within the Hadoop ecosystem that can be used by data scientists to solve their specific problems.

## Hadoop Data Science Methods

A number of Hadoop technologies and methods can be applied to data science. MapReduce can be sufficient to solve certain classes of problems. Pig and Hive can be used for set-based problems, although they are not good choices when iterative algorithms are needed.

Apache Hama and Apache Spark are specifically designed to solve problems that fall into the data science category, as well as many others. In addition, R can be used to interactively analyze large datasets using the Hadoop platform. The rmr2 package supports MapReduce and HDFS operations as well as visualization and all the capabilities of R. Each of these technologies is introduced in this chapter, and several examples are provided to show how these technologies can be used. The examples are chosen for their ease of understanding, so they do not necessarily represent optimal solutions to specific real-world problems. More information on methods and technologies can be found online and in published literature.

# Apache Hama

Apache Hama is a distributed computing framework based on a bulk synchronous parallel (BSP) model. Hama, which is short for "Hadoop Matrix," was inspired by the Google Pregel framework. Hama was developed by Edward J. Yoon and is now a top-level Apache project. It was designed to efficiently solve a class of large-scale problems that require iteration. Like all parallel models, it solves many problems, but not all. Still, Hama is effective and relatively easy to use compared with MapReduce.

## Bulk Synchronous Parallel Model

BSP is a parallel computing model that was developed in the 1980s by Leslie Valiant at Harvard. Nodes in a BSP computing model operate in an iterative sequence of supersteps that contain three phases:

1. Concurrent computation
2. Interprocess communication (data exchange)
3. Barrier synchronization

Figure 15-1 illustrates the BSP execution model.

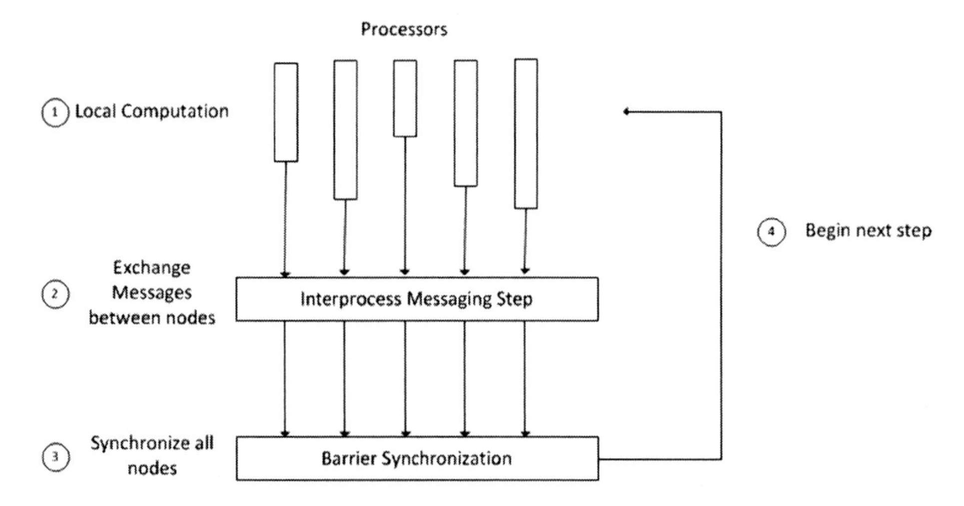


***Figure 15-1.*** *BSP model*

The processes in Hama are called peers. Prior to barrier synchronization, each peer performs local computations using its own data. This computation is wait-free, so performance is limited only by local I/O and CPU bounds. Peers can send messages to other peers.

Finally, peers enter the barrier synchronization. At this phase, each peer has completed its superstep and is ready to accept messages from other peers. During the synchronization, all messages are exchanged reliably, and all peers wait for others to enter the barrier. After coming out of the barrier, each peer can work independently on messages sent to it from other peers, and the cycle repeats.

One peer can act as the master task, which aggregates data from all other peers, maintains global state, and broadcasts global state to all other peers.

## Hama Hello World!

One of the key advantages of Hama is its ease of use compared with MapReduce. To illustrate the simplicity of the model, Listing 15-1 provides a basic Hello World! Hama job.

***Listing 15-1.*** Simple Hama Program

```
package org.apress.prohadoop.c15;

import java.io.IOException;

import org.apache.hadoop.fs.FileSystem;
import org.apache.hadoop.fs.Path;
import org.apache.hadoop.io.NullWritable;
import org.apache.hadoop.io.Text;
import org.apache.hama.HamaConfiguration;
import org.apache.hama.bsp.BSP;
import org.apache.hama.bsp.BSPJob;
import org.apache.hama.bsp.BSPJobClient;
import org.apache.hama.bsp.BSPPeer;
import org.apache.hama.bsp.ClusterStatus;
import org.apache.hama.bsp.FileOutputFormat;
import org.apache.hama.bsp.NullInputFormat;
import org.apache.hama.bsp.TextOutputFormat;
import org.apache.hama.bsp.sync.SyncException;

public class DemoBSP extends BSP<NullWritable, NullWritable, Text, Text, Text> {
    private static Path TMP_OUTPUT = new Path("/tmp/demo-" + System.currentTimeMillis());

    @Override
    public void setup(BSPPeer<NullWritable, NullWritable, Text, Text, Text> peer)
            throws IOException, SyncException, InterruptedException {
//        System.out.printf("Setup: %d, %s\n", peer.getPeerIndex(), peer.getPeerName());
    }

    @Override
    public void bsp(BSPPeer<NullWritable, NullWritable, Text, Text, Text> peer)
            throws IOException, SyncException, InterruptedException {
        String message = String.format("Hello from %s!", peer.getPeerName());
        for (int i = 0; i < peer.getNumPeers(); i++) {
            int next = (peer.getPeerIndex() + 1) % peer.getNumPeers();
            String nextName = String.format("local:%d", next);
            peer.send(nextName, new Text(message));
            peer.sync();

            Text currentMessage = peer.getCurrentMessage();
            System.out.printf("bsp: %s, %s\n", peer.getPeerName(), currentMessage);
            message = currentMessage.toString();
        }
    }
```

```
    @Override
    public void cleanup(
            BSPPeer<NullWritable, NullWritable, Text, Text, Text> peer)
            throws IOException {
//        System.out.printf("Cleanup: %d, %s\n", peer.getPeerIndex(), peer.getPeerName());
    }

    public static void main(String[] args) throws Exception {
        HamaConfiguration conf = new HamaConfiguration();

        BSPJob bsp = new BSPJob(conf, DemoBSP.class);
        bsp.setJobName("Demo BSP");
        bsp.setBspClass(DemoBSP.class);
        bsp.setInputFormat(NullInputFormat.class);
        bsp.setOutputKeyClass(Text.class);
        bsp.setOutputValueClass(Text.class);
        bsp.setOutputFormat(TextOutputFormat.class);
        FileOutputFormat.setOutputPath(bsp, TMP_OUTPUT);

        BSPJobClient jobClient = new BSPJobClient(conf);
        ClusterStatus cluster = jobClient.getClusterStatus(true);

        if (args.length > 0) {
            bsp.setNumBspTask(Integer.parseInt(args[0]));
        } else {
            bsp.setNumBspTask(cluster.getMaxTasks());
        }

        long startTime = System.currentTimeMillis();
        if (bsp.waitForCompletion(true)) {
            System.out.println("Job Finished in "
                    + (System.currentTimeMillis() - startTime) / 1000.0
                    + " seconds");
        }

        FileSystem fs = FileSystem.get(conf);
        fs.delete(TMP_OUTPUT, true);
    }
}
```

In this job, each peer sends a greeting message to the next peer in sequence; then the barrier synchronization is entered by calling `peer.sync()`. At this point, all messages are exchanged between peers. Upon returning from the barrier synchronization, each peer collects its message and prints it. The process repeats for the number of peers, so each peer has an opportunity to send and receive a message to and from every other peer.

The default Hama installation runs in local mode, just like Hadoop. To run this example in local mode, compile it using all JAR files in `$HAMA_HOME` and `$HAMA_HOME/lib`. This example has been packaged in a JAR file named `hama.jar`. To execute it, use the hama command and specify the number of peers as an argument on the command line:

```
hama jar demo.jar org.apress.prohadoop.c15.DemoBSP 3
```

For three peers, the output looks like this:

```
bsp: local:1, Hello from local:0!
bsp: local:2, Hello from local:1!
bsp: local:0, Hello from local:2!
bsp: local:1, Hello from local:2!
bsp: local:2, Hello from local:0!
bsp: local:0, Hello from local:1!
bsp: local:1, Hello from local:1!
bsp: local:2, Hello from local:2!
bsp: local:0, Hello from local:0!
```

Notice that there are 3 x 3 (9) output lines.

## Monte Carlo Methods

Monte Carlo methods are algorithms that rely on repeated random sampling to compute a result. The modern Monte Carlo method was developed at Los Alamos National Laboratory in 1946 and its initial problem involved computing the average distance of neutron travel in a material. Closed-form solutions were difficult to compute, so Stanislaw Ulam and Jon Von Neumann chose a statistical method. The project was classified at the time, so the code name Monte Carlo was chosen (after the famous Monaco casino).

The value of Pi can be estimated using Monte Carlo methods. It is not the most efficient way to do so, but it is easy to visualize and serves as an effective demonstration of the method.

The estimation works by inscribing a unit circle (circle with radius=1unit) within a perfect square. The area of the circle is $\pi r^2 = \pi$, and the area of the square is $4r^2 = 4$. If a large number of uniformly distributed random points are chosen within the square, the percentage of points falling within the circle is $\frac{\pi}{4}$. The value of $\pi$ is simply the proportion of points falling within the circle multiplied by 4. Points falling within the circle must satisfy the equation of a circle, $x^2 + y^2 = r^2$. This equation enables us to determine whether the point falls within the circle. From this information and a sufficiently large number of random points, it is possible to estimate the value of Pi.

Figure 15-2 shows the previous scheme. If points are randomly generated within the square, count the points falling in the shaded area and multiply them by 4 to get the value of Pi. For a circle with radius equal to 1 unit, its area is 0.785 times the area of the square it is inscribed into. This means, if 10000 points are uniformly generated inside the square approximately 7850 fall inside the circle. The value of $\pi$ is $4 \times 0.785 = 3.14$.

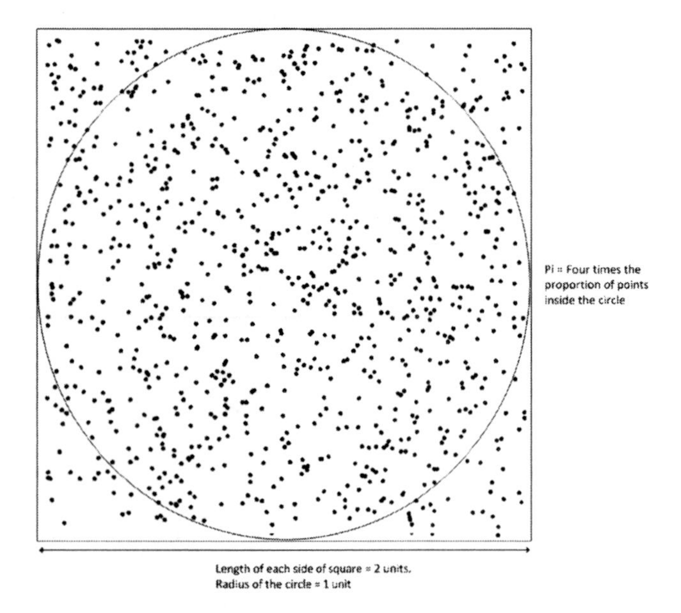


***Figure 15-2.*** *Calculating Pi using Monte Carlo*

To achieve high performance, the problem can be parallelized using Hama. Several peers can be created, each generating a large number of distinct random values. When they complete, they send a message to one master with their estimated value. The master then averages the values from all peers.

Listing 15-2 shows the Hama code to estimate Pi. One of the peers is arbitrarily chosen as the master in the `setup()` method, which runs prior to the `bsp()` method. The `bsp()` method for each peer computes its estimate, sends it to the master, and calls `sync`. In this case, only one superstep is used.

***Listing 15-2.*** Pi Estimation Using Hama

```
package org.apress.prohadoop.c15;

import java.io.IOException;

import org.apache.hadoop.fs.FSDataInputStream;
import org.apache.hadoop.fs.FileStatus;
import org.apache.hadoop.fs.FileSystem;
import org.apache.hadoop.fs.Path;
import org.apache.hadoop.io.DoubleWritable;
import org.apache.hadoop.io.IOUtils;
import org.apache.hadoop.io.NullWritable;
import org.apache.hadoop.io.Text;
import org.apache.hama.HamaConfiguration;
import org.apache.hama.bsp.BSP;
import org.apache.hama.bsp.BSPJob;
import org.apache.hama.bsp.BSPJobClient;
import org.apache.hama.bsp.BSPPeer;
import org.apache.hama.bsp.ClusterStatus;
import org.apache.hama.bsp.FileOutputFormat;
import org.apache.hama.bsp.NullInputFormat;
```

```
import org.apache.hama.bsp.TextOutputFormat;
import org.apache.hama.bsp.sync.SyncException;

public class BSPPiEstimator extends
        BSP<NullWritable, NullWritable, Text, DoubleWritable, DoubleWritable> {
    private String masterTask;
    private static final int ITERATIONS = 10000;
    private static Path TMP_OUTPUT = new Path("/tmp/pi-" + System.currentTimeMillis());

    @Override
    public void setup(
            BSPPeer<NullWritable, NullWritable, Text, DoubleWritable, DoubleWritable> peer)
            throws IOException {
        // Choose one as a master
        this.masterTask = peer.getPeerName(peer.getNumPeers() / 2);
    }

    @Override
    public void bsp(
            BSPPeer<NullWritable, NullWritable, Text, DoubleWritable, DoubleWritable> peer)
            throws IOException, SyncException, InterruptedException {

        int in = 0;
        for (int i = 0; i < ITERATIONS; i++) {
            double x = 2.0 * Math.random() - 1.0, y = 2.0 * Math.random() - 1.0;
            if ((Math.sqrt(x * x + y * y) < 1.0)) {
                in++;
            }
        }

        double data = 4.0 * in / ITERATIONS;

        peer.send(masterTask, new DoubleWritable(data));
        peer.sync();
    }

    @Override
    public void cleanup(
            BSPPeer<NullWritable, NullWritable, Text, DoubleWritable, DoubleWritable> peer)
            throws IOException {
        if (peer.getPeerName().equals(masterTask)) {
            double pi = 0.0;
            int numPeers = peer.getNumCurrentMessages();
            DoubleWritable received;
            while ((received = peer.getCurrentMessage()) != null) {
                pi += received.get();
            }

            pi = pi / numPeers;
            peer.write(new Text("Estimated value of PI is"),
                    new DoubleWritable(pi));
        }
    }
```

```
    public static void main(String[] args) throws Exception {
        HamaConfiguration conf = new HamaConfiguration();
        BSPJob bsp = new BSPJob(conf, BSPPiEstimator.class);
        bsp.setJobName("Pi Estimation Example");
        bsp.setBspClass(BSPPiEstimator.class);
        bsp.setInputFormat(NullInputFormat.class);
        bsp.setOutputKeyClass(Text.class);
        bsp.setOutputValueClass(DoubleWritable.class);
        bsp.setOutputFormat(TextOutputFormat.class);
        FileOutputFormat.setOutputPath(bsp, TMP_OUTPUT);

        BSPJobClient jobClient = new BSPJobClient(conf);
        ClusterStatus cluster = jobClient.getClusterStatus(true);

        int numTasks = cluster.getMaxTasks();
        if (args.length > 0) {
            numTasks = Integer.parseInt(args[0]);
        }
        System.out.printf("Using %d tasks\n", numTasks);
        bsp.setNumBspTask(numTasks);

        long startTime = System.currentTimeMillis();
        if (bsp.waitForCompletion(true)) {
            FileSystem fs = FileSystem.get(conf);
            FileStatus[] files = fs.listStatus(TMP_OUTPUT);
            for (int i = 0; i < files.length; i++) {
                if (files[i].getLen() > 0) {
                    FSDataInputStream in = fs.open(files[i].getPath());
                    IOUtils.copyBytes(in, System.out, conf, false);
                    in.close();
                    break;
                }
            }

            fs.delete(TMP_OUTPUT, true);
            System.out.println("Job Finished in "
                    + (System.currentTimeMillis() - startTime) / 1000.0
                    + " seconds");
        }
    }
}
```

The output for 10,000 iterations is as follows:

```
Pi is roughly 3.1401200000
```

The value is not particularly accurate, but it is within the ballpark of the real value. When executed locally, Hama supports ten tasks by default. That result is a total of 100,000 iterations to estimate Pi.

## K-Means Clustering

K-means clustering is an example of unsupervised machine learning: it is an attempt to automatically find structure in a set of unlabeled data. Given a set of n observations, the goal is to find a set of k clusters such that each observation belongs to the cluster with the nearest mean. Figure 15-3 is an image of a set of possible real-world observations plotted in two dimensions. The casual human observer easily notices three clusters. The goal of k-means clustering is to develop computer algorithms to do what humans consider simple.

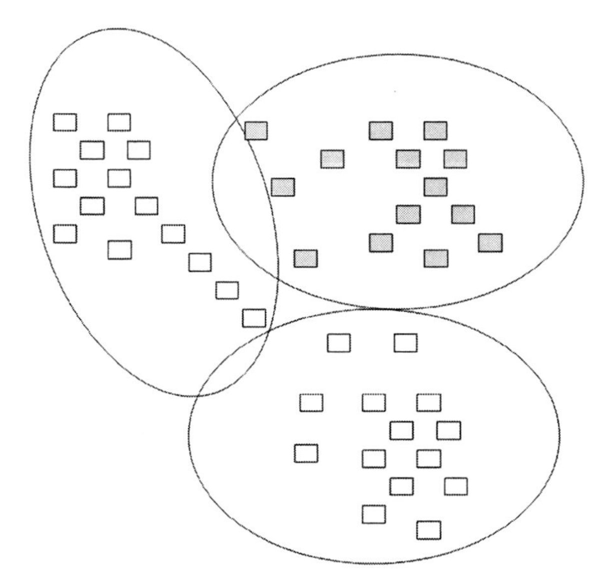

***Figure 15-3.*** *K-means clustering*

Computing the optimum solution is known to be NP-hard, meaning that it represents one of the most difficult problems from a computational complexity perspective. Fortunately, heuristic methods exist that yield satisfactory results and are not difficult to implement. The typical method is performed in two steps: an assignment step and an iterative update step. In the assignment step, it is common to choose k random observations as the prototype centers for each cluster. During the update step, the mean distance between the centers and each observation is computed, and observations are assigned to a cluster. The center of each cluster is recalculated. The center calculation is repeated until it no longer changes.

Although this method usually produces good results, it might not converge on a global optimum. It is common to run the algorithm multiple times with different initial centers. Certain sets of observations require exponential time to converge, but they tend not to occur in practice.

Choosing the number of clusters, k, is a problem different from the clustering algorithm proper. The choice of k is a trade-off because the number of clusters can always be increased to improve accuracy. Choosing a value of k close to the number of observations offers little value. One rule of thumb is to choose $K \approx \sqrt{\left(\frac{n}{2}\right)}$. The elbow method is another approach that attempts to reduce variance as the k is increased. More information on clustering algorithms and selecting an appropriate value of k is widely available online and in published literature.

As an example, consider the following observations, or input dataset:

```
foo     5.1     3.5
foo     4.9     3.0
foo     4.7     3.2
bar     9.1     3.5
bar     9.9     3.0
bar     9.7     3.2
baz     9.1     60.5
baz     9.9     60.0
baz     9.7     60.2
```

It is easy to see that there are three distinct clusters. The data was chosen so that the labels (foo, bar, and baz) correspond to the three clusters. Real-world observations usually do not contain such obvious clusters; this simplified data is used for illustration purposes only,

Hama includes a convenient class for k-means clustering, KMeansBSP, which contains methods analogous to the assignment and update steps described previously. The class is relatively easy to use. It expects a path to the input dataset, which can be a file or directory. Each line in the input represents one observation, and each observation contains multiple columns delimited by the TAB character. The first column for each observation is an arbitrary label that does not affect the cluster computation. The remaining columns must be numeric values.

There can be many numeric values for each observation. KMeansBSP uses Euclidean distance to determine the means. The data in the previous example can be visualized on a two-dimensional plane. Other datasets, such as the classic iris flower dataset collected by Edgar Anderson, contain 4 features from 50 samples of 3 species of iris. The features are the length and width of petals and sepals in centimeters.

The source code for this example is shown in Listing 15-3.

***Listing 15-3.*** KMeans Using BSP

```
package org.apress.prohadoop.c15;

import java.util.HashMap;
import java.util.List;
import java.util.Set;

import org.apache.hadoop.fs.FileSystem;
import org.apache.hadoop.fs.Path;
import org.apache.hama.HamaConfiguration;
import org.apache.hama.bsp.BSPJob;
import org.apache.hama.commons.math.DoubleVector;
import org.apache.hama.ml.kmeans.KMeansBSP;

public class BSPKMeansExample {
    public static void main(String[] args) throws Exception {
        if (args.length != 4) {
            System.out.println("Usage: <input> <output> <iterations> <K (number of centers)>");
            return;
        }
        HamaConfiguration conf = new HamaConfiguration();

        Path in = new Path(args[0]);
        Path out = new Path(args[1]);
```

```
        int iterations = Integer.parseInt(args[2]);
        int k = Integer.parseInt(args[3]);

        FileSystem fs = FileSystem.get(conf);
        Path centerIn;
        if (fs.isFile(in)) {
            centerIn = new Path(in.getParent(), "center/center.seq");
        } else {
            centerIn = new Path(in, "center/center.seq");
        }
        Path centerOut = new Path(out, "center/center_output.seq");

        conf.set(KMeansBSP.CENTER_IN_PATH, centerIn.toString());
        conf.set(KMeansBSP.CENTER_OUT_PATH, centerOut.toString());
        conf.setInt(KMeansBSP.MAX_ITERATIONS_KEY, iterations);

        in = KMeansBSP.prepareInputText(k, conf, in, centerIn, out, fs, true);

        BSPJob job = KMeansBSP.createJob(conf, in, out, true);

        job.waitForCompletion(true);

        System.out.println("Cluster results:");
        List<String> results = KMeansBSP.readOutput(conf, out, fs, 100);
        for (String line : results) {
            System.out.println(line);
        }

        System.out.println("\nCluster centers:");
       HashMap<Integer, DoubleVector> centers = KMeansBSP.readClusterCenters(conf, out, centerOut, fs);
        Set<Integer> keys = centers.keySet();
        for (Integer n : keys) {
            System.out.printf("Center %d, vector: %s\n", n, centers.get(n));
        }
    }
}
```

The example is executed locally with the following command. KMeansBSP also expects the output directory, number of iterations, and number of clusters (k):

```
hama jar hama.jar org.apress.prohadoop.c15.BSPKMeansExample simple_kmeans.txt /tmp/result 10 3
```

The output looks like this:

```
Cluster results:
foo: [5.1, 3.5] belongs to cluster 2
foo: [4.9, 3.0] belongs to cluster 2
foo: [4.7, 3.2] belongs to cluster 2
bar: [9.1, 3.5] belongs to cluster 1
bar: [9.9, 3.0] belongs to cluster 1
bar: [9.7, 3.2] belongs to cluster 1
baz: [9.1, 60.5] belongs to cluster 0
```

```
baz: [9.9, 60.0] belongs to cluster 0
baz: [9.7, 60.2] belongs to cluster 0

Cluster centers:
Center 0, vector: [9.566666666666666, 60.233333333333333]
Center 1, vector: [9.566666666666666, 3.233333333333333]
Center 2, vector: [4.8999999999999995, 3.233333333333333]
```

Clearly, the algorithm converges toward expected values, even with ten iterations. The observations are sufficiently far apart so that clusters are easy to find, which might not be the case with other real-world datasets.

# Apache Spark

Apache Spark is another distributed computing model that can run on both Hadoop and Apache Mesos. Spark was developed at the UC Berkeley AMPLab and became a top-level Apache project in February 2014. The big advantage of Spark over other parallel models is performance and ease of use. Spark is advertised as running up to 30 times faster than MapReduce-based workloads. Much of this performance is achieved through a more efficient runtime model that leverages memory where possible and disk when necessary.

Spark can access datasets stored in the HDFS and Amazon S3 buckets as well as Hadoop file formats such as sequence files, which makes Spark an easy transition from other Hadoop ecosystem solutions.

Spark is written primarily in Scala and supports Scala, Java, and Python APIs. Scala, another language that runs on the Java Virtual Machine (JVM), is a statically typed, object-functional language. Some of the appeal of Scala is its concise syntax: in many cases, it uses type inference to avoid explicit variable type declarations. Scala supports first-class functions and local functions, which can often be convenient. It also supports interactive execution, similar to Python, but achieves performance close or equal to Java.

## Resilient Distributed Datasets (RDDs)

One of the goals of Spark is to support distributed collection operations as if they are local. For example, Scala has a built-in sequence trait that includes many high-level methods (as well as common low-level collection methods). A Scala trait is similar to a Java interface, although traits can be partially implemented.

Table 15-1 lists some of the methods available in the Scala Seq trait. A Seq trait represents sequences, which are a kind of iterable that has a length and whose elements have fixed index positions starting from 0.

***Table 15-1.** Scala Methods Available in Scala Seq Traits*

| Method | Description |
| --- | --- |
| `map(f: T => U): Seq[U]` | Applies function `f` to every element in a sequence |
| `flatMap(f: T => Seq[U]): Seq[U]` | Applies function `f` to every element, producing multiple objects |
| `filter(f: T => Boolean): Seq[T]` | Filters out any element for which function `f` returns `false` |
| `exists(f: T => Boolean): Boolean` | Returns `true` if function `f` applied to any element returns `true` |
| `forall(f: T => Boolean): Boolean True` | Returns `true` if function `f` applied to all elements returns `true` |
| `reduce(f: (T, T) => T): T` | Applies function `f` to all elements (two at a time) and returns a single object |
| `groupBy(f: T => K): Map[K,List[T]]` | Groups elements by function `f` |
| `sortBy(f: T => K): Seq[T]` | Sorts elements by function `f` |

The key concept of Spark is the Resilient Distributed Dataset (RDD). RDDs are like immutable local Scala sequences, but they operate in parallel on a cluster. RDDs are powerful abstractions that support many high-level operations. RDDs contain partitions, which are similar to blocks within a Map or Reduce task in Hadoop MapReduce. The number of partitions in an RDD is based on the number of blocks in an HDFS path or the number of nodes in the case of in-memory data.

RDDs are reliable, which means that node failure on a cluster is not catastrophic. RDD reliability is achieved through a lineage, and RDD operations are recorded in their lineage. When a node containing an RDD partition fails, that partition is reconstructed using information in its lineage. The technique, which is similar to a database transaction log or journaling filesystem, is quite efficient. In many ways, RDDs are more powerful, flexible, and extensible than Map and Reduce tasks; in fact, RDD `flatMap()` and `reduceByKey()` methods are conceptually similar to the `map()` and `reduce()` methods in Hadoop MapReduce. By this measure, Spark can be considered a superset of MapReduce.

RDD methods fall into two categories: transformations and actions. Transformations create new RDDs from other RDDs—for example, `map()`, `filter()`, and `groupBy()`. Transformations are lazy; they are not executed until needed. Actions return results or save output to storage—for example, `collect()`, `count()`, and `save()`.

## Monte Carlo with Spark

The prior Pi estimation example can be performed in Spark. Although the example is written in Java, it could also have been written in Scala or Python. The `SparkPiEstimator` class creates a list containing `NUM_SAMPLES` integers. The values in the list are not important; the list is used to parallelize the computation on multiple nodes using the `parallelize()` method in `SparkContext`.

The example counts the number of random points that fall into the quarter circle, as in the Hama BSP example. The `parallelize()` and `filter()` methods are transformations (they create RDDs), and the `count()` method is an action. The final value is computed by dividing by the number of total samples and multiplying by 4.0.

The source code appears in Listing 15-4.

***Listing 15-4.*** Pi Estimation Using Spark

```
package org.apress.prohadoop.c15;

import java.util.ArrayList;
import java.util.List;

import org.apache.log4j.Level;
import org.apache.log4j.LogManager;
import org.apache.spark.api.java.JavaSparkContext;
import org.apache.spark.api.java.function.Function;

public class SparkPiEstimator {
    public final static int NUM_SAMPLES = 100000;

    public static void main(String[] args) throws Exception {
        LogManager.getRootLogger().setLevel(Level.ERROR);
        JavaSparkContext sc = new JavaSparkContext("local", "Pi Estimator");

        List<Integer> list = new ArrayList<>();
        for (int i = 0; i < NUM_SAMPLES; i++) {
            list.add(i);
        }
```

```
        double count = sc.parallelize(list)
                .filter(new Function<Integer, Boolean>() {
                    public Boolean call(Integer i) {
                        double x = Math.random();
                        double y = Math.random();
                        return x * x + y * y < 1;
                    }
                }).count();
        System.out.printf("Pi is roughly %12.10f\n", 4.0 * count / NUM_SAMPLES);
    }
}
```

The output is the following:

```
Pi is roughly 3.1401200000
```

When executed locally, parallelize() creates only one partition, so all calculation is performed in one task. When executed on multiple workers, such as a cluster, parallelize() divides up the list of integers evenly among the workers. To get similar results between Hama and Spark, it is necessary to specify 100,000 total values in the list.

Spark RDDs support a rich set of transformations and actions. It is possible to rewrite the preceding example as a MapReduce job in Spark, as shown in Listing 15-5.

***Listing 15-5.*** MapReduce using Spark

```
package org.apress.prohadoop.c15;

import java.util.ArrayList;
import java.util.List;

import org.apache.log4j.Level;
import org.apache.log4j.LogManager;
import org.apache.spark.api.java.JavaRDD;
import org.apache.spark.api.java.JavaSparkContext;
import org.apache.spark.api.java.function.Function;
import org.apache.spark.api.java.function.Function2;

public class SparkMRPiEstimator {

    public static void main(String[] args) {
        LogManager.getRootLogger().setLevel(Level.ERROR);
        JavaSparkContext jsc = new JavaSparkContext("local", "MR Pi Estimator");

        int slices = 10;
        int n = 10000 * slices;
        List<Integer> list = new ArrayList<Integer>(n);
        for (int i = 0; i < n; i++) {
            list.add(i);
        }

        JavaRDD<Integer> dataSet = jsc.parallelize(list, slices);
```

```
        int count = dataSet.map(new Function<Integer, Integer>() {
            @Override
            public Integer call(Integer integer) {
                double x = Math.random() * 2 - 1;
                double y = Math.random() * 2 - 1;
                return (x * x + y * y < 1) ? 1 : 0;
            }
        }).reduce(new Function2<Integer, Integer, Integer>() {
            @Override
            public Integer call(Integer integer, Integer integer2) {
                return integer + integer2;
            }
        });

        System.out.printf("Pi is roughly %12.10f\n", 4.0 * count / n);
    }
}
```

The output is the following:

```
Pi is roughly 3.1371600000
```

Notice that the value is slightly different from previous examples because of the statistical nature of the Monte Carlo methods. To obtain identical results, it is necessary to seed the random number generators with a consistent value.

## KMeans with Spark

Similar to Hama, Spark includes a KMeans class. In this case, the data is read in using the textFile() method defined in SparkContext. textFile() returns an RDD (a JavaRDD in this case). The number partitions in the RDD depend on the number of input blocks; in this case, it contains only one partition because it is a small file.

The same input data is used as in the Hama k-means example. The Spark KMeans class expects an array of doubles only as input, so the first columns containing the label are ignored in the map() step. The analogous assignment step is performed in the train() method of the KMeans class, which accepts the input RDD, the number of clusters, and the number of iterations. The train() method returns a KMeansModel that is used in the update step. The center of each cluster can be obtained from the KMeansModel.

The source code for this example is shown in Listing 15-6.

***Listing 15-6.*** KMeans Using Spark

```
package org.apress.prohadoop.c15;

import org.apache.log4j.Level;
import org.apache.log4j.LogManager;
import org.apache.spark.api.java.JavaRDD;
import org.apache.spark.api.java.JavaSparkContext;
import org.apache.spark.api.java.function.Function;
import org.apache.spark.mllib.clustering.KMeans;
import org.apache.spark.mllib.clustering.KMeansModel;
```

```
public class SparkKMeans {
    public static void main(String[] args) {
        LogManager.getRootLogger().setLevel(Level.ERROR);
        JavaSparkContext sc = new JavaSparkContext("local", "Spark K Means");

        JavaRDD<String> input = sc.textFile("simple_kmeans.txt");
        JavaRDD<double[]> parsedData = input.map(new Function<String, double[]>() {
            @Override
            public double[] call(String in) throws Exception {
                String[] columns = in.split("\t");
                double[] values = new double[columns.length-1];
                for (int i = 0; i < values.length; i++) {
                    values[i] = Double.parseDouble(columns[i+1]);
                }
                return values;
            }
        });
        int k = 3;
        int iterations = 10;
        KMeansModel clusters = KMeans.train(parsedData.rdd(), k, iterations);
        double[][] centers = clusters.clusterCenters();
        for (int i=0; i<centers.length; i++) {
            double[] ds = centers[i];
            System.out.printf("Center %d, vector: [%5.2f, %5.2f]\n", i, ds[0], ds[1]);
        }
    }
}
```

As expected, the output compares well with the Hama BSP example:

```
Center 0, vector: [ 9.57,  3.23]
Center 1, vector: [ 9.57, 60.23]
Center 2, vector: [ 4.90,  3.23]
```

The center order is not the same as the Hama example due to implementation differences.

The `clusters` object also contains a `predict()` method to determine which cluster contains a new input vector.

It is interesting to compare the Scala version of the same code:

```
import org.apache.spark.mllib.clustering.KMeans

val input = sc.textFile("simple_kmeans_no_key.txt")
val parsedData = input.map( _.split('\t').map(_.toDouble))

val k = 3
val iterations = 10
val clusters = KMeans.train(parsedData, k, iterations)
```

Scala uses type inference to reduce the need to explicitly define variable types, which creates more concise code. The preceding example uses an input file that does not include the key in the first column. This shortcut simplifies the file parsing logic.

The Scala version can be executed within the interactive Spark shell, which is convenient for exploratory analytics.

# RHadoop

R, which is a powerful language and runtime environment for statistical analysis, interactively analyzes local datasets of modest size. R also supports convenient visualization through graphical plotting as well as a large set of packages that can extend the capabilities of R. R is a GNU project, so it is available as free software. There are binary versions with easy-to-use installers for Windows, Mac OS X, and UNIX platforms.

Although R is powerful, it is not well-suited for very large datasets out of the box. Fortunately, Revolution Analytics has developed RHadoop, which is an R package that integrates R with Hadoop clusters. The RHadoop package contains several libraries that support MapReduce execution on a remote cluster as well as HDFS and HBase access from within the R graphical environment.

The R language and environment is another large topic that can fill a book. Such material is widely available online and in published literature. To get started, a brief summary is described here. The R application must be installed first, which is a straightforward download and install. Then the RHadoop package can be downloaded as a zip file. From within the RGui, select `menu item Packages | Install package(s) from local zip files` and choose the downloaded zip file.

The rmr2 library requires several other R packages, which are listed here:

- `Rcpp`
- `RJSONIO`
- `bitops`
- `digest`
- `functional`
- `reshape2`
- `stringr`
- `plyr`
- `caTools`

Fortunately, these packages are available using the `Packages | Install Package(s)` menu item.

The rmr library can be run locally for development and testing purposes by setting the following:

```
rmr.options(backend = "local")
```

This code disables Hadoop cluster integration and uses local MapReduce support. For cluster integration, consult the RHadoop online documentation.

Figure 15-4 shows a screenshot after executing R code. It was contributed by Eric Lecoutre, the winner of the R Homepage graphics competition, and it illustrates some of the computation and graphical capabilities of R.

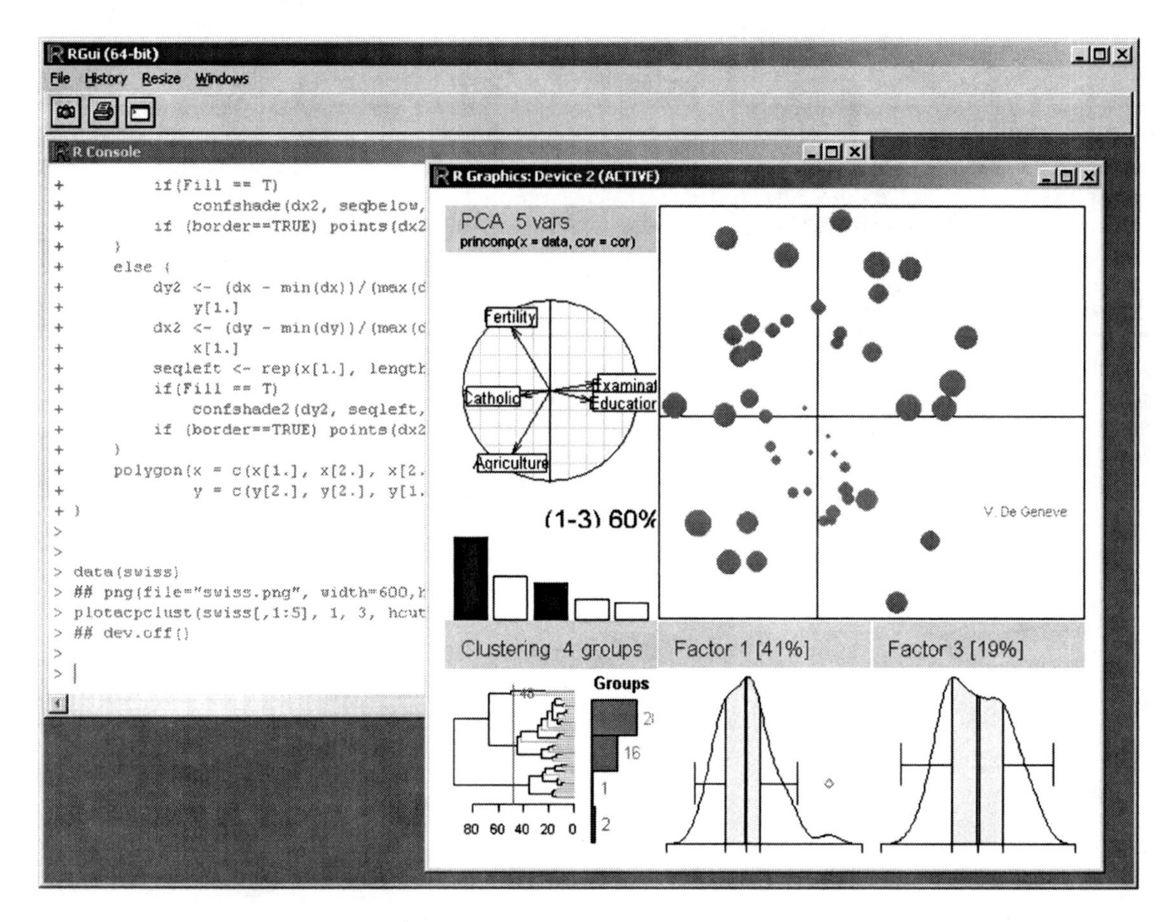


*Figure 15-4.* *Graphical capabilities of R*

Just like Hama and Spark, RHadoop can be used to implement many of the solutions described in this chapter. RHadoop relies on MapReduce for much of its computation, so it is constrained by the batch-oriented nature of MapReduce. An R package for Spark is under development by AMPLab.

# Summary

Data science is a growing field of study due to the abundance of Big Data and powerful analysis frameworks and tools. Becoming an experienced data scientist requires interdisciplinary skills such as domain knowledge, mathematics, and adaptation to the changing landscape of Big Data.

A few technologies and techniques have been described in this chapter to show what is possible. We explored the BSP model and Hama. You learned how to use the BSP model to perform Monte Carlo–style computations.

We discussed Spark, which is becoming very popular, and all vendors are putting their weight behind it. We explored Spark in the same context as BSP. Finally we explored RHadoop.

It is a well-known concept in data mining that more data trumps better models. The traditional methods applied data mining techniques on sampled data because most data mining applications used a single system view of the data. This method limited the amount of data that could be handled using data mining software. The techniques described in this chapter enable data mining algorithms to be applied on large scales of data on a cluster of commodity machines, which will change the landscape considerably in the coming years.

CHAPTER 16

# Hadoop in the Cloud

Hadoop requires commodity cluster hardware to operate. One solution is to design a cluster, procure hardware, select a distribution, install Hadoop, and administer the cluster in-house. Some vendors deliver a completely configured cluster based on customer specifications, but the jobs of administration, maintenance, and upgrading remain. Installing and maintaining a cluster can be an effective solution, but it requires significant initial investment; an experienced administration and maintenance staff; and data archival, backup, and restore facilities.

An alternative is to use a cloud solution. Numerous vendors offer Hadoop cloud solutions in which the physical cluster is located in data centers all over the world. Customers can provision their clusters with specific software images, log in remotely, load data from external sources, and run Hadoop jobs. Cloud vendors offer several options, from bare machines to a menu of installed packages. Cloud solutions enjoy economies of scale, but introduce additional challenges such as security, data ingress/egress, and retention time.

This chapter discusses some of the benefits and challenges associated with Hadoop in the cloud. We also explore specific solutions offered by three well-known cloud vendors. Finally, we step through the process of provisioning clusters using Amazon Web Services (AWS).

## Economics

The first question that is often asked when considering a cloud solution is whether it makes economic sense. For cloud solutions to be competitive, they must offer significant economic advantages to offset challenges not found in self-hosted clusters. Data center cost models vary significantly, depending on many factors. For the purposes of a cloud versus self-hosted comparison, we use an estimated total cost of ownership (TCO) model. Our model offers a rough idea of costs for a nominal, 20-node cluster operated over 5 years. More accurate models must consider regional factors such as energy and labor costs.

### Self-Hosted Cluster

At a minimum, the total cost of a self-hosted cluster must consider the following line items:

- Hardware procurement
- Real estate
- Power, environmental, and fire-suppression controls
- Internet connections
- Operations and administration

For our 20-node cluster, a total of 25 physical machines might be required, including data nodes, NameNodes, and edge nodes. A wide variety of hardware configurations can be used, and the per-node cost can vary significantly, depending on CPU choice, memory configuration, drive capacity, and so on. For expediency, we choose a nominal price of $1,000/node. An additional $2,000 is added for network switches, racks, and cabling (we ignore battery backup for this estimate). This brings the hardware procurement cost to roughly $27,000.

Our cluster requires physical real estate, power, environmental, and possibly fire-suppression controls. Two 19-inch racks could easily hold all the nodes and network switches, so the real estate cost is not significant. Power and environmental costs vary by region, but for simplicity, we choose a nominal cost of $0.10 per kilowatt hour. We assume 24-hour, 365 days per year of uptime. Estimating no more than 500 watts per node, the total power and environmental costs over 5 years come to roughly $54,750.

Internet connection costs also vary widely by region, bandwidth, and other factors. Let us choose a reliable symmetric bandwidth fiber Internet connection. We also choose a nominal $500/month cost for Internet connectivity. This brings the 5-year Internet cost to roughly $30,000.

Self-hosted clusters require maintenance and administration from time to time. The exact staff cost can vary widely, depending on hardware stability, upgrade schedule, level of support, and labor rates. Let us estimate that no more than 1 full-time administrator spending an average of 3 hours per week is sufficient to manage the cluster. Labor rates vary widely, so we choose a nominal $100/hour, fully loaded labor rate. Over 5 years, administrative labor costs come to roughly $78,000.

Based on these gross estimates, the total cost of ownership for a 20-node cluster is $189,750. The average hourly cost for the cluster is $4.33. or about $0.17 per node. Clearly, the cost could be significantly higher if operational costs increase due to hardware failures, software bugs and patches, or additional support. Some cost reductions are possible if the cluster is located in a region with reduced electrical, environmental, and Internet costs.

## Cloud-Hosted Cluster

For a rough comparison, Amazon offers several on-demand cloud instances. A typical on-demand, m3.large Linux instance with 7.5 GiB of memory and 32 TB of SSD storage costs $0.14/hour, and there are instances that are priced as low as $0.04/hour. So cloud solutions can be very cost competitive, and more importantly, cloud costs are more predictable and flexible.

Compute power has become a commodity, so consolidation has advantages. Standardized hardware, highly trained administrative staff, and high-quality remote management tools all contribute to efficiency. Cloud solutions can offer lower cost due to economies of scale.

## Elasticity

Over the course of 5 years, it is very likely that compute demand will change significantly. Procuring, provisioning, and managing more self-hosted hardware increase costs. Reduced demand does not significantly reduce costs because self-hosted hardware is a sunk capital cost.

The most cost-effective cluster is sized for particular workloads. In self-hosted clusters, trade-offs are made to address the most common workloads. Workloads that are poorly matched to a self-hosted cluster can lead to increased operating costs.

Cloud vendors offer elastic clusters that can be sized based on demand. Cluster size can be tuned to specific workload and minimize overall cost. As data volume and demand increase over time, nodes can be easily added without costly procurement and installation. Elasticity alone is economically very appealing.

## On Demand

On demand is a simple pricing model in which instances are billed according to a fixed schedule. It is very similar to an Internet hosting model with full machine access.

On-demand instances can remain running indefinitely. Cloud vendors offer high reliability and maintenance as part of the package. To further reduce costs, it is possible to run instances as needed and shut them down when the computation is complete. In this pay-as-you-go model, developers are encouraged to write efficient algorithms that require fewer resources for shorter periods of time. In the cloud, time is money.

## Bid Pricing

Bid pricing is a low-cost alternative to on-demand pricing. Customers can bid for compute resources by specifying \a bid price, which is usually much lower than on-demand pricing.

Bid pricing can be used at any time, but it works best during times of low usage. It is possible that no resources will be available at a given price, in which case a bid might fail. One strategy is to bid at the on-demand price, which all but guarantees availability but adds the benefit of potentially acquiring instances at a lower price.

## Hybrid Cloud

Many organizations are opting for hybrid solutions. Self-hosted clusters can be used for sensitive workloads, whereas cloud solutions are used for less-sensitive workloads and elasticity.

# Logistics

Working with a cluster in the cloud presents different logistical challenges. Data must be moved in and out of a provider's data center, perhaps at a geographically distant location.

Transferring large data volumes over the Internet can saturate wide area network (WAN) connections, so proactive planning is required. Fortunately, cloud vendors offer several options for data ingress/egress and retention.

## Ingress/Egress

Data transfer rate in and out of the HDFS can be a significant challenge. Self-hosted clusters can be connected to low-cost, high-speed local area networks (LANs). Additional LAN capacity can be added for a modest cost as needed.

In cloud-hosted clusters, data must be transferred over WANs such as the Internet or private networks. Although WAN bandwidth has increased significantly, it is still significantly more expensive than LANs. Adding Internet capacity to satisfy peak demand is not cost effective.

To meet this challenge, cloud vendors offer import/export services using physical media. Large volumes of data can be packaged in portable disk drives and shipped via mail or a shipping company. Hundreds of terabytes can be compressed, stored on several hard drives, and physically shipped overnight at a manageable cost.

## Data Retention

In self-hosted servers, large volumes of data can be stored indefinitely at a reasonable cost. Persistent, on-demand, cloud-hosted clusters can also store data long term, but this is not the most cost-effective option.

In many cases, years of data retention might be required to satisfy legal or other regulatory compliance. To meet this challenge, cloud vendors offer low-cost, long-term data storage. Tools are available to move data in and out of this long-term storage, as well as efficient transfer in and out of cloud-hosted clusters.

A popular operating model is to save large volumes of data in low-cost storage. When processing is required, an appropriately sized cluster can be provisioned in the cloud in a matter of minutes. Data is transferred from long-term storage into the HDFS in the cluster. Hadoop jobs can be executed as needed, and results can then be transferred out of the HDFS back into long-term storage. At this point, data in the cluster is no longer needed, so it can be shut down. This model optimizes compute and storage resources and minimizes overall cost.

# Security

Self-hosted clusters enjoy the advantage of physical on-premises security. Many organizations are reluctant to transfer their sensitive data to storage outside of their physical control, so this can be a significant area of concern.

No system is perfectly secure. A motivated attacker can break any system, given enough time and/or resources. Attackers often use back doors, social engineering, or even threat of physical harm if the value of data is significantly greater than the cost of a successful attack. Fortunately, the bad guys are generally opportunists that perform some level of cost-benefit analysis. Increasing the cost of a successful attack beyond the value of the data is usually a sufficient deterrent. When evaluating the security of any system, it is necessary to determine the value to a potential intruder and erect enough barriers to make a successful attack economically disadvantageous.

To manage security risks and make objective decisions, it is necessary to formally understand security concepts. Secure systems must address the following concerns:

- Authentication
- Authorization
- Confidentiality
- Integrity
- Nonrepudiation

To meet these challenges, cloud providers offer multiple layers of security. Authentication usually requires multiple factors such as user IDs/passphrases and public key cryptography. Each user can create a unique key pair that is required for any data access or remote system interaction.

Fine-grained access policies can be applied to users and groups by group administrators and ensure that users are authorized to perform only a limited set of operations. Access to long-term storage can be restricted as well.

Confidentiality is achieved through multiple layers. First, all data transferred over the Internet is encrypted, either through Secure Sockets Layout (SSL) or a virtual private network (VPN). Second, the data can be stored in encrypted form in long-term storage. This encryption at rest attempts to achieve a high degree of confidentiality, integrity, and nonrepudiation.

# Cloud Usage Models

Job execution in a self-hosted cluster is straightforward: data is loaded in to the HDFS, jobs are executed, and results are collected. In most organizations, many jobs compete for cluster resources at any given time, so jobs are scheduled based on configured job queues.

Given the flexibility of cloud solutions, other usage models are possible. The simplest approach is to create a cluster sized for average demand and submit jobs just like a self-hosted cluster. This is known as a persistent cluster. Persistent clusters are preferred when frequent jobs benefit from data stored in the HDFS. Jobs might have I/O dependencies, so keeping data in the HDFS after a job has ended reduces overall processing time and cost.

Another alternative is to run jobs in a transient cluster, which is operated through a series of steps listed here:

1. The cluster is configured and started.
2. Data is copied from external durable storage into the HDFS.
3. One or more jobs are executed.
4. Processed data is copied from the HDFS back to durable storage.
5. The cluster is shut down, and all resources are released.

Transient clusters are a more cost-effective option when there is significant dead time between jobs. In a self-hosted cluster, dead time is like excess inventory, which is a source of organizational waste. In a cloud-hosted transient cluster, resources can be returned to the cloud when not needed, which reduces overall cost and improves organizational efficiency.

The choice of a persistent or transient cluster depends on a job's execution rate, average execution time, cluster startup time, and any data ingress/egress time. Cloud vendors offer guidelines to determine the best cluster-usage model for given workloads. These guidelines are available on their web sites.

# Cloud Providers

Many cloud providers offer a wide variety of services, so it is difficult to compare the exact value that each provides. For the purposes of completeness, we discuss services offered by three of the most well-known cloud providers.

## Amazon Web Services

Amazon offers a wide range of remote storage, compute, and naming services in the cloud, which are collectively known as Amazon Web Services (AWS). All services can be configured and monitored from a single web-based console. A screenshot of the AWS console is shown in Figure 16-1.

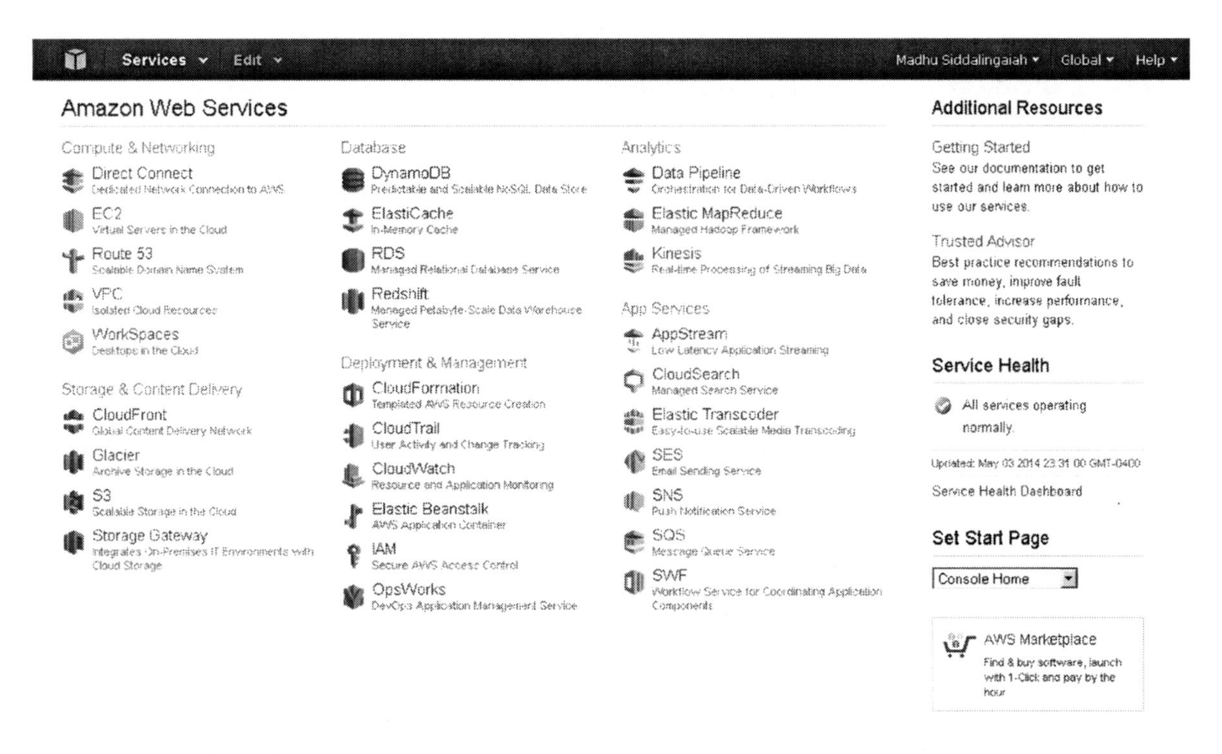


***Figure 16-1.** AWS console*

Although Amazon offers many services, the most relevant services for Hadoop developers are Amazon Simple Storage Service (S3), Elastic Compute Cloud (EC2), and Elastic MapReduce (EMR).

## Simple Storage Service

S3 is a low-cost, durable cloud storage service that is capable of storing large volumes of data, with additional features such as life cycle management and security. The life cycle can be controlled with rules. These rules can specify when it should be archived or expired. Access to S3 data can be restricted or encrypted if required.

S3 is highly scalable, so it is not necessary to be concerned with ultimate capacity. Clusters can access data in S3 efficiently, which makes transient clusters a viable option, even for large datasets.

S3 is like a key/value store. The fundamental unit of storage is an object, and objects contain data and metadata. The metadata contains key/value pairs that describe the object, including default metadata such as object last modification date. Objects are stored in a named collection called a bucket, and an object within a bucket is uniquely identified by a key and a version. Any object within S3 can be identified with a bucket name, key, and version.

Bucket names must be unique for all of S3. Bucket names are like domain names, meaning that there can be only one owner of a bucket. Buckets can appear as subdomains of `s3.amazonaws.com`.

S3 objects can be accessed from web browsers, programmatic APIs, and desktop tools. For example, it is possible to access a bucket named `madhu.com` from a web browser using the URL `http://madhu.com.s3.amazonaws.com`. Access to the bucket and its objects depends on security settings. By default, a bucket with no configured security returns the web page shown in Figure 16-2.

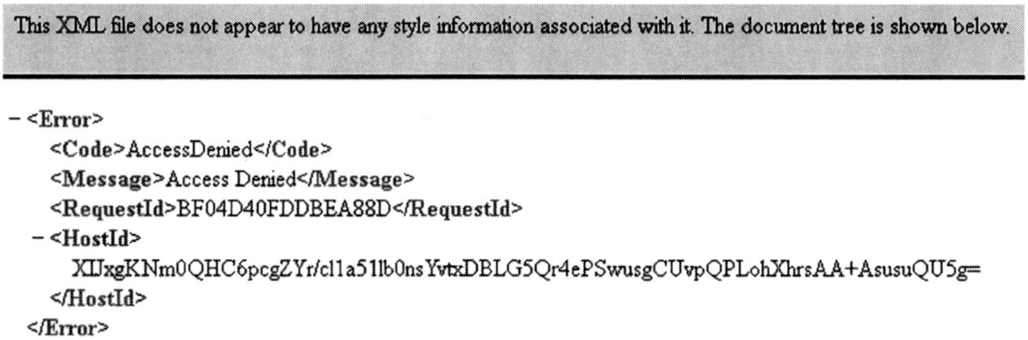

This XML file does not appear to have any style information associated with it. The document tree is shown below.

```
- <Error>
    <Code>AccessDenied</Code>
    <Message>Access Denied</Message>
    <RequestId>BF04D40FDDBEA88D</RequestId>
  - <HostId>
      XIJxgKNm0QHC6pcgZYr/cl1a51lb0nsYvtxDBLG5Qr4ePSwusgCUvpQPLohXhrsAA+AsusuQU5g=
    </HostId>
  </Error>
```

***Figure 16-2.*** *Result page for a bucket with no security*

## Elastic MapReduce

EMR is a web service that supports Apache Hadoop and support tools for accessing other Amazon services, such as S3. EMR offers easy access to configured Hadoop clusters. Users can select a Hadoop distribution, such as Amazon's own distribution or MapR, and a Hadoop version. The number of nodes and the instance type (machine size) can also be specified. More details on EMR are presented in the case study toward the end of this chapter.

## Elastic Compute Cloud

EC2 is a web service that offers scalable compute capacity. Users can create machine instances with an Amazon Machine Image (AMI) that can include an operating system and other software. Many third-party vendors prepare AMIs that support common functions, such as web servers, Hadoop distributions, or other services. More details on EC2 are presented in the case study toward the end of this chapter.

## Google Cloud Platform

Google offers a number of high- and low-level services, although not as comprehensive as Amazon (see Figure 16-3).

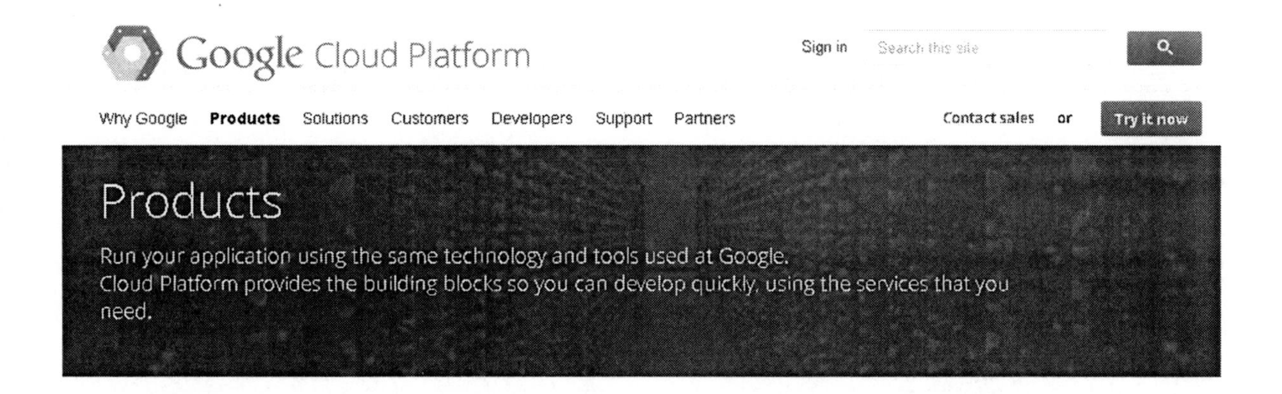


Compute

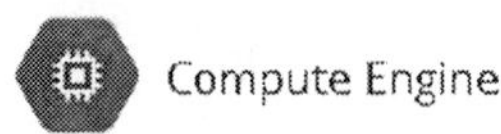


Compute Engine is Google's Infrastructure-as-a-Service (IaaS). Run large-scale workloads on virtual machines hosted on Google's infrastructure. Choose a VM that fits your needs and gain the performance and consistency of Google's worldwide fiber network. With per-minute billing, you pay only for what you use.

App Engine is Google's Platform-as-a-Service (PaaS). Develop your application easily using built-in services that make you more productive. Deploy to a fully-managed platform and let Google carry the pager. Just download the SDK and start building immediately for free with no credit card required.

***Figure 16-3.*** *Google Cloud Platform*

Table 16-1 summarizes some of the products offered on the Google Cloud Platform that are of interest to Hadoop developers.

***Table 16-1.*** *Google Cloud Platform Services Summary*

| Product | Description |
| --- | --- |
| Compute Engine | Networked virtual machines with per-minute billing |
| Cloud Storage | Durable, long-term object storage with high availability and efficient caching for high performance |
| Cloud Datastore | Scalable, transactional, NoSQL store with SQL-like capabilities |
| BigQuery | Analytic engine capable of executing SQL-like queries against terabytes of data |
| Management and Developer Tools | Easy-to-use cloud resource management tools and APIs |

Google offers some services that are very similar to those of Amazon. One difference is that Google does not currently offer a packaged Hadoop solution such as Amazon EMR. Still, it is possible to provision a cluster with Hadoop using Google Compute Engine.

Both Google and Amazon compete aggressively on price, which is a great benefit to customers in the cloud marketplace.

## Microsoft Azure

Microsoft Azure offers a variety of cloud services, including VMs, servers, storage, directory services, and network services (see Figure 16-4).

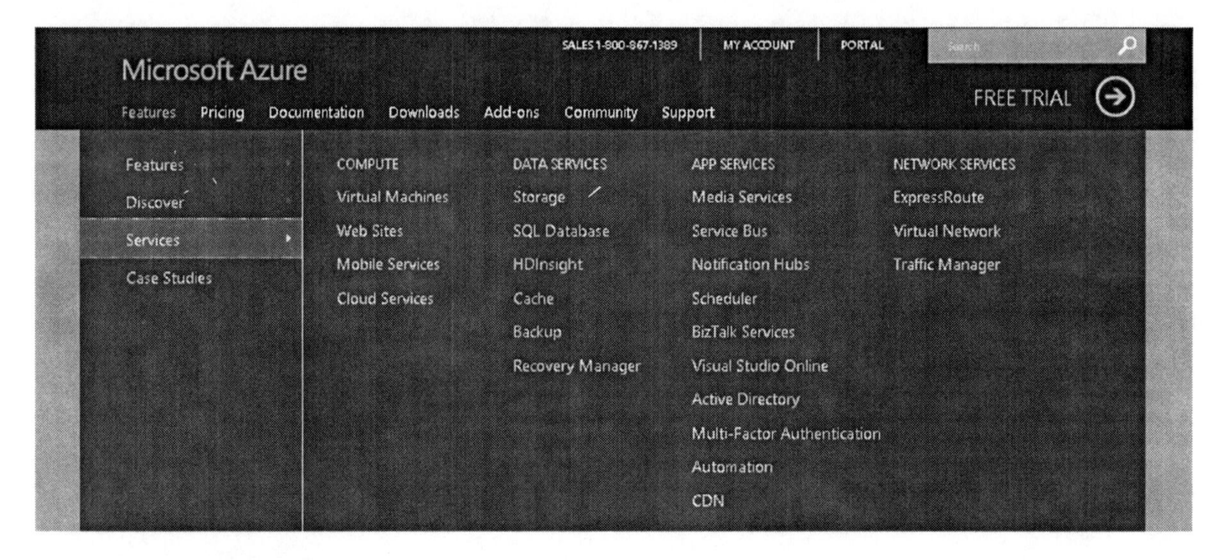


***Figure 16-4.*** *Microsoft Azure*

Historically, Microsoft has promoted Microsoft Windows exclusively and downplayed the value of open-source software. With Azure, Microsoft has fully embraced both Linux and Hadoop. The Azure VM can be provisioned with either Windows or Linux operating systems. Azure supports HDInsight a Hadoop implementation which integrates with Microsoft Business Intelligence (BI) tools.

For storage, Microsoft also offers block blobs, similar to Amazon S3. Microsoft tables and queues are NoSQL stores for structured and semistructured data. Azure can be a good choice for Microsoft-centric organizations that need a Big Data strategy.

## Choosing a Cloud Vendor

Given all the choices of cloud vendors, which one is best? The answer depends on the types of available services, cost, and many other factors. Cloud services are a highly competitive landscape. Vendors continue to innovate and offer greater value at lower prices.

Many organizations hedge their bets and select two or more vendors for their cloud solutions. This strategy avoids vendor lock-in and minimizes risk in case one vendor declines in popularity or fails to offer value. Given the popularity and availability of open-source software, migrating applications from one cloud vendor to another is a manageable exercise.

# Case Study: Amazon Web Services

This case focuses on EC2 and EMR. To get started, it is necessary to create an AWS account, which requires a credit card for billing and identity verification. Expired credit card information can result in authentication failure under certain conditions, so it is important to keep that information up to date.

The AWS account is like a superuser with unlimited access to AWS resources. In general, it is preferable to create Identity Access Management (IAM) users and restrict access through policies. Amazon documents IAM best practices here: `http://docs.aws.amazon.com/IAM/latest/UserGuide/IAMBestPractices.html`.

To create IAM users, visit `https://console.aws.amazon.com/iam/#users` and click Create New Users. When created, IAM users have the following credentials:

- User name (defined by AWS account owner)
- Access key ID (AWS-generated alphanumeric string)
- Secret access key (AWS-generated alphanumeric string)
- RSA private key name (defined by AWS account owner)
- Base64-encoded RSA private key (AWS generated `.pem` file)

All these credentials are needed by the user to access AWS services. Like any other secret credentials, confidentiality is critical, so they must be kept in a safe place.

Access to specific services requires IAM policies, and adding policies is straightforward from the IAM console. Simply select a user, click Attach User Policy, and select a policy template from the available list. Advanced users can customize policies if needed. For this case study, we need two policies: Elastic MapReduce and Elastic Compute Cloud. Without these policies, users receive an error message stating that access to a service is not authorized.

## Elastic MapReduce

EMR clusters can be launched from the Amazon web console, command line, or programmatically through several programming language APIs. To create an EMR cluster from the AWS console, click Elastic MapReduce. Figure 16-5 shows the form that displays.

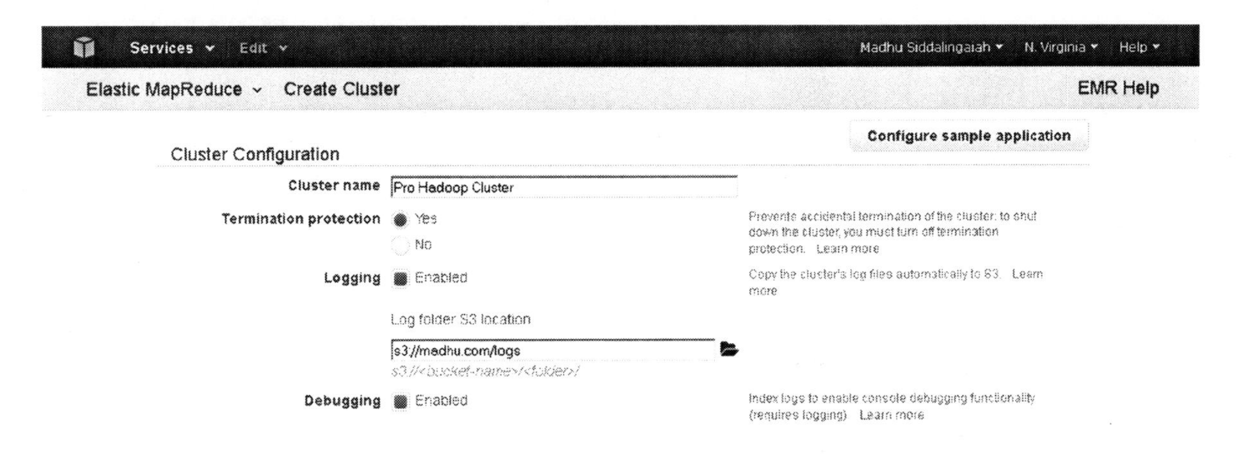


***Figure 16-5.*** *Amazon Elastic MapReduce console*

The cluster name is needed to identify a cluster in the console. Termination protection prevents accidental termination and loss of all data within the cluster, including the HDFS. Log output can be sent to an S3 bucket accessible by the user, which allows logs to be viewed outside of the cluster in case it fails to start properly or after it has been shut down.

EMR instances can be tagged with key/value pairs to categorize or add metadata to a cluster. Tags are optional, so it is not necessary to add any. Specific software configurations can be selected, such as the version of Hadoop, distribution (Amazon or MapR), and additional applications (Hive, Pig, HBase etc.). The Hadoop version and distribution is defined by an AMI, which is a binary image containing the operating system and related software for an instance.

Farther down, the cluster hardware can be configured, as shown in Figure 16-6.

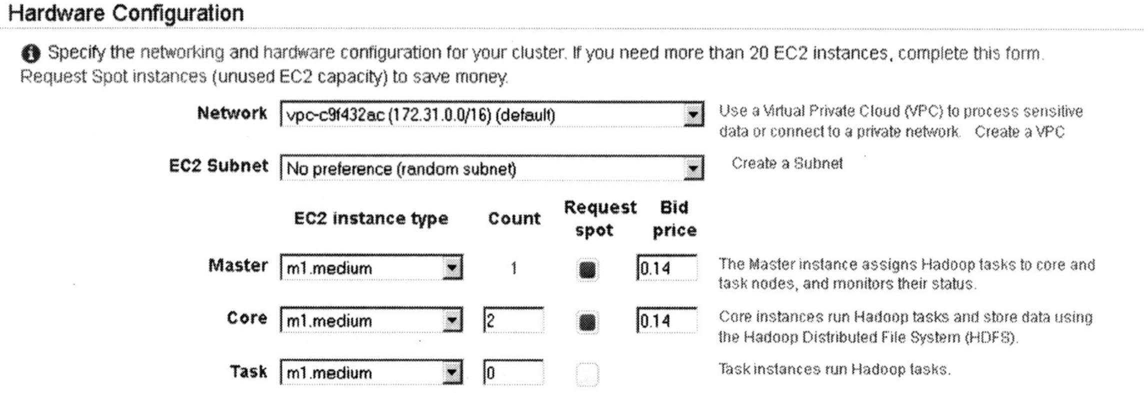

Hardware Configuration

Specify the networking and hardware configuration for your cluster. If you need more than 20 EC2 instances, complete this form. Request Spot instances (unused EC2 capacity) to save money.

Network: vpc-c9f432ac (172.31.0.0/16) (default) — Use a Virtual Private Cloud (VPC) to process sensitive data or connect to a private network. Create a VPC

EC2 Subnet: No preference (random subnet) — Create a Subnet

| | EC2 instance type | Count | Request spot | Bid price | |
|---|---|---|---|---|---|
| Master | m1.medium | 1 | ■ | 0.14 | The Master instance assigns Hadoop tasks to core and task nodes, and monitors their status. |
| Core | m1.medium | 2 | ■ | 0.14 | Core instances run Hadoop tasks and store data using the Hadoop Distributed File System (HDFS). |
| Task | m1.medium | 0 | | | Task instances run Hadoop tasks. |

***Figure 16-6.** Amazon Elastic MapReduce Hardware Configuration screen*

For this example, we choose an m1.medium instance type. The instance type defines the number of CPUs, memory configuration, and hard drive size. AWS documentation lists all these details for reference. If Request Spot is selected, a bid price can be entered; if it is not selected, on-demand pricing is in effect. We select a cluster with one master node and two core (slave) nodes and a maximum bid price of $0.14.

A user key pair can be selected to log in to the master node without a password. The user's SSH keys are automatically installed in the master node if this option is selected. Finally, bootstrap actions and execution steps can be added. (Bootstrap actions are scripts executed before Hadoop is started in the cluster that install additional software or perform other configurations.)

Execution steps, which are programs or scripts that can be executed after Hadoop starts, can include one or more jobs that should be executed. After all steps are completed, the cluster can be automatically shut down. This configuration enables jobs to be executed in a transient cluster. Source data can be copied from S3 to the HDFS in the transient cluster, jobs executed, and output data copied back to S3 before the cluster is shut down.

Once the cluster has been provisioned and configured, the EMR web console displays the master public DNS for the master node (see Figure 16-7). This DNS name can be used to log in to the cluster using SSH on Linux or PuTTY on Windows.

*Figure 16-7. Amazon Elastic MapReduce public DNS settings screen*

If you log in to the cluster through PuTTY, you have to convert the AWS-supplied `.pem` private key file to a `.ppk` file by using the `PuTTYgen.exe` utility available from the Internet. The `.ppk` file can be specified in PuTTY under the Connection | SSH | Auth configuration setting.

The login username is hadoop. If the IAM user private key is specified, no password is required. Figure 16-8 is a screenshot of a login to the master node through PuTTY.

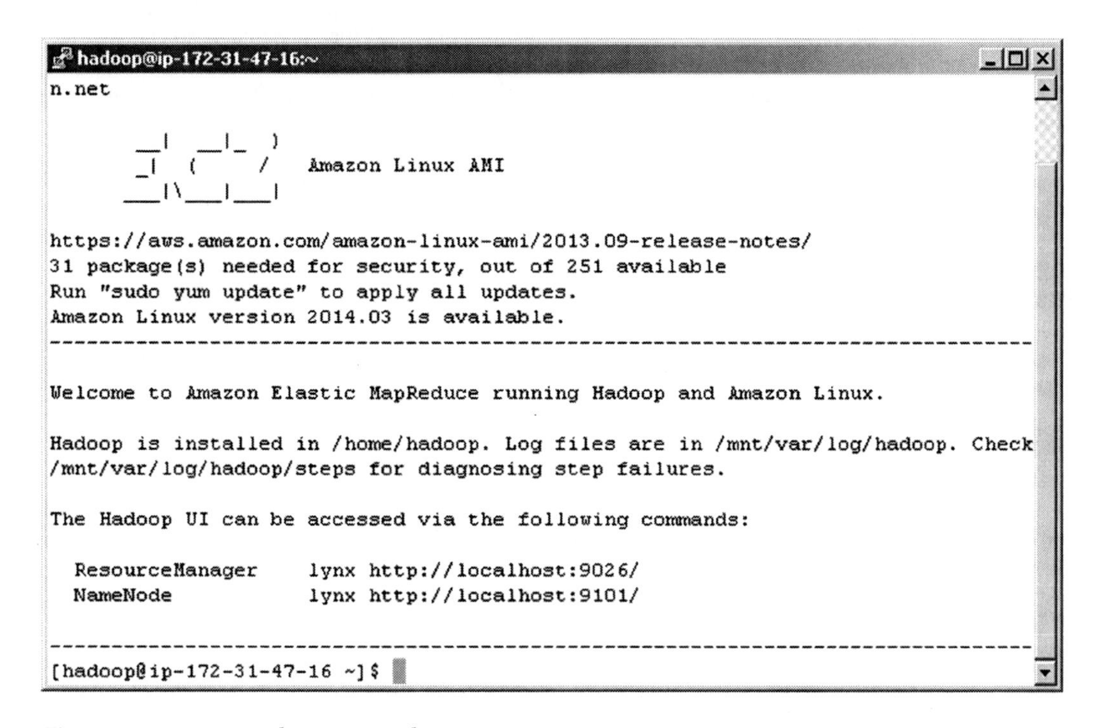


*Figure 16-8. Amazon Elastic MapReduce using PuTTY to start a remote session*

The cluster is now ready. From within the cluster, it is possible to start MapReduce jobs, access the Internet, or access Amazon S3. Once work is completed, the cluster can be terminated from the EMR web console.

You can also start a cluster through the Amazon Command Line Interface (CLI), which is a Ruby-based application that is available as a free download from the AWS web site. The CLI requires a specific Ruby version as documented online. Once Ruby and the CLI are installed, a similar cluster (as before) can be started using the following command:

```
ruby elastic-mapreduce --create --alive --name "Pro Hadoop Cluster" --log-uri s3://madhu.com/logs
--instance-group master --instance-type m1.small --instance-count 1 --bid-price 0.14 --instance-
group core --instance-type m1.small --instance-count 2 --bid-price 0.14
```

The CLI includes many options, such as bootstrap actions, steps, and so on. Refer to the online Amazon CLI documentation for details.

## Elastic Compute Cloud

EC2 differs from EMR in that Hadoop is not automatically installed. Installing Hadoop can be performed manually after instances are running or programmatically through an API. Amazon offers many machine images to choose from, including several Linux distributions and Windows operating systems.

This example provisions a Spark/Shark cluster using the `spark-ec2` Python script found in the Spark distribution. This script connects to AWS, selects an AMI, installs Spark/Shark (as well as other software), and installs SSH keys for easy access. The `spark-ec2` script requires the SSH command to connect to the cluster, which is generally not available on Windows operating systems. For this reason, it is easiest to run this script from a local Linux distribution.

To use the `spark-ec2` script, it is necessary to set the following environment variables:

```
export AWS_ACCESS_KEY_ID=<redacted AWS supplied access ID>
export AWS_SECRET_ACCESS_KEY=<redacted AWS supplied access key>
```

The `spark-ec2.py` script is found in the `ec2` directory of the Spark distribution. The following command starts an EC2 Spark cluster with two worker instances and a bid price of $0.14:

```
./spark-ec2 -k <redacted key name> -i <redacted keyfile.pem> -s 2 launch ProHadoopSpark --spot-
price=0.14
```

The startup takes a bit of time, but it is interesting to follow the steps as they progress. The Listing 16.1 demonstrate the starting up of Spark on EC2.

***Listing 16-1.*** Starting Spark on EC2 using spark-ec2.py

```
Setting up security groups...
Creating security group ProHadoopSpark-master
Creating security group ProHadoopSpark-slaves
Searching for existing cluster ProHadoopSpark...
Spark AMI: ami-5bb18832
Launching instances...
Requesting 2 slaves as spot instances with price $0.140
Waiting for spot instances to be granted...
0 of 2 slaves granted, waiting longer
...
0 of 2 slaves granted, waiting longer
```

```
All 2 slaves granted
Launched master in us-east-1c, regid = r-ab2c9988
Waiting for instances to start up...
Waiting 120 more seconds...
Generating cluster's SSH key on master...
...
Warning: Permanently added 'ec2-54-86-127-162.compute-1.amazonaws.com,54.86.127.162' (RSA)
to the list of known hosts.
Connection to ec2-54-86-127-162.compute-1.amazonaws.com closed.
Transferring cluster's SSH key to slaves...
ec2-54-86-40-33.compute-1.amazonaws.com
Warning: Permanently added 'ec2-54-86-40-33.compute-1.amazonaws.com,54.86.40.33' (RSA)
to the list of known hosts.
ec2-54-86-113-189.compute-1.amazonaws.com
Warning: Permanently added 'ec2-54-86-113-189.compute-1.amazonaws.com,54.86.113.189' (RSA)
to the list of known hosts.
Cloning into 'spark-ec2'...
...
Unpacking Scala
...
Unpacking Spark
...
Unpacking Shark
...
Unpacking Hadoop
...
Unpacking Tachyon
...
Initializing ganglia
Connection to ec2-54-86-40-33.compute-1.amazonaws.com closed.
Connection to ec2-54-86-113-189.compute-1.amazonaws.com closed.
Creating local config files...
..
Configuring /etc/httpd/conf/httpd.conf
Configuring /root/mapreduce/hadoop.version
Configuring /root/mapreduce/conf/core-site.xml
Configuring /root/mapreduce/conf/slaves
Configuring /root/mapreduce/conf/mapred-site.xml
Configuring /root/mapreduce/conf/hdfs-site.xml
...
Setting up shark
...
Formatting ephemeral HDFS namenode...
...
starting org.apache.spark.deploy.master.Master, logging to /root/spark/sbin/../logs/spark-root-org.
apache.spark.deploy.master.Master-1-ip-172-31-43-187.ec2.internal.out
ec2-54-86-113-189.compute-1.amazonaws.com: starting org.apache.spark.deploy.worker.Worker, logging
to /root/spark/sbin/../logs/spark-root-org.apache.spark.deploy.worker.Worker-1-ip-172-31-38-253.ec2.
internal.out
ec2-54-86-40-33.compute-1.amazonaws.com: starting org.apache.spark.deploy.worker.Worker, logging to
/root/spark/sbin/../logs/spark-root-org.apache.spark.deploy.worker.Worker-1-ip-172-31-39-144.ec2.
internal.out
```

```
Setting up tachyon
...
Starting master @ ec2-54-86-127-162.compute-1.amazonaws.com
...
Setting up ganglia
...
Starting httpd:                                          [  OK  ]
...
Spark standalone cluster started at http://ec2-54-86-127-162.compute-1.amazonaws.com:8080
Ganglia started at http://ec2-54-86-127-162.compute-1.amazonaws.com:5080/ganglia
Done!
```

Now that the cluster is provisioned, we can log in to the master node using SSH and specify the IAM user's private key with the `-i` command-line option:

```
$ ssh root@ec2-54-86-127-162.compute-1.amazonaws.com -i ~/AWS/<radacted keyfile.pem>

       __|  __|_  )
       _|  (     /   Amazon Linux AMI
      ___|\___|___|

https://aws.amazon.com/amazon-linux-ami/2013.03-release-notes/
There are 42 security update(s) out of 257 total update(s) available
Run "sudo yum update" to apply all updates.
Amazon Linux version 2014.03 is available.
root@ip-172-31-43-187 ~]$ ls
ephemeral-hdfs  mapreduce        scala  spark      tachyon
hadoop-native   persistent-hdfs  shark  spark-ec2
```

The EC2 instances can be shut down from the AWS console, but it is more convenient to do it with the `spark-ec2` script:

```
$ ./spark-ec2 -k <redacted keyname> -i <radacted keyfile.pem> destroy ProHadoopSpark
Are you sure you want to destroy the cluster ProHadoopSpark?
ALL DATA ON ALL NODES WILL BE LOST!!
Destroy cluster ProHadoopSpark (y/N): y
Searching for existing cluster ProHadoopSpark...
Found 1 master(s), 2 slaves
Terminating master...
Terminating slaves...
```

# Summary

Hadoop hosted in the cloud is a cost-effective solution with many benefits. Several reputable cloud vendors offer a range of flexible elastic solutions that can scale easily as workload demand increases. Although moving sensitive data into the cloud might seem risky, vendors have made every effort to ensure confidentiality and reliability.

Standing up a Hadoop cluster in the cloud is as simple as filling out a web form and pressing submit. In a matter of minutes, a highly scalable, customized cluster is available in a persistent or transient model. Given the economies of scale and competition in the cloud marketplace, cloud solutions will very likely continue to offer greater value at lower costs.

# CHAPTER 17

# Building a YARN Application

Hadoop 2.0 allows a developer to plug in to other frameworks. A large number of frameworks have developed around data stored in the HDFS, and some have been covered in this book (HAMA and Spark, for example). Some of these frameworks were developed to overcome the limitations of using MapReduce for all types of problems. For example, the key limitation of MapReduce is that each MapReduce phase reads and writes data to the HDFS, so iterative algorithms run several times slower in MapReduce. Each iteration is a separate MapReduce job that reads the output of the earlier iteration's MapReduce job from the HDFS. Frameworks such as HAMA and Spark were developed to address these limitations. They leveraged the HDFS, but existed as separate frameworks that had their own resource management capabilities. YARN or Hadoop 2.0 allows these frameworks to be integrated into the larger Hadoop framework, which centralizes resource management across all types of jobs in the cluster.

As discussed in Chapter 2, each type of job has its own Application Master in YARN, and the MapReduce framework also uses its own Application Master. Incorporating another framework requires developing an Application Master for the framework.

This chapter illustrates how to develop an Application Master using a simple real-world example. This example also shows how YARN can be used as a general-purpose framework to develop a distributed application.

## YARN: A General-Purpose Distributed System

We describe developing a custom Application Master by building a practical distributed application. The author has developed a faceted search application for the US Patents dataset. The entire patent grant text data can be downloaded from the following website:

```
https://www.google.com/googlebooks/uspto-patents-grants-text.html
```

The United States Patent and Trademark Office (USPTO) releases patent documents every Tuesday for the patents granted that week. Each file is a compressed zip file with text for patents published that week, so there are as many zip files per year as there are weeks in each year.

The entire list of URLs for each zip file from 1976 through 2014 is provided in the `src/,main/resources/input/patents/PatentList.txt` file in the book's source code (there are 2051 URLs). Sample lines from this file are shown in Listing 17-1.

***Listing 17-1.*** PatentList.txt

```
http://storage.googleapis.com/patents/grant_full_text/2014/ipg140107.zip
http://storage.googleapis.com/patents/grant_full_text/2014/ipg140114.zip
...
http://storage.googleapis.com/patents/grant_full_text/1986/pftaps19860225_wk08.zip
http://storage.googleapis.com/patents/grant_full_text/1986/pftaps19860304_wk09.zip
http://storage.googleapis.com/patents/grant_full_text/1986/pftaps19860311_wk10.zip
...
http://storage.googleapis.com/patents/grant_full_text/1976/pftaps19761221_wk51.zip
http://storage.googleapis.com/patents/grant_full_text/1976/pftaps19761228_wk52.zip
```

Each file is between 10 MB and 100 MB, which is a 100–150 GB download that can take a long time on a single machine. Downloading with a multithreaded program does not help much, however; this approach eventually saturates the network bandwidth. You want to distribute the process of downloading files to multiple machines, but track the progress from a central location. In other words, you want to build a distributed download service.

Developing distributed services that are failure resilient is a challenging task. In this chapter, we develop a distributed download service using the YARN framework. YARN provides all the plumbing necessary to develop such a distributed service, which enables us to focus on the key logic to download files and copy them to distributed file storage.

Although in our example we use the HDFS as our distributed storage, we can also use network attached storage (NAS) if it is visible from all the Hadoop slave nodes in which the containers tasks execute.

Figure 17-1 illustrates the high-level requirements of the distributed download service.

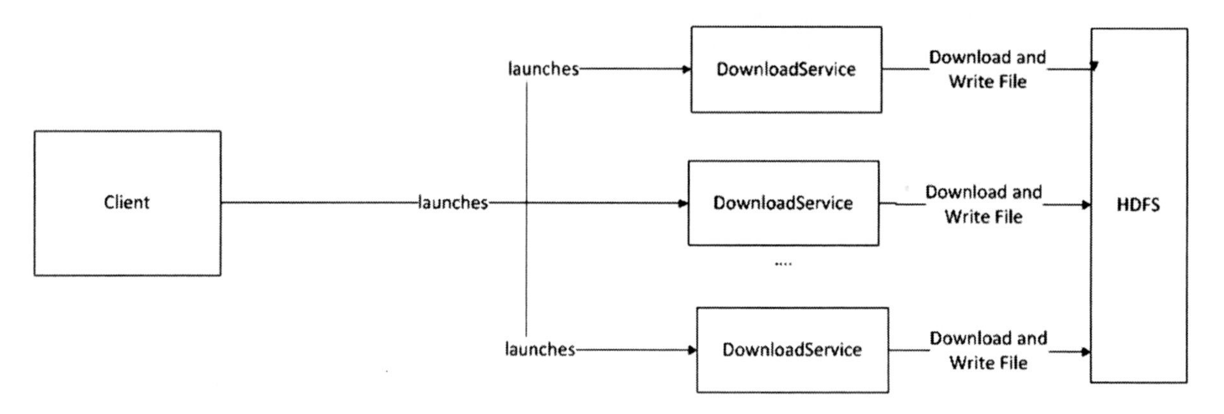
Client
launches
launches
launches
DownloadService
DownloadService
DownloadService
Download and Write File
Download and Write File
Download and Write File
HDFS
Each DownloadService instance runs on its own node
Each DownloadService instance receives a single URL to download
Each DownloadService instance writes the file it downloads to a folder in common folder in HDFS

***Figure 17-1.*** *Requirements for the distributed download service*

Our goal is to develop a distributed download service in which each file is downloaded from its own worker task that executes in a node on the cluster.

# YARN: A Quick Review

In Chapter 2 of this book, we discussed how a YARN application is submitted. Each YARN application has an Application Master instance managing its life cycle. Each type of application has its own Application Master class used to manage the given type of applications. For instance, MapReduce has its own Application Master, and a separate instance of this Application Master manages each MapReduce job.

The Resource Manager is responsible for managing resources. An Application Master is responsible for acquiring resources from the Resource Manager to execute worker tasks (Mappers and Reducers in MapReduce). A Resource Manager allocates resources (in the form of containers on the task nodes) to each Application Master. The Application Master launches tasks in the containers on the task nodes. The job of launching containers in a task node is performed by the Node Manager. The Application Master tracks the status of worker tasks and monitors progress.

At a high level, a YARN application contains the following Hadoop Framework components:

- *Resource Manager (RM)*: A central agent responsible for allocating cluster resources.
- *Node Manager (NM)*: An agent that runs on each node of the cluster. Its task is to enforce Node resource allocations.

A YARN application has the following custom components that we will write:

- *Client*: Runs on a client node (also referred to as an edge node) in the cluster. Its task is to launch the application. A client communicates with the Resource Manager to obtain container resources to launch the custom Application Master.
- *Application Master (AM)*: Runs on a remote node. Its task is to negotiate resources from the Resource Manager to execute the worker tasks in remote containers. It also monitors the progress of each launched container.
- *Worker task*: Implements the application logic. In our example, it downloads files from a remote URL and saves them to the HDFS.

Our goal is to develop a YARN application that meets the requirements shown in Figure 17-1. The application executes as shown in Figure 17-2.

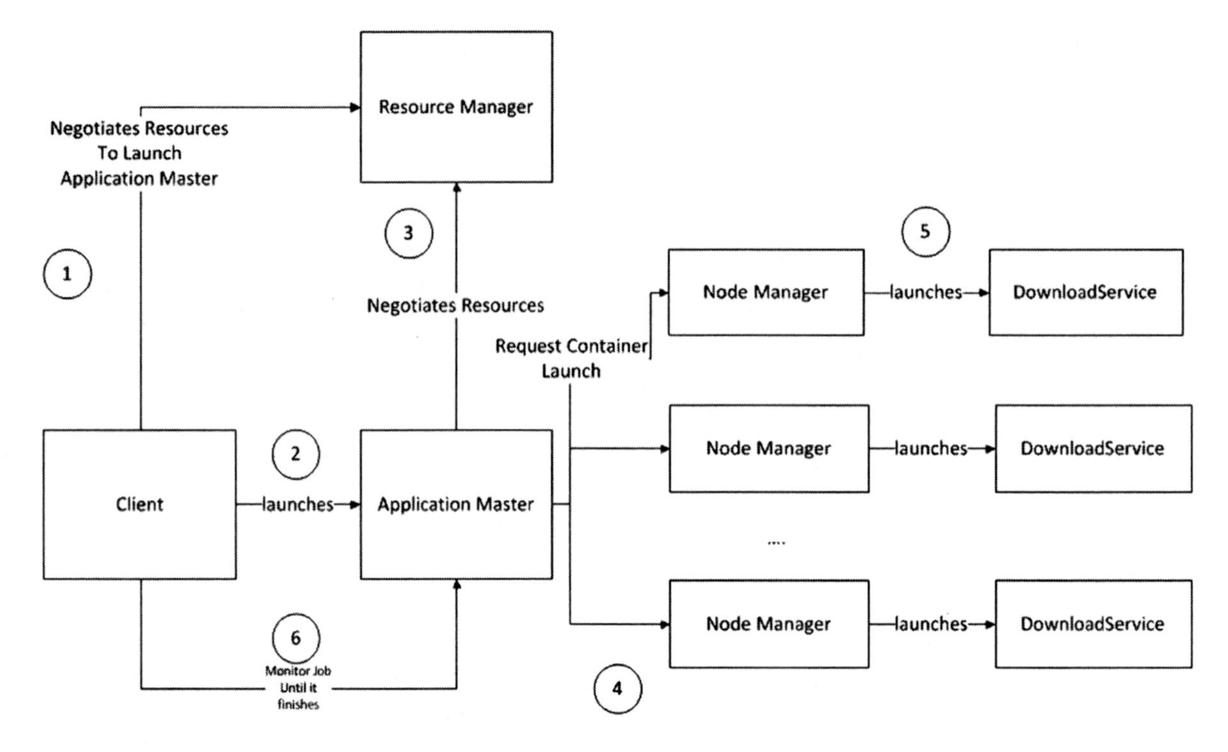


*Figure 17-2. DownloadService Execution as a YARN Application*

At a high level, the flow of the distributed download service implemented as a YARN application is as follows:

1. We decide to develop a custom Application Master program to be launched by the client.
2. The list of URLs for the files to be downloaded is placed in a file in the HDFS.
3. The client requests a container from the Resource Manager to launch the custom Application Master.
4. The client program launches the Application Master using the container resource negotiated from the Resource Manager. The client launches the Application Master with a list of two parameters:
    a. Input file path in the HDFS that contains the list of URLs
    b. Output directory folder in the HDFS in which the downloaded files will be placed
5. The Application Master obtains the list of download URLs from the input file path provided by the client. The Application Master requests as many containers as there are files to be downloaded. These containers are requested from the Resource Manager.
6. When the container requests are granted, the Application Master launches the containers with the `DownloadService` task, which takes two parameters:
    a. The URL of the file to the download
    b. The destination HDFS folder into which the downloaded file will be placed

7. The Application Master monitors the containers. (In our simplified example, we do not handle failure and restarts.)

8. When all the containers return, the Application Master unregisters with the Resource Manager.

9. The client program is notified that the Application Master terminated and it returns.

A YARN application uses three protocols, as illustrated in Figure 17-3.

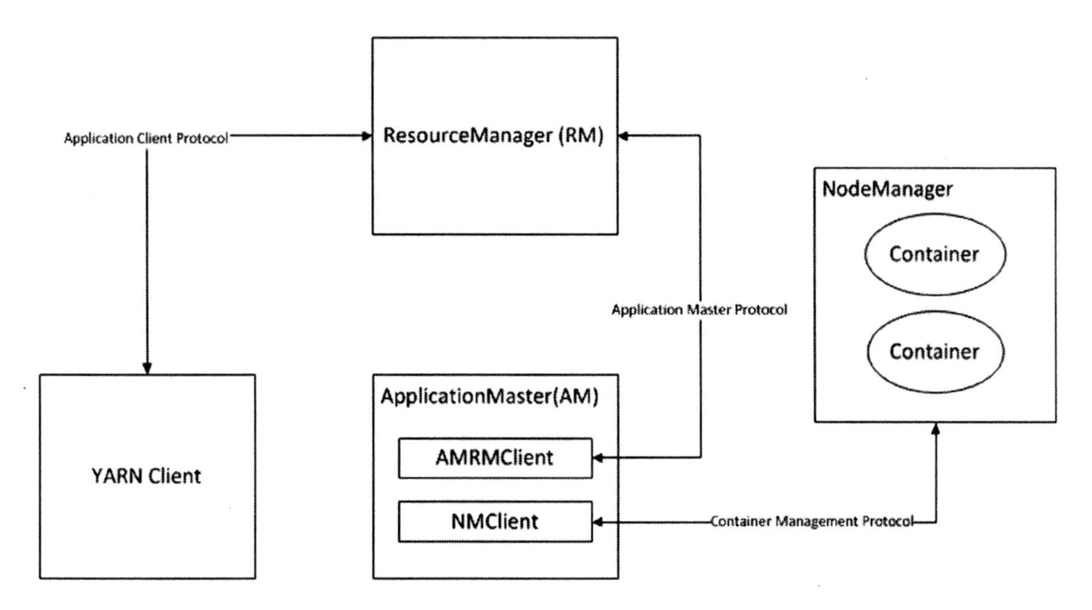


***Figure 17-3.*** *YARN application protocols*

- *Application Client Protocol*: Client to Resource Manager to launch the Application Master. The client uses this protocol to request resources in the cluster to launch the Application Master.
- *Application Master Protocol*: Application Master to Resource Manager to request container resources to execute worker tasks. The Application Master uses this protocol to request resources in the cluster to launch the worker tasks.
- *Container Management Protocol*: Application Master to Node Manager to launch containers to execute the tasks in the launched containers. The Application Master uses this protocol to launch the worker tasks through the Node Manager once the containers are allocated by the Resource Manager.

# Creating a YARN Application

Let us begin developing a distributed download service as a YARN application. The source code in this chapter has been adapted from the source code by Hortonworks published at this GitHub location: `https://github.com/hortonworks/simple-yarn-app`.

This is a simple YARN application to run *n* copies of a UNIX command. The source code is kept very simple with minimalistic error handling, which simplifies its use for the purpose of teaching the main concepts. We follow the same approach.

Our custom YARN application is defined using the following set of custom classes:

- `DownloadService.java`: The task launched on the container nodes to download the remote files from the Google bulk download site for USPTO patents. It takes two parameters to store the downloaded file: the URL of the file to download and the root HDFS directory.
- `Client.java`: The class the user launches. Its responsibility is to launch the Application Master and consequently the download application.
- `ApplicationMaster.java`: The custom Application Master that launches the download file task (`DownloadService`) on the container nodes.

## POM Configuration

Before we start developing our YARN application, we need to configure the Maven configuration file to ensure that the YARN client libraries are made available. The following entries in our `pom.xml` file ensure this:

```
<dependency>
   <groupId>org.apache.hadoop</groupId>
   <artifactId>hadoop-yarn-client</artifactId>
   <version>2.2.0</version>
</dependency>
<dependency>
   <groupId>org.apache.hadoop</groupId>
   <artifactId>hadoop-common</artifactId>
   <version>2.2.0</version>
</dependency>
```

# DownloadService.java Class

Let us first look at the class that contains the custom logic to download the files. The `DownloadService` class is responsible for downloading remote files in a task node. A separate instance of `DownloadService` runs in a container on a task node. It takes two parameters:

- The URL of the file to download. An example URL is this: `http://storage.googleapis.com/patents/grant_full_text/2014/ipg140107.zip`.
- The HDFS path in which all the download files are downloaded. For example, if this parameter value is `/user/hdfs/patents/`, the file `ipg140107.zip` is downloaded in the `/user/hdfs/patents/` folder.

The command line to execute the `DownLoadService.java` class is shown here:

```
java DownloadService <DOWNLOAD_URL> <HDFS_PATH_TO_DOWNLOAD>
```

The `DownloadService.main` method is shown in Listing 17-2.

***Listing 17-2.*** DownloadService.main Method

```
public static void main(String[] args) throws Exception {
   DownloadFileService service = null;
   try{
      String url = args[0].trim();
      String rootHDFSPath = args[1];
      service = new DownloadFileService(url,rootHDFSPath);
      service.performDownloadSteps();
   }
   finally{
      service.closeAll();
   }
}
```

The key method is `performDownloadSteps()`, which is shown in Listing 17-3. This method first checks to see whether the file is already downloaded in the HDFS folder. If it exists, the process returns.

***Listing 17-3.*** DownloadService.performDownloadSteps Method

```
public void performDownloadSteps() throws Exception
{
   //If File already exists simply exit
   if(DownloadUtilities.doesFileExists(this.fs, this.outFile)){
     System.out.println(this.fileURL +
                      " file is already downloaded");
     return;
   }
   else{ //Else download and save file
       this.initializeDownload();
       this.downloadAndSaveFileFromUrl();
   }
}
```

If the file does not exist in the HDFS, the process initializes the download by performing the following steps:

1. It opens an input stream to the remote URL.
2. It opens an output stream to the HDFS file.
3. It downloads the file from the remote URL and saves it to the HDFS folder.

First let us examine the attributes of the `DownloadService` class and the constructor. This code is shown in Listing 17-4.

***Listing 17-4.*** DownloadService Attributes and Constructor

```
public class DownloadFileService {

    private String fileURL = null;//URL to download from
    private String hdfsRootPath = null;//HDFS Root folder to save file

    private FileSystem fs = null;//HDFS File System
    private Path outFile = null;//Full path of output fle
```

```
    private BufferedInputStream in = null;//Input stream to remote URL
    private FSDataOutputStream out = null;//Outout stream to file

    public DownloadFileService(String fileURL, String hdfsRootPath)
                                                throws Exception {
        this.fileURL = fileURL;
        this.hdfsRootPath = hdfsRootPath;
        String p = DownloadUtilities.
                    getFilePathFromURL(hdfsRootPath,this.fileURL);
        this.outFile = new Path(p);
        Configuration conf = new Configuration();
        this.fs = FileSystem.get(conf);
    }
......
}
```

The DownloadUtilities utility class provides a method that converts the download URL to a fully qualified path in which the file is downloaded in the HDFS. Listing 17-5 shows this method.

***Listing 17-5.*** DownloadUtilities.getFilePathFromURL

```
public static String getFilePathFromURL(String rootPath,
                                        String fileUrl){
    String filePath =  rootPath + "/" +
        fileUrl.substring(fileUrl.lastIndexOf("/")+1);
    return filePath;
}
```

The listing for the methods that actually download and save the file are shown in Listing 17-6 and Listing 17-7.

***Listing 17-6.*** DownloadService.initializeDownload method

```
public void initializeDownload() throws Exception{
   this.in = new BufferedInputStream(
                 new URL(this.fileURL).openStream());
   this.out = fs.create(outFile);
}
```

***Listing 17-7.*** DownloadService.downloadAndSaveFileFromUrl Method

```
public void downloadAndSaveFileFromUrl()
                                   throws Exception {
   try {
      byte data[] = new byte[1024];
      int count;
      while ((count = in.read(data, 0, 1024)) != -1) {
          out.write(data, 0, count);
      }
   } finally {
      this.closeAll();
   }
}
```

Finally, we close all the streams. Listing 17-8 shows the closeAll() method.

***Listing 17-8.*** DownloadService.downloadAndSaveFileFromUrl Method

```
public void closeAll() throws Exception {
   if (in != null)
      in.close();
   if (out != null)
      out.close();
   if (fs != null)
      fs.close();
}
```

# Client.java

At a high level, the client communicates with the Resource Manager to launch the Application Master. The source code for the client can be found in the Java class `org.apress.prohadoop.c17.Client`.

The client program takes the following three parameters:

- The HDFS path to `PatentList.txt` in Listing 17-1
- The HDFS path to the destination folder in which the downloaded zip files will be stored
- The HDFS path to the JAR file that is used to launch both the Application Master and instances of `DownloadService`

The following lines of code in the client program configure these three input parameters:

```
final String hdfsPathToPatentFileList = args[0];
final String hdfsOutFolder = args[1];
final Path jarPath = new Path(args[2]);
```

## Steps to Launch the Application Master from the Client

This section describes how the Application Master is launched from the client node. Figure 17-4 demonstrates these steps.

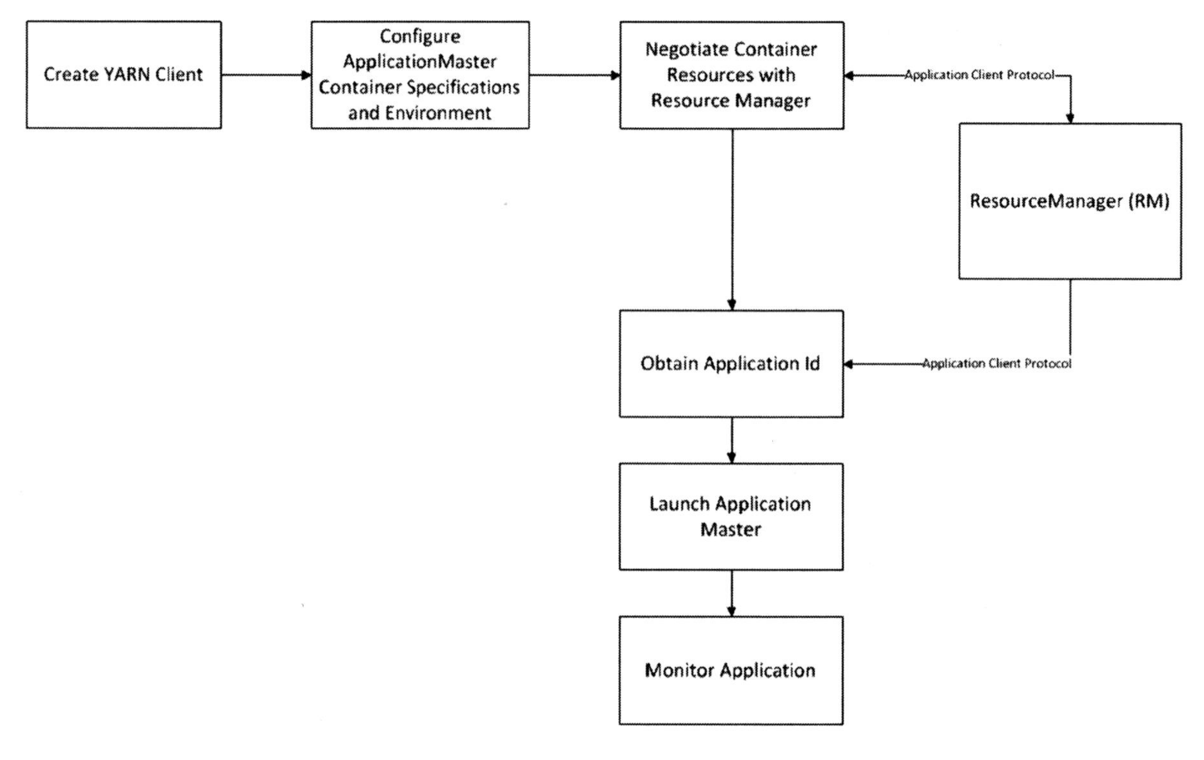


***Figure 17-4.*** *Launching an Application Master*

The client performs the following steps to launch the Application Master:

1. Creates a `YarnClient` instance and initializes it with the `YarnConfiguration` instance.
2. Uses the `YarnClient` instance to create and configure the remote container environment in which the custom Application Master will be launched. At this point, the application is simply configured and not launched. This process includes the following steps:
    a. Sets up the application JAR in the local working directory of the container that launches the Application Master.
    b. Configures the container classpath variable so that all the Hadoop JAR files are available along with the application JAR.
    c. Negotiates the container resources for launching the Application Master with the Resource Manager, including negotiating for memory and CPU cores.
3. Obtains an application ID from the Resource Manager using the Application Client Protocol illustrated in Figure 17-3 and launches the Application Master.
4. The client monitors the application every 100 milliseconds and returns when it finishes (succeeds or fails).

## Creating the YarnClient

The lines of code used to configure the YarnClient are as follows:

```
// Create yarnClient
YarnConfiguration conf = new YarnConfiguration();
YarnClient yarnClient = YarnClient.createYarnClient();
yarnClient.init(conf);
yarnClient.start();
```

The YarnClient is configured using the YarnConfiguration instance.

## Configuring the Application

In the next few lines, we configure the application. First, we create a YarnClientApplication instance:

```
YarnClientApplication app = yarnClient.createApplication();
```

Next, an instance of a ContainerLaunchContext is created. This is a very important class that contains all the information needed by the Node Manager to launch the container. It includes details such as the following:

- The command to launch the container.
- The container ID of the container.
- Resource instances allocated to this container. As discussed in Chapter 2, they include the amount of memory and virtual cores allocated to the container.
- The user to whom the container is allocated.
- Any security tokens if security is turned on.
- Any LocalResource instance needed by the container to execute the tasks, including all the binary files, JAR files, or any other files needed to execute the tasks.
- Environment variables needed to execute the container, including variables such as a classpath or other path variables.

We instantiate a ContainerLaunchContext instance to configure the container to launch the ApplicationMaster instance:

```
ContainerLaunchContext amContainer =
        Records.newRecord(ContainerLaunchContext.class);
```

We configure the command to launch the Application Master:

```
amContainer.setCommands(
        Collections.singletonList(
            " java " +
            " -Xmx128M" +
            " org.apress.prohadoop.c17.ApplicationMaster" +
            " " + hdfsPathToPatentFileList +
            " " + hdfsOutFolder
            )
        );
```

Next, we configure the LocalResource to launch the Application Master. This is the JAR file location in the HDFS that contains all the classes needed to execute the Application Master instance. The Application Master uses this:

```
LocalResource appMasterJar = setupAppMasterJar(jarPath); amContainer.setLocalResources(
       Collections.singletonMap("downloadapp.jar", appMasterJar));
```

The listing for setupAppMasterJar is shown in Listing 17-9. This code section configures the JAR file resource for the container that launches the Application Master. It configures the JAR file in the HDFS to be copied into a container local folder.

***Listing 17-9.*** Configuring the LocalResource with the JAR File

```
private LocalResource setupAppMasterJar(Path jarPath)
           throws IOException {
   LocalResource appMasterJar = Records.
                              newRecord(LocalResource.class);
   FileStatus jarStat = FileSystem.get(conf).
                                 getFileStatus(jarPath);
   appMasterJar.setType(LocalResourceType.FILE);
   appMasterJar.setVisibility(
                    LocalResourceVisibility.APPLICATION);
   appMasterJar.setResource(
              ConverterUtils.getYarnUrlFromPath(jarPath));
   appMasterJar.setSize(jarStat.getLen());
   appMasterJar.setTimestamp(jarStat.getModificationTime());
   return appMasterJar;
}
```

This method configures the JAR file so that the file is visible only to the instance of the launched Application Master. Other parameters specified are as follows:

- Whether the resource is a file or an archive. An archive resource is exploded on the container node. We indicate that our resource is a file.
- The timestamp and the length are specified to ensure some sanity checks. The framework makes sure that the local resource being used is the same resource the client intended to be used with the application.
- The HDFS location of the resource to be copied over into the working directory is set using the setResource call.

Finally, the following calls set up the configuration to configure a symbolic link called "downloadapp.jar" in the working directory that is linked back to the actual file. The remote environment references the actual file using the symbolic link reference that is local to the local working directory.

```
LocalResource appMasterJar = setupAppMasterJar(jarPath); amContainer.setLocalResources(
          Collections.singletonMap("appmaster.jar",
                                appMasterJar));
```

Now we configure the classpath for the Application Master. Listing 17-10 shows the source code for setAppMasterEnv.

***Listing 17-10.*** Configuring the ApplicationMaster Classpath

```
private void setupAppMasterEnv(Map<String, String> appMasterEnv) {
   for (String c : conf.getStrings(
        YarnConfiguration.YARN_APPLICATION_CLASSPATH,
        YarnConfiguration.DEFAULT_YARN_APPLICATION_CLASSPATH)) {
      Apps.addToEnvironment(appMasterEnv,
                            Environment.CLASSPATH.name(),
                            c.trim());
   }
   Apps.addToEnvironment(appMasterEnv,
                         Environment.CLASSPATH.name(),
                         Environment.PWD.$() + File.separator + "*")
}
```

This configuration is executed on the remote container where the Application Master is launched. The purpose is to the launch the Application Master instance on the remote node after configuring the environment on the remote node. (See the highlighted line in Listing 17-10.) The listing achieves the following

- All the Hadoop jar files are configured to the remote classpath environment.
- The highlighted portion configures all the files in the local working directory of the container that will be launched to run the Application Master. It usually refers to the path "`./*`".

The `ApplicationMaster` container classpath is configured using the following lines of code that invoke the method illustrated in Listing 17-10:

```
Map<String, String> appMasterEnv = new HashMap<String, String>();
setupAppMasterEnv(appMasterEnv);
amContainer.setEnvironment(appMasterEnv);
```

The resource requirements for the container that will host the Application Master are configured next. We configure the container to use 256 MB of memory and 1 core:

```
Resource capability = Records.newRecord(Resource.class);
capability.setMemory(256);
capability.setVirtualCores(1);
```

Finally, we configure the `ApplicationSubmissionContext` instance, which is an instance that contains all the information that the Resource Manager needs to launch the Application Master instance:

```
ApplicationSubmissionContext appContext =
    app.getApplicationSubmissionContext();
//set Application name
appContext.setApplicationName("download-yarn-app");
//Configure Container Specifications
appContext.setAMContainerSpec(amContainer);
//Provide Resource Requirements for the Container
appContext.setResource(capability);
//Configure Queue to use to launch the ApplicationMaster
appContext.setQueue("default");
```

## Launching the Application Master

ApplicationSubmissionContext is now used to launch the application. The client first requests an application identifier from the Resource Manager. This is a blocking call. The client code cannot proceed until the application identifier is returned by the Resource Manager. Finally, the application is submitted.

```
ApplicationId appId = appContext.getApplicationId();
System.out.println("Submitting application " + appId);
yarnClient.submitApplication(appContext);
```

## Monitoring the Application

The application is monitored using the following call:

```
appReport = yarnClient.getApplicationReport(appId);
YarnApplicationState appState = appReport.getYarnApplicationState();
```

The YarnApplicationState instance indicates the current status of the application that could include one of the following states:

- FINISHED
- RUNNING
- FAILED

The status is monitored every 100 milliseconds:

```
while (appState != YarnApplicationState.FINISHED &&
           appState != YarnApplicationState.KILLED &&
           appState != YarnApplicationState.FAILED) {
   Thread.sleep(100);
   appReport = yarnClient.getApplicationReport(appId);
   appState = appReport.getYarnApplicationState();
}
```

After emerging from the previous loop, the final status of the application and additional information is reported:

```
System.out.println("Application with " + appId + " finished with" +
                       " status " + appState + "." +
                       " The application started at " +
                    appReport.getStartTime() +"." +
                       " The application ended at  " +
                    appReport.getFinishTime());
```

Listing 17-11 shows the full source code for Client.java.

***Listing 17-11.*** Client.java Source Code

```
/*Package and Import declarations*/

public class Client {

  Configuration conf = new YarnConfiguration();

  public void run(String[] args) throws Exception {
    final String hdfsPathToPatentFileList = args[0];
    final String hdfsOutFolder = args[1];
    final Path jarPath = new Path(args[2]);

    // Create yarnClient
    YarnConfiguration conf = new YarnConfiguration();
    YarnClient yarnClient = YarnClient.createYarnClient();
    yarnClient.init(conf);
    yarnClient.start();

    // Create application via yarnClient
    YarnClientApplication app = yarnClient.createApplication();

    // Set up the container launch context for the application master
    ContainerLaunchContext amContainer =
        Records.newRecord(ContainerLaunchContext.class);
    amContainer.setCommands(
        Collections.singletonList(
            " java " +
            " -Xmx128M" +
            " org.apress.prohadoop.c17.ApplicationMaster" +
            " " + hdfsPathToPatentFileList +
            " " + hdfsOutFolder
            )
        );

    // Setup jar for ApplicationMaster
    LocalResource appMasterJar = Records.newRecord(LocalResource.class);
    setupAppMasterJar(jarPath, appMasterJar);
    amContainer.setLocalResources(
        Collections.singletonMap("downloadapp.jar", appMasterJar));

    // Setup CLASSPATH for ApplicationMaster
    Map<String, String> appMasterEnv = new HashMap<String, String>();
    setupAppMasterEnv(appMasterEnv);
    amContainer.setEnvironment(appMasterEnv);

    // Set up resource type requirements for ApplicationMaster
    Resource capability = Records.newRecord(Resource.class);
    capability.setMemory(256);
    capability.setVirtualCores(1);
```

```
    // Finally, set-up ApplicationSubmissionContext for the application
    ApplicationSubmissionContext appContext =
    app.getApplicationSubmissionContext();
    appContext.setApplicationName("download-yarn-app"); // application name
    appContext.setAMContainerSpec(amContainer);
    appContext.setResource(capability);
    appContext.setQueue("default"); // queue

    // Submit application
    ApplicationId appId = appContext.getApplicationId();
    System.out.println("Submitting application " + appId);
    yarnClient.submitApplication(appContext);

    ApplicationReport appReport = yarnClient.getApplicationReport(appId);
    YarnApplicationState appState = appReport.getYarnApplicationState();
    while (appState != YarnApplicationState.FINISHED &&
           appState != YarnApplicationState.KILLED &&
           appState != YarnApplicationState.FAILED) {
      Thread.sleep(100);
      appReport = yarnClient.getApplicationReport(appId);
      appState = appReport.getYarnApplicationState();
    }

    System.out.println(
        "Application " + appId + " finished with" +
                " state " + appState + "." +
                " The application started at " + appReport.getStartTime() +"." +
                " The application ended at  " + appReport.getFinishTime());

  }

  private void setupAppMasterJar(Path jarPath, LocalResource appMasterJar) throws IOException {
    /*Code already covered inline text*/
  }

  private void setupAppMasterEnv(Map<String, String> appMasterEnv) {
    /*Code already covered inline text*/
  }

  public static void main(String[] args) throws Exception {
    Client c = new Client();
    c.run(args);
  }
}
```

# ApplicationMaster.java

This is our custom `ApplicationMaster` class. As noted in the last section, the Application Master is launched with two parameters:

- The HDFS path to the patent list file from Listing 17-1
- The HDFS path in which the downloaded zip files will be stored

The first parameter is converted into a list of URLS with the following call:

```
List<String> files = DownloadUtilities.
                     getFileListing(hdfsPathToPatentFiles);
//Each entry in the list will be a line from the hdfsPathToPatentFiles
```

The preceding call converts the file in Listing 17-1 into a list of URL strings. The source code for `DownloadUtilities` is available with the book source code (we do not describe this method here).

## Communication Protocol between Application Master and Resource Manager: Application Master Protocol

Now we introduce a very critical class that defines the communication protocol between the Application Master and Resource Manager. Figure 17-3 demonstrates the Application Master Protocol, which is the primary interface through which the Application Master communicates with the Resource Manager to request resources for launching the work tasks. Node Managers take instructions from Resource Managers to manage resources available on a single node. Thus, when the Application Master requests resources from the Resource Manager, the Resource Manager in turn uses the information it gains continuously from each Node Manager to allocate resources to a node with available resources to execute the containers requested by the Application Master.

The class used to define the Application Master Protocol is this:

```
org.apache.hadoop.yarn.api.records.AMRMClient
```

An asynchronous version of this class also exists: `org.apache.hadoop.yarn.api.records.AMRMClientAsync`. Use the asynchronous version if you need to interact with the Resource Manager using multiple threads.

## Node Manager Communication Protocol: Container Management Protocol

The Container Management Protocol shown in Figure 17-3 launches containers on remote nodes in communication with the Node Manager only after the resources are negotiated with the Resource Manager.

The class used to communicate with the Node Manager is `org.apache.hadoop.yarn.client.api.NMClient`.

The asynchronous version of the class is `org.apache.hadoop.yarn.client.api.NMClientAsync`.

The Application Master in our example uses the `NMClient` instance to launch the containers obtained from the Resource Manager.

## Steps to Launch the Worker Tasks

This section discusses the list of activities that occur when the Application Master starts up and begins execution. Figure 17-5 illustrates these activities.

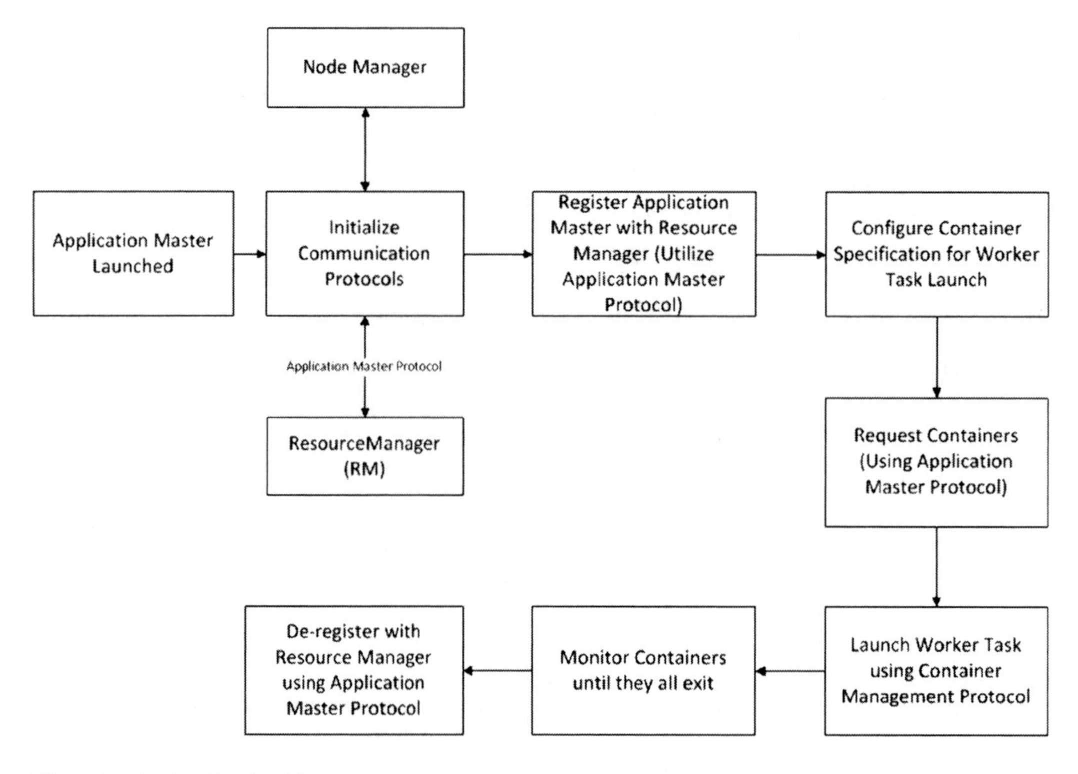


***Figure 17-5.*** *Application Master execution*

The Application Master launches the worker tasks by executing the following steps:

1. Initializes the Application Master Protocol and Container Management Protocol.
2. Registers the Application Master with the Resource Manager using the Application Master Protocol.
3. Configures the specifications for the containers that launch the worker tasks.
4. Requests containers using the Application Master Protocol. This step also obtains the reference to the allocated containers using the Application Master Protocol.
5. Launches the worker tasks using the Container Management Protocol.
6. Waits for containers to complete their tasks.
7. After all containers have returned, unregisters the Application Master with the Resource Manager.

For simplicity, we assume that we will start as many containers as we have patent files. Each container executes the `DownloadService` instance to download one file.

## Initializing the Application Manager Protocol and the Container Management Protocol

The following lines of code establish the connections between the Application Master and the Resource Manager as well as the Application Master and the Node Manager:

```
Configuration conf = new YarnConfiguration();

AMRMClient<ContainerRequest> rmClient = AMRMClient
                                          .createAMRMClient();
rmClient.init(conf);
rmClient.start();

NMClient nmClient = NMClient.createNMClient();
nmClient.init(conf);
nmClient.start();
```

## Registering Application Master with the Resource Manager

The Application Master is not registered with the Resource Manager using the following call:

```
rmClient.registerApplicationMaster("", 0, "");
```

## Configuring Container Specifications

Next we configure the specifications of the container. As mentioned before, we request as many containers as we have URLs to download:

```
Resource capability = Records.newRecord(Resource.class);
capability.setMemory(128);
capability.setVirtualCores(1);
```

## Requesting Containers from the Resource Manager

Having configured the containers, we now request them from the Resource Manager:

```
for (int i = 0; i < files.size(); ++i) {
      ContainerRequest containerAsk = new
           ContainerRequest(capability, null, null, priority);
      rmClient.addContainerRequest(containerAsk);
}
```

## Launching Containers on the Task Nodes

Finally, we launch the containers on the task node. Each container is configured with the actual command it executes:

```
int allocatedContainers = 0;
while (allocatedContainers < files.size()) {
   AllocateResponse response = rmClient.allocate(0);
```

```
    for (Container container : response.getAllocatedContainers()) {
    //Command to execute to download url to HDFS
    command = "java org.apress.prohadoop.c17.DownloadFileService"
                   +" " + files.get(allocatedContainers) + " " +
                     destHdfsFolder;

    // Launch container by creating a ContainerLaunchContext
    ContainerLaunchContext ctx =
            Records.newRecord(ContainerLaunchContext.class);
    ctx.setCommands(Collections.singletonList(command));
    nmClient.startContainer(container, ctx);
     ++allocatedContainers;
   }
   Thread.sleep(100);
}
```

The actual call that requests containers and receives container allocations is this:

```
AllocateResponse response = rmClient.allocate(0);
```

With this call, new containers assigned to the Application Master are retrieved. The status of completed containers and node health updates are also retrieved. It also doubles up as a heartbeat to the Resource Manager and must be made periodically. The call might not always return any new allocations of containers. An Application Master must be careful not to make concurrent allocate requests because it can cause request loss. The parameter passed to the `allocate()` method is just a progress parameter that indicates the amount of progress made by the Application Master. We simply pass 0, indicating that the Application Master is making its first call to allocate containers.

The key class with respect to launching containers that has been discussed in the `Client.java` section is `org.apache.hadoop.yarn.api.records.ContainerLaunchContext`.

## Waiting for Containers to Finish the Worker Tasks

Finally, we wait for all containers to complete. Once again, we use the call to `allocate()` to discover the container progress:

```
int completedContainers = 0;
while (completedContainers < files.size()) {
   AllocateResponse response = rmClient
         .allocate(completedContainers/files.size());
   for (ContainerStatus status :
      response.getCompletedContainersStatuses()) {
      ++completedContainers;
   }
   Thread.sleep(100);
}
```

## Unregistering Application Master from the Resource Manager

Once all the containers launched by the Application Master have returned, we unregister the Application Master from the Resource Manager. The application life cycle is now complete and it exits:

```
rmClient.unregisterApplicationMaster(
        FinalApplicationStatus.SUCCEEDED, "", "");
```

Listing 17-12 shows the full source code for ApplicationMaster.java.

***Listing 17-12.*** ApplicationMaster.java Full Source Code

```
/*Package and Import declarations*/
public class ApplicationMaster {
  public static void main(String[] args) throws Exception {
    final String hdfsPathToPatentFiles = args[0];
    final String destHdfsFolder = args[1];
    List<String> files = DownloadUtilities.getFileListing(hdfsPathToPatentFiles);
    String command = "";
    // Initialize clients to ResourceManager and NodeManagers
    Configuration conf = new YarnConfiguration();
    AMRMClient<ContainerRequest> rmClient = AMRMClient.createAMRMClient();
    rmClient.init(conf);
    rmClient.start();

    NMClient nmClient = NMClient.createNMClient();
    nmClient.init(conf);
    nmClient.start();

    // Register with ResourceManager
    rmClient.registerApplicationMaster("", 0, "");
    // Priority for worker containers - priorities are intra-application
    Priority priority = Records.newRecord(Priority.class);
    priority.setPriority(0);

    // Resource requirements for worker containers
    Resource capability = Records.newRecord(Resource.class);
    capability.setMemory(128);
    capability.setVirtualCores(1);

    // Make container requests to ResourceManager
    for (int i = 0; i < files.size(); ++i) {
      ContainerRequest containerAsk = new ContainerRequest(capability, null, null, priority);
      rmClient.addContainerRequest(containerAsk);
    }

    // Obtain allocated containers and launch
    int allocatedContainers = 0;
    while (allocatedContainers < files.size()) {
      AllocateResponse response = rmClient.allocate(0);
```

```
      for (Container container : response.getAllocatedContainers()) {
        //Command to execute to download url to HDFS
        command = "java org.apress.prohadoop.c17.DownloadFileService" +" "
                  + files.get(allocatedContainers) + " " +  destHdfsFolder;
        // Launch container by create ContainerLaunchContext
        ContainerLaunchContext ctx =
            Records.newRecord(ContainerLaunchContext.class);
        ctx.setCommands(Collections.singletonList(command));
        nmClient.startContainer(container, ctx);
        ++allocatedContainers;
      }
      Thread.sleep(100);
    }

    // Now wait for containers to complete
    int completedContainers = 0;
    while (completedContainers < files.size()) {
      AllocateResponse response = rmClient.allocate(completedContainers/files.size());
      for (ContainerStatus status : response.getCompletedContainersStatuses()) {
        ++completedContainers;
      }
      Thread.sleep(100);
    }

    // Un-register with ResourceManager
    rmClient.unregisterApplicationMaster(
        FinalApplicationStatus.SUCCEEDED, "", "");
  }
}
```

# Executing the Application Master

Finally, we are ready to execute the Application Master. Our custom application can be launched in the unmanaged and managed mode. The former is used to launch the Application Master on the development machine to simplify debugging. The latter is used to launch the application on the cluster. We perform the following steps:

1. Build the application. We get the JAR file `prohadoop-0.0.1-SNAPSHOT.jar`.
2. Copy the `PatentList.txt` from Listing 17-1 to an HDFS folder: `/user/hdfs/c17/PatentList.txt`. We refer to this entire path as `$HDFS_PATH_PATENT_LIST_FILE`.
3. Create a `/user/hdfs/c17/patents/` folder as the output HDFS folder in which we will save our downloaded patents. We refer to this entire path as `$HDFS_PATENT_ROOT_DEST`.
4. Copy the `prohadoop-0.0.1-SNAPSHOT.jar` to the HDFS in the `/user/hdfs/c17/jar` folder. We refer to this path as `$HDFS_JAR_ROOT`.

## Launch the Application in Un-Managed Mode

To launch our application using the un-managed mode use the following command after building the application.

```
hadoop jar $HADOOP_YARN_JAR_FILE_LOC/ hadoop-yarn-applications-unmanaged-am-launcher-2.2.0 Client \
-classpath prohadoop-0.0.1-SNAPSHOT.jar \
-cmd "java org.apress.prohadoop.c17.ApplicationMaster $HDFS_PATH_PATENT_LIST_FILE $HDFS_PATENT_ROOT_DEST"
```

Once the command returns, all our patents files should be downloaded into the `$HDFS_PATENT_ROOT_DEST` folder.

## Launch the Application in Managed Mode

To launch the application in the cluster (managed mode), we use the following command after building the application:

```
hadoop jar prohadoop-0.0.1-SNAPSHOT.jar org.apress.prohadoop.c17.Client $HDFS_PATH_PATENT_LIST_FILE
$HDFS_PATENT_ROOT_DEST $HDFS_JAR_ROOT/prohadoop-0.0.1-SNAPSHOT.jar
```

Once the application completes, all the patent files should be downloaded into the `$HDFS_PATENT_ROOT_DEST` folder.

# Summary

This chapter explored the key feature of Hadoop 2.0 that makes it distinct from Hadoop 1.0: the capability to create distributed applications on top of Hadoop while using the Hadoop Framework capabilities to manage resources. Using a simple yet practical example, we demonstrated how a distributed application can be created using the YARN framework. We adapted the simple application provided by Hortonworks to achieve our goals.

We discussed the key classes that define the communication protocols between the client and the Resource Manager to launch the Application Master. We also explored the protocol classes that are used to communicate with the Node Manager. They are used by the Application Master to launch worker tasks.

You should be able to use this basic introduction to create more-complex applications. Vendors are already integrating various frameworks into YARN. From a typical reader's perspective, YARN provides a developer to build custom distributed applications using a consistent framework that centralizes cluster resource management.

# APPENDIX A

# Installing Hadoop

This appendix discusses installation of the Hadoop Framework in a local and pseudo-cluster mode on Windows and Linux. Hadoop is an evolving software, so its installation is very complex. The steps discussed in this appendix must be seen as a set of general guidelines for installation; your mileage may vary.

## Installing Hadoop 2.2.0 on Windows

Windows is a preferred development environment for many developers. The good news for Windows-based developers is that Hadoop 2.2.0 supports running on Microsoft Windows. However, the downloaded files from the Apache web site do not contain some of the Windows native components (such as `winutils.exe`, `hadoop.dll`, etc.). So running Hadoop on Windows requires that the Hadoop distribution be built from source.

### Preparing the Installation Environment

First, you need to prepare your Windows environment. Following are some steps to use to get the right files and directories in place, and to otherwise prepare for the actual compiling and building:

1. Ensure that JDK 1.6 is or higher is installed. We assume that it is installed in the `c:/MyApps/jdk16/` folder, which should have a `bin` subfolder.
2. Download the `hadoop-2.2.x-src.tar.gz` files (2.2.0 at the time of this writing) from the download section of the Apache web site for Hadoop. The download link is as follows: `http://www.apache.org/dist/hadoop/core/hadoop-2.2.0/hadoop-2.2.0-src.tar.gz`.
3. Explode the `tar` file into a directory. For the purpose of this section, we assume that the directory is called `c:/myapps/hadoop/`.
4. If using Visual Studio, use Visual Studio 2010 Professional (not 2012). Do not use Visual Studio Express because it does not support compiling for 64-bit, which presents problems if running on a 64-bit system. Alternatively, download the Microsoft Windows SDK v7.1. At the time of writing, the link is the following:

   `http://www.microsoft.com/en-us/download/confirmation.aspx?id=8279`.
5. Place the SDK in the folder with full privileges. For this section, we assume that the folder is `c:/myapps/winsdk71/`.

6. Download and install the Linux-like environment for Windows named Cygwin. We assume that Cygwin home folder is `c:/myapps/cygwin/`. Ensure that the following UNIX command-line tools are installed: `sh`, `mkdir`, `rm`, `cp`, `tar`, and `gzip`. The download location for Cygwin is `http://cygwin.com/install.html`.

7. Download and install Maven 3.1.1 or a compatible version. Installation consists of simply exploding the `tar` bundle into a Windows folder. We assume that the home folder for Maven download is `c:/MyApps/apache-maven-3.1.1/`. Verify that this folder has the `bin` subfolder.

8. Download and install Protocol Buffers 2.5.0. At the time of writing, the link is `http://protobuf.googlecode.com/files/protoc-2.5.0-win32.zip`.

9. Simply unzip the zip file into a folder (assumed to be `c:/MyApps/protobuf/`). Verify that the installation is correct by ensuring that the folder contains the `protoc.exe` file.

10. Install ZLIB from `http://zlib.net/zlib128-dll.zip` and explode the zip file in the `c:/zlib-1.2.8` folder. At runtime, this folder must be accessible from the `PATH` variable.

11. Add the environment variables shown in Table A-1.

***Table A-1.*** *Environment Variables for Windows*

| Environment Variable | Value |
|---|---|
| JAVA_HOME | c:/MyApps/jdk16/ |
| M2_HOME | c:/MyApps/apache-maven-3.1.1/ |
| Platform | x64 (or Win32 when building on the 32-bit system) |
| CYGWIN_HOME | c:/myapps/cygwin/ |
| PROTOBUF_HOME | c:/myapps/protobuf/ |
| ZLIB_HOME | c:/zlib-1.2.8 |

12. Add the following to the Windows `PATH` variable:

   - `%JAVA_HOME%/bin/`
   - `%CYGWIN_HOME%/bin/`
   - `%M2_HOME%/bin/`
   - `%PROTOBUF_HOME%/bin/`
   - `%ZLIB_HOME`

---

■ **Note** The `Platform` variable mentioned previously is case-sensitive. Ensure that it is not spelled *PLATFORM* or *platform*. Although Windows variables are case-insensitive, MAVEN is case-sensitive for variable names. Failure to set this variable correctly causes `msbuild` to fail while building the native code in `hadoop-common`. This code is needed to execute a Hadoop job in the Windows environment.

---

You now are ready to move on to the build step and actually compile Hadoop to run under Windows.

## Building Hadoop 2.2.0 for Windows

Similar to painting a house, most of the work has been in the preparation. Now it is time to compile Hadoop 2.2.0 from source. Here are the steps to follow:

1. Go to the Windows Start ➤ All Programs ➤ Microsoft Windows SDK v7.1 and start the Windows SDK 7.1 command prompt in Run As Administrator mode.
2. Change the folder to `c:/MyApps/hadoop`. Recall that this is the folder in which the Hadoop source was placed.
3. Execute the following command mvn package with the `-Pdist,native-win -DskipTests -Dtar` options. The exact command is this:

   ```
   mvn package -Pdist,native-win -DskipTests -Dtar
   ```
4. If there are no errors, the `hadoop-2.2.0.tar.gz` file is created in the `c:/myapps/hadoop/hadoop-dist/hadoop-2.2.0/` folder.

## Installing Hadoop 2.2.0 for Windows

The previous section compiled the Hadoop software from source. In this section, we describe how to install the binary files created in the previous steps and how to configure the system to complete the installation. The key steps in the installation of the Hadoop software are as follows:

1. Explode the `hadoop-2.2.0.tar.gz` file in the `c:/hdp/` folder. Always use very short folder names to avoid running into problems related to limitations on the Windows maximum path length.
2. Add the `HADOOP_HOME` environment variable with the value `c:/hdp/`.
3. Add `%HADOOP_HOME%/bin` to the `PATH` environment variable.

## Configuring Hadoop 2.2.0

You can edit the Hadoop configuration files to set up the Hadoop installation to work on a local machine. There are several files to edit, and each is described in the following subsections.

## core-site.xml configuration

The core-site.xml file enables you to overwrite a number of default values that control the core of the Hadoop install. The location of this file is %HADOOP_HOME%/etc/hadoop/core-site.xml.

On my system, the HADOOP_HOME environment variable is set to c:/myapps/Hadoop/. So the path to the core-site.xml path works out to be c:/hdp/etc/Hadoop/core-site.xml.

Following is a copy of core-site.xml from my own system:

```
<configuration>
   <property>
      <name>fs.defaultFS</name>
      <value>hdfs://localhost:9000</value>
   </property>
</configuration>
```

Table A-2 describes in more detail the property modified in this listing. In addition, the full list of properties for core-site.xml can be found here: http://hadoop.apache.org/docs/current/hadoop-project-dist/hadoop-common/core-default.xml.

***Table A-2.** fs.defaultFS Property*

| Property | Documentation |
|---|---|
| fs.defaultFS | The scheme, host name, and port number of the default file system used by the Hadoop system. This value indicates that the HDFS file system is used and that NameNode runs on the localhost on port 9000. |

## hdfs-site.xml configuration

The hdfs-site.xml file enables you to overwrite a number of default values that control the HDFS. The location of this file is %HADOOP_HOME%/etc/hadoop/hdfs-site.xml. On my system, the path works out to be c:/hdp/etc/Hadoop/hdfs-site.xml.

Following is a copy of hdfs-site.xml from my own system. The lines in bold show where I have modified my configuration to specify the path prefix for the file system client.

```
<configuration>
   <property>
      <name>dfs.replication</name>
      <value>1</value>
   </property>
   <property>
      <name>dfs.namenode.name.dir</name>
      <value>file:/hdp/data/dfs/namenode</value>
   </property>
   <property>
     <name>dfs.datanode.data.dir</name>
     <value>file:/hdp/data/dfs/datanode</value>
   </property>
</configuration>
```

In the preceding example, the directories mentioned under the dfs.namenode.name.dir and dfs.datanode.data.dir properties are created under the root directory c:.

Table A-3 lists the properties updated previously. The full list of properties for `core-site.xml` can be found here: `http://hadoop.apache.org/docs/current/hadoop-project-dist/hadoop-hdfs/hdfs-default.xml`.

***Table A-3.*** *hdfs-site.xml Properties*

| Property | Documentation |
|---|---|
| `dfs.replication` | The default replication factor to use for each block of a file. Hadoop splits a file into blocks (the default size is 128 MB) and replicates them for redundancy. This parameter specifies the number of times each block is replicated. Because you are operating in the pseudo-distributed mode, specify the replication factor of 1. In a multinode production cluster, the typical (default) value is 3. |
| `dfs.namenode.name.dir` | The NameNode stores the metadata of the file system (similar to the File Allocation Table in the old Windows days) in the local file system in the specified directory. |
| `dfs.datanode.data.dir` | The DataNode stores the actual blocks of file data in the local file system in the specified directory. |

## yarn-site.xml configuration

The `yarn-site.xml` file enables you to overwrite a number of default values controlling the YARN components. The location of this file is `%HADOOP_HOME%/etc/hadoop/yarn-site.xml`. On my system, the path works out to be `c:/hdp/etc/Hadoop/yarn-site.xml`.

Following is a copy of `yarn-site.xml` from my own system:

```
<configuration>
<property>
   <name>yarn.nodemanager.aux-services</name>
   <value>mapreduce_shuffle</value>
</property>
<property>
   <name>
      yarn.nodemanager.aux-services.mapreduce_shuffle.class
   </name>
   <value>org.apache.hadoop.mapred.ShuffleHandler</value>
</property>
</configuration>
```

Table A-4 lists the properties updated previously. The full list of properties for `yarn-site.xml` can be found here: `http://hadoop.apache.org/docs/current/hadoop-yarn/hadoop-yarn-common/yarn-default.xml`.

***Table A-4.*** *yarn-site.xml Properties*

| Property | Documentation |
|---|---|
| `yarn.nodemanager.aux-services` | The name of an auxiliary service being added to the Node Manager. In Hadoop 2.x, Shuffle/Sort (See Chapter 5,6,7 for details on Shuffle/Sort) is an auxiliary service. The name of the service we specified is `mapreduce_shuffle`. The class implementing this service is specified in the property described next. If more than one service is specified, these service names must be comma-separated. |
| `yarn.nodemanager.aux-services.mapreduce_shuffle.class` | Implements the `mapreduce_shuffle` service. Note that the name of the property must always be `yarn.nodemanager.aux-service[service_name].class`. In the example, `service_name` is `mapreduce_shuffle`, as specified in the `yarn.nodemanager.aux-services` property. |

## mapred-site.xml configuration

The mapred-site.xml file enables you to overwrite a number of default values controlling the MapReduce job execution. The location of this file is %HADOOP_HOME%/etc/hadoop/mapred-site.xml. On my system, the path works out to be c:/hdp/etc/Hadoop/mapred-site.xml.

Following is a copy of mapred-site.xml from my own system:

```
<configuration>
    <property>
       <name>mapreduce.framework.name</name>
       <value>yarn</value>
    </property>
</configuration>
```

Table A-5 lists the properties updated previously. The full list of properties for mapred-site.xml can be found here: http://hadoop.apache.org/docs/stable/hadoop-mapreduce-client/hadoop-mapreduce-client-core/mapred-default.xml.

***Table A-5.*** *mapred-site.xml Properties Documentation*

| Property | Documentation |
|---|---|
| mapreduce.framework.name | The runtime framework that is used to execute MapReduce jobs. We have specified YARN, implying that the job is submitted to the YARN cluster. Setting the value to local would cause the LocalJobRunner to be used. This setting is appropriate in a development environment in which a job is executed in a single JVM. The yarn setting configures the cluster in a pseudo-distributed mode. |

The preceding property determines whether you are running MapReduce framework in local mode, classic (MapReduce v1) mode, or yarn ((MapReduce v2) mode. The local mode indicates that the job is run locally using the LocalJobRunner. If set to YARN, the job is submitted and executed via the YARN cluster.

## Preparing the Hadoop Cluster

Prepare to start the Hadoop cluster by formatting the NameNode, which needs to be done only once. Execute the following steps in sequence in the Windows SDK 7.1 command prompt:

1. Go to Windows Start ➤ All Programs ➤ Microsoft Windows SDK v7.1 and start the Windows SDK 7.1 command prompt in Run As Administrator mode.

2. Run the following set of commands in sequence from the command prompt window opened in the prior step:

```
cd c:/%HADOOP_HOME%/bin
hdfs namenode -format
```

## Starting HDFS

You are now ready to start the Hadoop system. The first step is to start the HDFS, which in turn entails starting the NameNode and DataNode by executing the following steps in sequence:

1. Go to Windows Start ➤ All Programs ➤ Microsoft Windows SDK v7.1 and start the Windows SDK 7.1 command prompt in Run As Administrator mode.
2. Run the following set of commands in sequence from the command prompt window opened in the prior step:

```
cd c:/%HADOOP_HOME%/sbin
start-dfs
```

Executing the `start` command causes two command prompt windows to be opened on your behalf by the `start` command to run the NameNode and DataNode services. Do not close these windows.

## Starting MapReduce (YARN)

In this step, you start the Resource Manager and Node Manager by executing the following steps in sequence:

1. Go to the Windows Start ➤ All Programs ➤ Microsoft Windows SDK v7.1 and start the Windows SDK 7.1 command prompt in Run As Administrator mode.
2. Run the following set of commands in sequence:

```
cd c:/%HADOOP_HOME%/sbin
start-yarn
```

Two command prompt windows should be automatically opened to run the Resource Manager and the Node Manager. Leave these windows open.

## Verifying that the Cluster Is Running

Verify that the cluster is running by opening the administrator pages for the Resource Manager and Node Manager, as well as for the NameNode. Type the following two URLs into your browser:

- `http://localhost:8042`: This is the URL for the Node Manager and Resource Manager
- `http://localhost:50070`: This is the URL for the NameNode

## Testing the Cluster

Test the cluster to verify correct operation. Execute the following in sequence. This sequence creates a new folder in the HDFS:

1. Open a Windows command prompt.
2. Execute the following command: `hadoop dfs -mkdir /test`. It creates a folder called `test` in the root of the HDFS.
3. Get a listing of the root HDFS folder by executing this command: `hadoop dfs -ls /`.

# Installing Hadoop 2.2.0 on Linux

The installation of Hadoop 2.x on Linux follows similar steps. Because the steps for Windows 7 were discussed in considerable detail, we quickly go through the Linux installation.

Some of the prerequisites of Linux-based installation are as follows:

- Java 6 is installed, and the `$JDK_HOME` environment variable points to the JDK 1.6 installation. Also, `$JDK_HOME/bin` is part of the `$PATH` variable.
- SSH is configured on the machine.
- There is a dedicated user for Hadoop. Let's call this user `hduser` and assume that the home directory of this user is `/home/hduser/` and is exported as the variable `$HOME`.

The installation steps are as follows:

1. Download the `hadoop-2.2.x tar` ball from one of the Apache Hadoop download mirrors. The name of the file is `hadoop-2.2.0.tar.gz`.
2. Explode the preceding `tar` file in the location `$HOME/hadoop`. Make the `hduser` the owner of the folder (and subfolders of) `$HOME/hadoop`.
3. Set up the following environment variables and add them to your startup scripts to ensure that you do not have to set them up each time:

```
$ export HADOOP_HOME=$HOME/hadoop/hadoop-2.2.2
$ export HADOOP_MAPRED_HOME=$HOME/hadoop/hadoop-2.2.2
$ export HADOOP_COMMON_HOME=$HOME/hadoop/hadoop-2.2.2
$ export HADOOP_HDFS_HOME=$HOME/hadoop/hadoop-2.2.2
$ export YARN_HOME=$HOME/hadoop/hadoop-2.2.2
$ export HADOOP_CONF_DIR=$HOME/hadoop/hadoop-2.2.2/etc/
$ export PATH=$PATH:$HADOOP_HOME/bin
```

4. Create the NameNode and DataNode data folders:

```
mkdir -p $HOME/hadoopdata/ hdfs/namenode
mkdir -p $HOME/hadoopdata/ hdfs/datanode
```

5. Set up the configuration files as you did for Windows.

- `core-site.xml`

```
    <configuration>
<property>
<name>fs.default.name</name>
<value>hdfs://localhost:9000</value>
</property>
    </configuration>
```

- hdfs-site.xml

```
    <configuration>
<property>
  <name>dfs.replication</name>
  <value>1</value>
</property>
<property>
  <name>dfs.namenode.name.dir</name>
  <value>
    file:/home/hduser/hadoopdata/hdfs/namenode
  </value>
</property>
<property>
  <name>dfs.datanode.data.dir</name>
  <value>
      file:/home/hduser/hadoopdata/hdfs/datanode
  </value>
</property>
   </configuration>
```

- mapred-site.xml

```
    <configuration>
<property>
  <name>mapreduce.framework.name</name>
  <value>yarn</value>
</property>
   </configuration>
```

- yarn-xml

```
    <configuration>
<property>
  <name>yarn.nodemanager.aux-services</name>
  <value>mapreduce_shuffle</value>
</property>
<property>
  <name>
    yarn.nodemanager.aux-services.mapreduce.shuffle.class
  </name>
  <value>
   org.apache.hadoop.mapred.ShuffleHandler
</value>
</property>
   </configuration>
```

6. Format the NameNode: `$HADOOP_HOME/bin/hadoop namenode -format`
7. Start the HDFS processes:
   a. Start the NameNode:

```
$HADOOP_HOME/sbin/hadoop-daemon.sh start namenode
```

b. Start the DataNode:

```
$HADOOP_HOME/sbin/hadoop-daemon.sh start datanode
```

8. Start the Map-Reduce processes.

a. Start the Resource Manager:

```
$HADOOP_HOME/sbin/hadoop-daemon.sh start resourcemanager
```

b. Start the Node Manager:

```
$HADOOP_HOME/sbin/hadoop-daemon.sh start nodemanager
```

c. Start the Job History Server:

```
$HADOOP_HOME/sbin/hadoop-daemon.sh start historyserver
```

9. Verify that the cluster is running by checking the following URLs:

- `http://localhost:8042`: The URL for the Node Manager and Resource Manager
- `http://localhost:50070`: The URL for the NameNode

10. Finally, you can stop the processes:

```
$HADOOP_HOME/sbin/hadoop-daemon.sh stop namenode
$HADOOP_HOME/sbin/hadoop-daemon.sh stop datanode
$HADOOP_HOME/sbin/hadoop-daemon.sh stop resourcemanager
$HADOOP_HOME/sbin/hadoop-daemon.sh stop nodemanager
$HADOOP_HOME/sbin/hadoop-daemon.sh stop historyserver
```

APPENDIX B

# Using Maven with Eclipse

The goal of this appendix is to show you how to use Maven with Eclipse. We do not delve into Maven details, but instead provide you with a list of references that explain Maven succinctly.

Although Maven is becoming a *de facto* tool of choice in all development environments, you might not have not worked with it. Several projects still use ANT as a build tool.

This appendix is provided for programmers who are not familiar with Maven because this book's examples use Maven. This appendix describes what Maven is, how to configure it, and how to use it with the Eclipse IDE. It should be an adequate introduction to get started with Java development using Maven.

## A Quick Introduction to Maven

Programmers who have not used Maven before are typically overwhelmed by the notion of using it for the first time. Maven can feel complex—and it is. However, it is not necessary to understand a lot of its details in order get started.

---

**Note** An excellent resource for getting started with Maven is "Maven in 5 Minutes," which can be found here: `http://maven.apache.org/guides/getting-started/maven-in-five-minutes.html`.

True to the spirit of this original link, several other 5-minute links on Maven are available. Here's another good one, especially if you are more familiar with ANT: `http://www.fiveminutes.eu/working-with-maven/`.

After you become familiar with Maven by perusing these sites, more details can be obtained from this link: `http://maven.apache.org/guides/getting-started/`.

---

### Creating a Maven Project

We discussed installing Maven in Appendix A. This section explores Maven by creating a Maven project from scratch. Follow these steps:

1. Create the folder in which you create the Maven project (for example, `C:/MyProjects/mymaven`).
2. Create a Maven project by executing the following command (you should be connected to the Internet when running this command):

```
mvn archetype:generate -DgroupId=com.apress.myapp -DartifactId=myapp
-DarchetypeArtifactId=maven-archetype-quickstart -DinteractiveMode=false
```

3. This command creates a subfolder called `myapp` in the `C:/MyProjects/mymaven` folder. The folder has the structure shown in Figure B-1. Note that there are separate source paths of application code and test code.

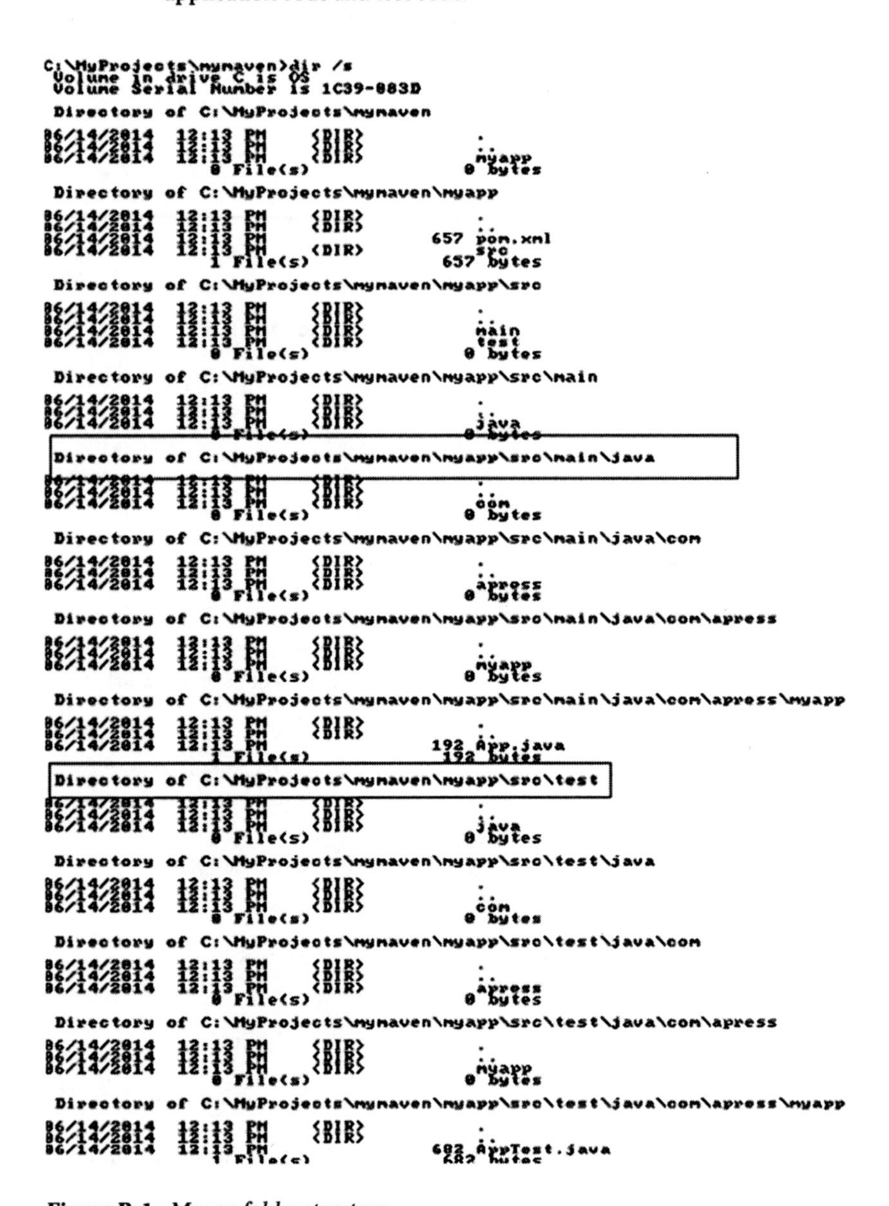


***Figure B-1.** Maven folder structure*

4. Open the `pom.xml` file, which can be found in the `C:/MyProjects/mymaven/app` folder. Note the dependencies section in the file as well as the dependency generated for `junit`. Maven handles dependency management transparently through a central Maven repository. An example global repository is mentioned in the following link (but you can create your own repository inside your company firewall):

   `http://repo.maven.apache.org/maven2/`

5. The link is not meant for access through the browser; the Maven process uses it internally to resolve dependencies. If you want to browse the Maven repository, try this link instead:

   `http://search.maven.org/#browse`

The role of dependency management is to manage the versioning. The version of the library you are using may depend on other libraries. But which versions should you use? What if those libraries need other libraries leading to transitive dependencies? Maven handles all this. All you need to mention in your `<dependency>` tag is the version number. Maven handles the dependency management transparently.

# Using Maven with Eclipse

Now let us explore how Maven can be used with the Eclipse IDE. Readers using another IDE should follow similar steps for their IDE.

## Installing the m2e Maven Eclipse Plug-in

First we need a Maven plug-in. We recommend the m2e plug-in for Maven integration with Eclipse. It can be accessed from the following update site for Eclipse: `https://eclipse.org/m2e/download/`.

---

▒ **Note** If you are not familiar with installing plug-ins using Eclipse from update sites, you can follow the instructions from several tutorials available online. Try doing a search for the term *Installing Plug-ins in the Eclipse Environment*. Some example links include the following:

- `http://agile.csc.ncsu.edu/SEMaterials/tutorials/install_plugin/index_v35.html`
- `https://www.youtube.com/watch?v=T7t5daTM-T8`

---

## Creating a Maven Project from Eclipse

To create a new Maven project in Eclipse follow the steps below:

1. Select File/New and click Other at the bottom of the submenu. The Select a Wizard dialog box displayed in Figure B-2 appears. Click Maven Project and then click the Next button.

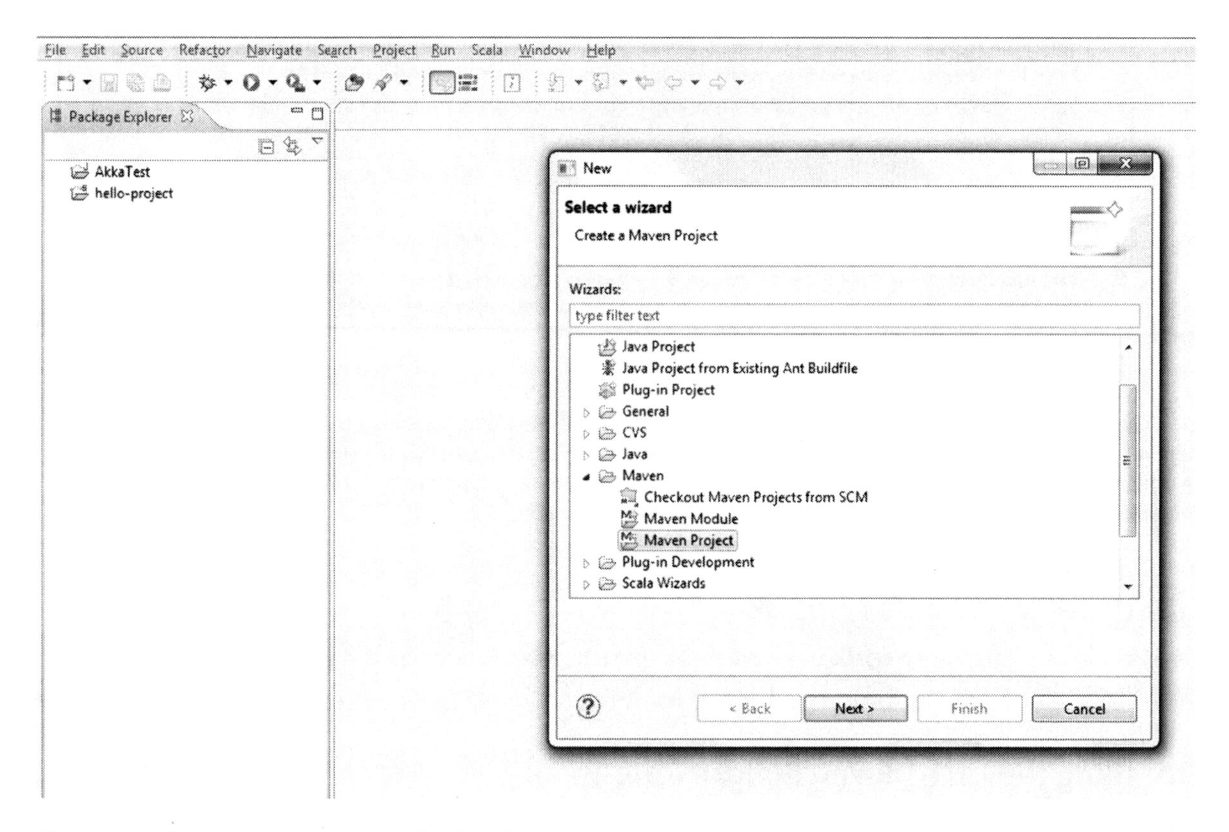


***Figure B-2.*** *Creating a new Maven project in Eclipse*

2. The New Maven Project dialog box displays (see Figure B-3).

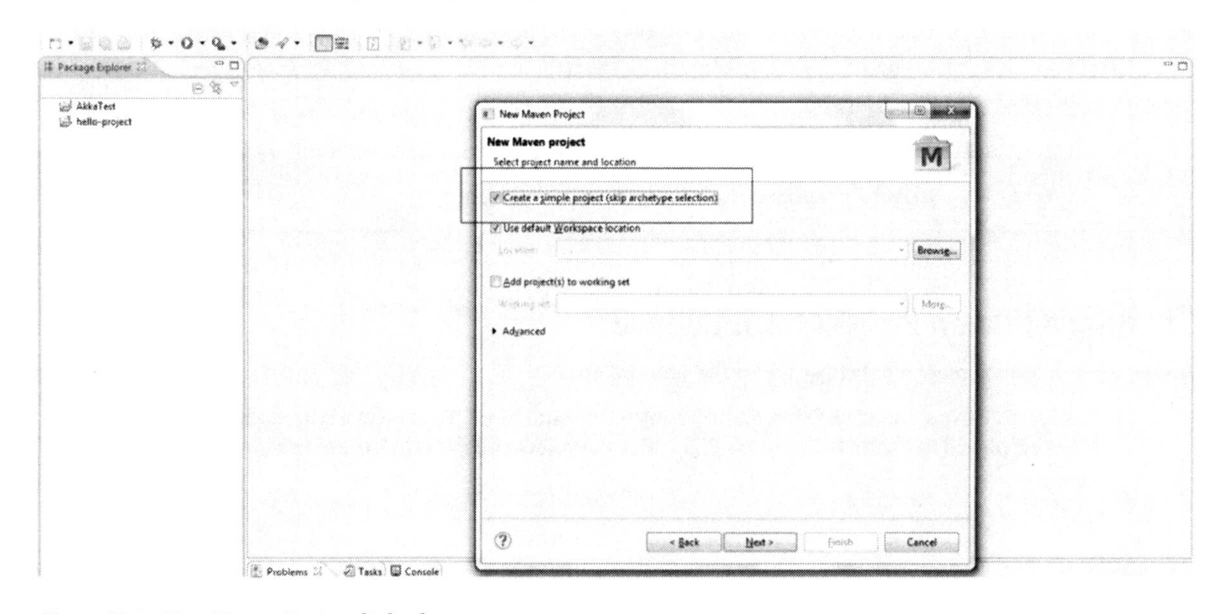


***Figure B-3.*** *New Maven Project dialog box*

3. Click Next, and the dialog box shown in Figure B-4 appears. Fill in the required details as highlighted in the boxed section. This step is similar to the `generate` command mentioned in the earlier section. We are creating our Maven project using this sequence of steps.

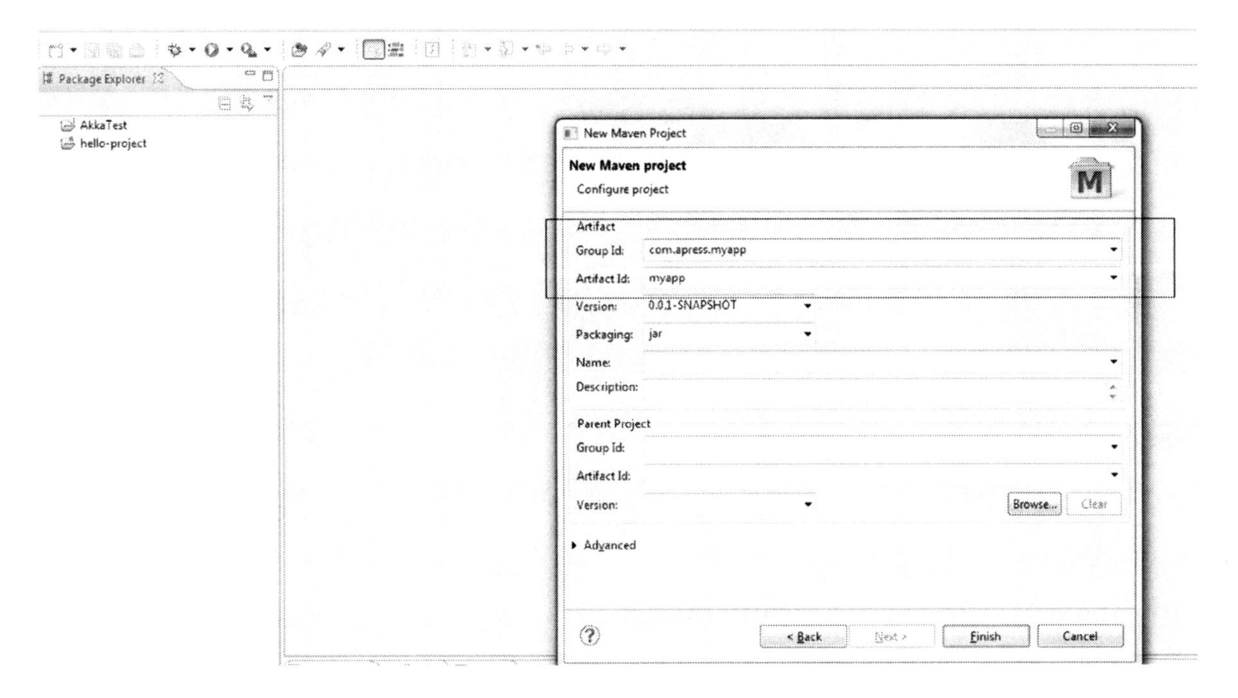


***Figure B-4.*** *New Maven Project (populate details)*

Click Finish, and the Maven project is created. The project and the corresponding folder structure created are shown in Figure B-5.

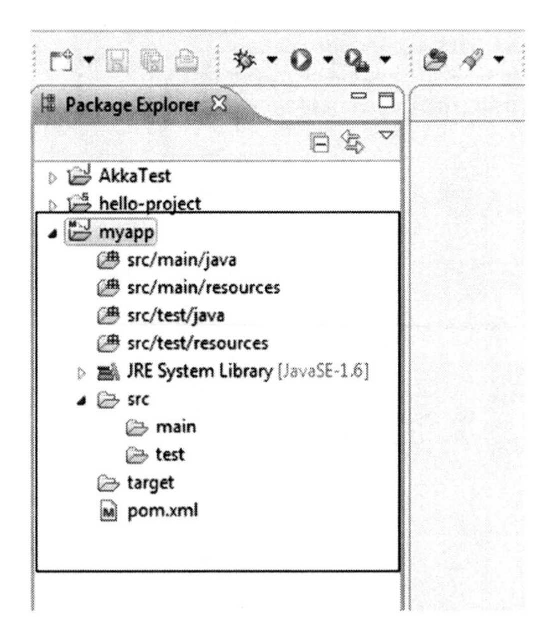


***Figure B-5.*** *myapp Maven project is created*

# Building a Maven Project from Eclipse

All the major commands for Maven can be executed from Eclipse directly. Right-click the project name in the Projects window and select Run As.

Your screen should look similar to Figure B-6.

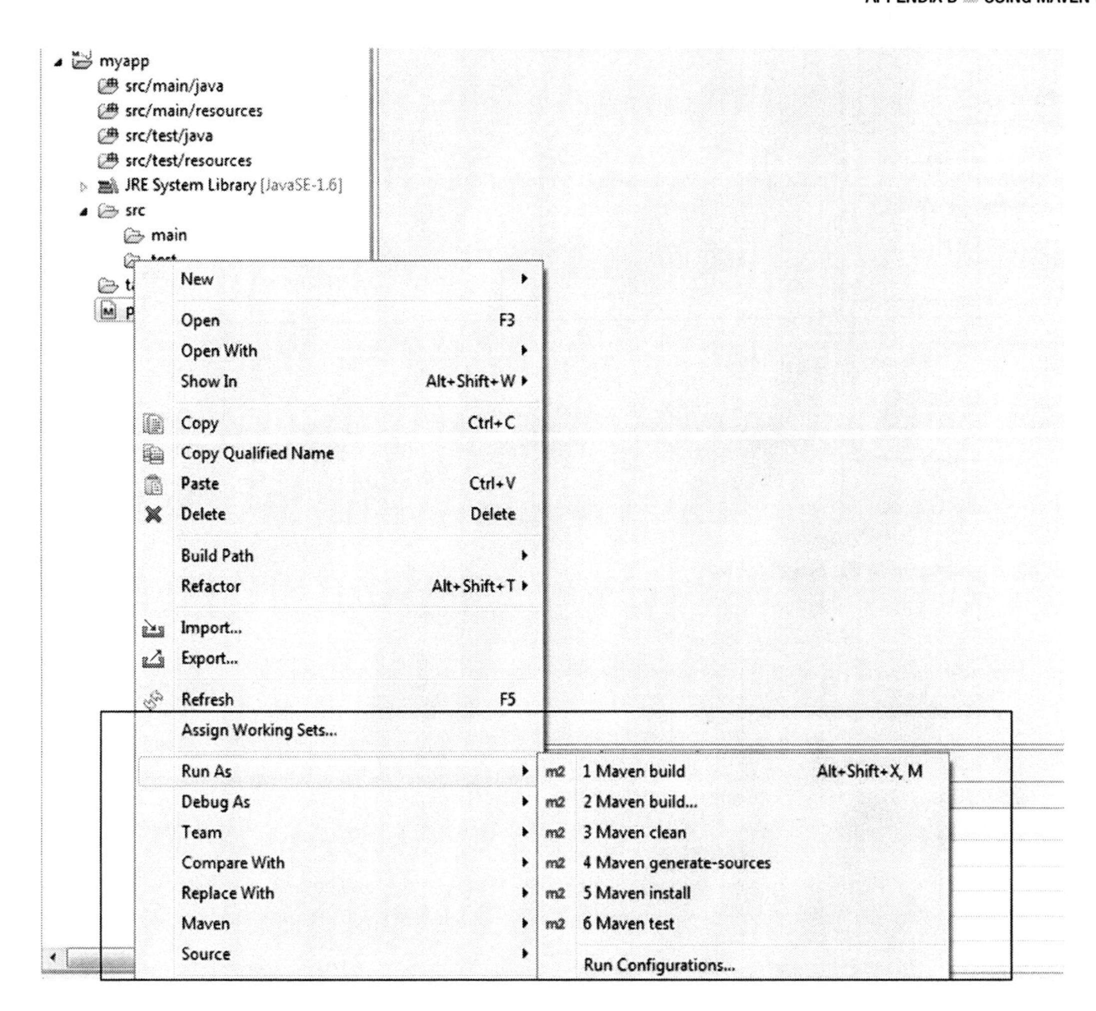


***Figure B-6.*** *Install step*

Click Maven Install, which starts the generation of the JAR file. It produces the JAR file in the target folder, as shown in Figure B-7. To clean it and rebuild again, run the Maven Clean command shown in the box to the lower right in Figure B-6 and then execute Maven Install again.

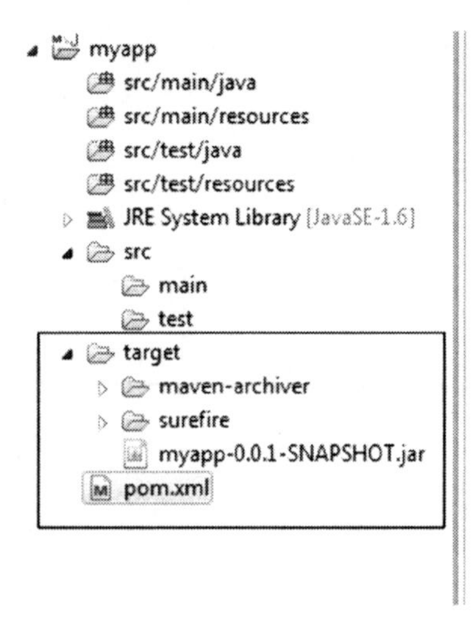


***Figure B-7.*** *JAR file is generated in the target folder*

APPENDIX C

# Apache Ambari

This appendix discusses Apache Ambari (refer to Chapter 9), which is an open-source tool for monitoring the Hadoop cluster. The purpose of this appendix is to describe the purpose of an external monitoring tool for Hadoop and the capabilities it provides. All major installations of Hadoop use similar tools. For example, Cloudera provides Cloudera Manager to perform functions similar to what Ambari provides. Ambari is fully open source and is discussed in this chapter as an illustration. You will encounter similar tools when you work on your Hadoop project. This appendix provides a flavor of what to expect when using such a tool.

## Hadoop Components Supported by Apache Ambari

Apache Ambari supports Hadoop administration through an intuitive web-based UI. The following set of Hadoop components are supported by Ambari:

- *HDFS*: Ambari allows management of NameNodes, Secondary NameNodes, and DataNodes.
- *MapReduce*: Ambari manages the default settings for MapReduce. Recall that in YARN, MapReduce jobs are run by a MapReduce Application Master that is started for each job. This Application Master uses these default settings when it starts a job unless they are overridden by the job submitter.
- *Hive*: A tool that provides data warehouse capabilities on top of Hadoop. Hive is discussed in Chapter 10.
- *Pig*: A tool for creating data flow programs in Hadoop using a high-level language called Pig Latin. Pig scripts are then compiled at runtime into MapReduce programs and executed in the cluster. Pig allows users to focus on data flows instead of the low-level details of MapReduce programs. Pig is discussed in Chapter 11.
- *HCatalog*: The table and storage management layer of HDFS that enables users who use MapReduce/Hive/Pig to be abstracted from the actual format in which data is stored. HCatalog presents the users with a relational view of the data, even though the actual format of the data can be RCFile, Text, or Sequence format. HCatlog is discussed in Chapter 12.
- *HBase*: A column-oriented database used for randomly accessing data stored in the Hadoop system. HBase is a multidimensional key value store. Users can read and write rows based on row key and column key values. HBase is discussed in Chapter 14.
- *ZooKeeper*: A centralized service for providing naming and distributed synchronization services for building customized distributed systems. It is currently used by HBase for managing region servers that stored HBase data. Given the YARN capabilities to plug in custom distributed frameworks, ZooKeeper will see a more expanded role in Hadoop 2.x.

- *Oozie*: A workflow engine used to support orchestration of Pig and MapReduce jobs in Hadoop.
- *Sqoop*: A component used to export/import data from external systems such as RDBMSs into HDFS. Sqoop is useful for ETL-type operations in which transactional data is aggregated into a data warehouse. Data is "sqooped" out of transactions databases into flat files; then MapReduce or other utilities such as Pig or Hive aggregate the data into flat files for the data warehouse. These flat files are then "sqooped" into the data warehouse. Ambari enables system administrators to perform the following range of activities through a centralized web interface.
- Hadoop cluster provisioning
    - Wizard-based approach to install Hadoop services on remote hosts. As noted in earlier chapters, installing Hadoop is a complex process, and it is even more challenging to add nodes to an existing Hadoop cluster. A wizard-based node provisioning service is very helpful for managing a Hadoop cluster.
    - Support for centralized configuration of remote Hadoop services. A large number of Hadoop configurations need to be applied on every data node, which can become very cumbersome.
- Hadoop cluster management
    - Support centralized start/stop/management of remote Hadoop services.
- Hadoop cluster monitoring
    - Ambari uses Ganglia for metrics collection. A Ganglia monitor (gmond) is installed on each remote host, which collects metrics and sends it to a central Ganglia collector component. Host metrics such as CPU utilization, I/O operations per second, average memory utilization, swap space utilization, and average network latency can be collected through the Ganglia monitors.
    - Ambari uses Nagios for notification. Nagios integrates with Hadoop-based APIs to provide alerts on service states. The system can be configured to provide new customized alerts
- REST-based APIs
    - REST-based management APIs are provided that extend the Hadoop basic REST-based APIs. These APIs are used by the web-based UI that serves as a dashboard. Application developers and system integrators can use it to integrate administration services in their services. A full set of APIs can be obtained by accessing this link:

      `https://github.com/apache/ambari/blob/trunk/ambari-server/docs/api/v1/index.md.`
    - Ambari Web is a visual dashboard for monitoring the health of the Hadoop cluster. It uses the REST APIs provided by Ambari Server.

# Installing Apache Ambari

Ambari currently supports 64-bit versions of the following operating systems:

- Redhat Enterprise Linux (RHEL) 5 and 6
- CentOS 5 and 6
- Oracle Enterprise Linux (OEL) 5 and 6
- SuSE Linux Enterprise Server (SLES) 11

The installation guide can be downloaded from this URL:

`http://docs.hortonworks.com/HDPDocuments/HDP1/HDP-1.2.0/bk_using_Ambari_book/content/index.html`.

Installation guides for all versions can be accessed from the Apache web site for Ambari:

`http://ambari.apache.org/`.

# Trying the Ambari Sandbox on Your OS

Although Ambari does not support Windows yet, you can easily try out Ambari through the VMs provided by Hortonworks. Simply follow these steps:

1. Download and install VirtualBox.
2. Download the Hortonworks Sandbox, which is the VM that comes installed with the Hortonworks distribution of Hadoop 2.x.
3. Import and start the Hortonworks Sandbox through your VirtualBox.
4. Go to `http://localhost:8888` and complete the registration process.
5. Click Enable Ambari.
6. Go to `http://localhost:8080` and enter the default username/password as admin/admin. You are now inside the Ambari UI.
7. Ambari monitors the Hadoop 2.x instance running in the sandbox.

# Index

## I



## J



## K



## L



## M

## N



## O



## P



## Q



## R



## S

## T



## U

## V



## W



## X



## Y, Z

# Contents